Scene Vision

Scene Vision
Making Sense of What We See

edited by Kestutis Kveraga and Moshe Bar

The MIT Press
Cambridge, Massachusetts
London, England

MIT Press books may be purchased at special quantity discounts for business or sales promotional use. For information, please email special_sales@mitpress.mit.

This book was set in Times New Roman MT Std 10/13pt by Toppan Best-set Premedia Limited, Hong Kong. Printed and bound in the United States of America.

Library of Congress Cataloging-in-Publication Data.

Scene vision : making sense of what we see / edited by Kestutis Kveraga and Moshe Bar.
pages cm
Includes bibliographical references and index.
ISBN 978-0-262-02785-4 (hardcover : alk. paper) 1. Visual perception. 2. Vision. I. Kveraga, Kestutis. II. Bar, Moshe (Neuroscientist)
BF241.S334 2014
152.14—dc23

2014007216

10 9 8 7 6 5 4 3 2 1

For Riki, Auris, and Indrius

Contents

Acknowledgments ix

The Current Scene 1
Moshe Bar

1 Visual Scene Representation: A Spatial-Cognitive Perspective 5
Helene Intraub

2 More Than Meets the Eye: The Active Selection of Diagnostic Information across Spatial Locations and Scales during Scene Categorization 27
George L. Malcolm and Philippe G. Schyns

3 The Constructive Nature of Scene Perception 45
Soojin Park and Marvin M. Chun

4 Deconstructing Scene Selectivity in Visual Cortex 73
Reza Rajimehr, Shahin Nasr, and Roger Tootell

5 The Neurophysiology of Attention and Object Recognition in Visual Scenes 85
Daniel I. Brooks, Heida Maria Sigurdardottir, and David L. Sheinberg

6 Neural Systems for Visual Scene Recognition 105
Russell A. Epstein

7 Putting Scenes in Context 135
Elissa M. Aminoff

8 Fast Visual Processing of "In Context" Objects 155
M. Fabre-Thorpe

9 Detecting and Remembering Briefly Presented Pictures 177
Mary C. Potter

10 Making Sense of Scenes with Spike-Based Processing 199
Simon Thorpe

11 A Statistical Modeling Framework for Investigating Visual Scene Processing in the Human Brain 225
Dustin E. Stansbury and Jack L. Gallant

12 On Aesthetics and Emotions in Scene Images: A Computational Perspective 241
Dhiraj Joshi, Ritendra Datta, Elena Fedorovskaya, Xin Lu, Quang-Tuan Luong, James Z. Wang, Jia Li, and Jiebo Luo

13 Emotion and Motivation in the Perceptual Processing of Natural Scenes 273
Margaret M. Bradley, Dean Sabatinelli, and Peter J. Lang

14 Threat Perception in Visual Scenes: Dimensions, Action, and Neural Dynamics 291
Kestutis Kveraga

Contributors 307
Index 311

Acknowledgments

We are grateful to our editor at MIT Press, Robert Prior, for his wisdom and patience along the way. We want to thank all the contributors to this volume for investing so much time and effort in making this volume a state-of-the-art review of current research in scene vision. We also would like to thank Jasmine Boshyan for her hard work in preparing the book materials for submission to the publisher and the editorial staff at the MIT Press—Christopher Eyer, Susan Clark, and Katherine Almeida—for their work preparing and promoting the book. Last, we would like to acknowledge the financial support that enabled our work on this volume: NIMH K01MH084011, MGH ECOR ISF grants and the Athinoula A. Martinos Center for Biomedical Imaging (to K.K.) and the Israeli Center of Research Excellence in Cognition (ICORE) grant 51/11 (to M.B.).

The Current Scene

Moshe Bar

For decades, visual recognition has been studied with objects rather than with scenes: single, clean, clear, and isolated objects presented to subjects at the center of the screen. This is the type of display you only see in a laboratory. In our real environment, objects do not appear so neatly. Our visual world is a stimulating scenery mess. Fragments, colors, occlusions, motion, eye movements, context, and distraction all have profound effects on perception. But except for a few brave researchers (with the seminal work of Irving Biederman from the 1960s and 1970s standing out), most have made the implicit or explicit decision that starting with objects, in spite of their meaningful lack of ecological validity, is complicated enough and will provide a good start.

This volume represents the collective contribution of modern pioneers of visual cognition, those who managed to build on solid foundations of past research to make the leap closer to real-world vision. Spanning issues of spatial vision to the study of context, from rapid perception to emotion, from attention to memory, from psychology to computational neuroscience, and from single neurons to the human brain, this book is expected to give the reader a reliable and stimulating snapshot of cutting-edge ideas on how we understand scenes and the visual world around us.

The problem begins with defining what constitutes a scene. Is a single object with background (a bird in the sky) a scene? Are two objects (a princess on a unicorn) a scene? As anyone who has tried to build a scene out of individual objects on a computer screen could attest, something is missing. Place ten objects in a circle, or even in realistic spatial relations to each other, and it does not feel right. But show a broken glass on the floor, and you know it is real. Real scenes have a "glue" that holds their elements together yet escapes definition.

A second problem is how to generalize what we have learned from rigorous and extensive research on individual object recognition to the realm of realistic, whole scenes. Those who had been following the literature may have strong views, for example, on whether individual objects are represented as a collection of 2D templates or as 3D structural descriptions, but those views seem to evaporate once we think of scenes, as if we are dealing with completely different topics. Similarly, size and viewpoint invariances have been central issues in object recognition but are hardly discussed, if at all, in the context of scenes. And the cortical pathways implicated in object recognition are generally treated as isolated from those that mediate scene understanding. The corresponding bodies of literature may give one the impression that scene perception and object recognition are nonoverlapping fields of research, although they are expected to be tightly related. There may be a gold mine of overarching links that could be made between what we know, and would like to know, about object and scene recognition, once we find the way.

One more concept whose discussion seems asymmetrical in that it is emphasized in one (scene understanding) but not the other (object recognition) is *gist*, a fascinating concept that also eludes definition. A glimpse at a scene is sufficient to extract so much information. Context, locations, presence of threat, and more are identifiable from mere tens of milliseconds of exposure. We can view a scene very briefly and, although we are not able to recognize its individual elements, we are left with a high-level conceptual gist impression such as "people having fun at a pool party." Although gist can also be defined for individual objects (e.g., by the use of low spatial frequencies), a gist of a scene tends to be richer and more complex. A scene's gist appears to involve more than information about the physical elements of the scene: it provides the basis for efficiently encapsulated memory (while sometimes giving rise to false recall), and it provides a powerful platform for the generation of predictions.

In addition to highlighting such important open questions, recent research on scene understanding has resulted in many interesting, and sometimes surprising, insights, as the chapters of this book demonstrate. Consider the following examples.

We know by now the speed limit on scene understanding and the minimal duration required for maintaining in memory what we have glimpsed, both for short and for long term. We know that observers extrapolate their memory of a scene beyond its presented boundaries, and this peculiar phenomenon tells us a lot about the spatial nature of scene representation. We also know that in spite of impressively quick and accurate scene understanding, we can miss information and not notice central changes made to the scene, and that those aspects that we are prone to miss typically do not affect the gist that is extracted from scenes.

The category of a scene (e.g., a street, a beach, an office) can be gleaned rapidly from selective spatial frequency bands in the image. Such sampling of spatial frequencies can be determined top-down and is modulated by task and the rudimentary diagnostics of the scene. This impressive ability may be based on relatively low-level visual characteristics that are common to members of each scene category. The accumulating evidence shows that in addition to the spatial layout of scenes, which is usually conveyed by the low spatial frequencies, a great deal of knowledge about content and category can be inferred from other statistics embedded in natural scenes.

Significant progress has also been made in recent years in charting the neural mechanisms subserving scene representation and processing. It has been shown repeatedly that scenes and contextual associations are represented and processed by a network that includes the parahippocampal cortex (in the medial temporal lobe), the retrosplenial complex, and parts of the medial prefrontal cortex. Scenes contain both spatial layout and information about contextual associations and, indeed, this network seems to mediate both such spatial and nonspatial information. Interestingly, this network shows a remarkable overlap with the brain's "default" network, an overlap that is taken to indicate that a lot of our thinking involves scenes and associations. This fits well with our subjective feeling of the extent to which our mental simulations, mind wandering, episodic memory, planning, and predictions rely on visual scenes and associations at their core.

We store our memories of scenes and their regularities in memory structures (e.g., "context frames") that are developed and fine-tuned with experience. These context frames are used not only for efficient memory storage but also for rapid recognition and for the generation of predictions about expected co-occurrences. Behavioral measures as well as physiological recordings (magnetoencephalography and electroencephalography) indicate that the benefits that context confers to recognition are exerted surprisingly early.

By now we know a great deal (although not enough) about the way the brain integrates multiple scene views across eye movements and differing levels of detail. Additionally, one of the main functions we deploy on scenes is searching for something (e.g., car keys, a friend). There are informative findings and insights on how search is guided by attention and depends on task requirements and stimulus characteristics. Search in scenes seems to be an iterative process that combines bottom-up and top-down processes regularly and rapidly. Many of these findings have also been reported in macaques and thus are supported by a wealth of neurophysiological data.

Finally, as creatures that seek to survive, we extract emotion and threat readily and continuously as an inherent part of our scene understanding. It turns out that we can discern different types of threats and for this rely on minute spatial and temporal properties in the scene. Furthermore, like many aspects of our environment, scenes are evaluated aesthetically. There have already been reports of some thought-provoking findings about what we find visually pleasing.

You can read about all this exciting research, and the fascinating questions that await, in the pages ahead. Enjoy.

1

Visual Scene Representation: A Spatial-Cognitive Perspective

Helene Intraub

Traditionally, scene perception has been conceptualized within the modality-centric framework of visual cognition. However, in the world, observers are spatially embedded within the scenes they perceive. Scenes are sampled through eye movements but also through movements of the head and body, guided by expectations about surrounding space. In this chapter, I will address the idea that scene representation is, at its core, a spatio-centric representation that incorporates multiple sources of information: sensory input, but also several sources of top-down information. Boundary extension (false memory beyond the edges of a view; Intraub 2010; Intraub & Richardson, 1989) provides a novel window onto the nature of scene representation because the remembered "extended" region has no corresponding sensory correlate. I will discuss behavioral, neuroimaging and neuropsychological research on boundary extension that supports a spatio-centric alternative to the traditional description of scene representation as a *visual* representation. I will suggest that this alternative view bears a relation to theories about memory and future planning.

The traditional modality-centric approach to scene representation continues to generate interesting questions and valuable research, but may unnecessarily constrain the way we think about scene perception and memory. A key motivation underlying much of the research on visual scene perception has been the mismatch between the phenomenology of vision (perception of a coherent, continuous visual world) and the striking limitations on visual input imposed by the physiology of vision (O'Regan, 1992). Put simply, the world is continuous, but visual sensory input is not. The visual field is spatially limited. To perceive our surroundings we must sample the world through successive eye fixations and movements of the head and body. Ballistic eye movements (saccades) shift the eyes' position between fixations, and during these eye movements, vision is suppressed (Volkmann, 1986). Thus, the currently available information during scene perception switches between visual sensory perception and transsaccadic memory (Irwin, 1991) as frequently as three times per second. Finally, each time the eyes land, our best visual acuity is limited to the tiny foveal region

(1° of visual angle) of each eye and declines outward into the large low-acuity periphery of vision (Rayner, 2009; Rayner & Pollatsek, 1992). How this piecemeal, inhomogeneous input comes to support the experience of a coherent visual world has been one of the classic mysteries of visual scene perception (Hochberg, 1986; Intraub, 1997; Irwin, 1991).

Rather than thinking of this problem in terms of the visual modality alone, an alternative approach is to consider that visual scene perception, even in the case of a 2D photograph, may be organized within the observer's spatial framework of surrounding space (the multisource model: Intraub, 2010, 2012; Intraub & Dickinson, 2008). Here, the underlying framework for scene perception is the observer's sense of space (e.g., "in front of me," "to the sides," "above," "below," and "behind me"; Tversky, 2009). This spatial framework acts as scaffolding that organizes not only the visual input but also rapidly available sources of information about the likely world from which the view was taken. These other sources of information include amodal completion of objects (Kanizsa, 1979) and amodal continuation of surfaces (Fantoni, Hilger, Gerbino, & Kellman, 2008; Yin, Kellman, & Shipley, 2000) that are cropped by the boundaries of the photograph; knowledge based on rapid scene classification (occurring within 100–150 ms of stimulus onset; Greene & Oliva, 2009; Potter, 1976; Thorpe, Fize, & Marlot, 1996); as well as object-to-context associations (Bar, 2004). The ability to rapidly identify objects and scenes provides early access to expectations (and constraints) about the likely layout and content of the surrounding world that a single view only partially reveals.

During day-to-day interactions with the world, the observer is embedded within a surrounding scene (e.g., standing in a kitchen) with online access to one view at a time. Scene representation, in this conceptualization, captures this fundamental reality. A single view (e.g., the first view on a scene, or the frozen view presented in a photograph) is thought to activate multiple brain areas that support a simulation (Barsalou, Kyle Simmons, Barbey, & Wilson, 2003) of the likely surrounding world that the view only partially reveals. In real-world perception as visual sampling continues, the representation increasingly reflects the specific details of the surrounding scene. What is suggested here is that the first fixation on a scene is sufficient to initiate a simulation that subsequent views can confirm or correct and embellish. According to the multisource model, just as the visual field itself is inhomogeneous, scene simulation too is inhomogeneous, shading from the highly detailed visual information in the current view to the periphery of vision, to amodal perception just beyond the boundaries, and to increasingly general and schematic expectations. A key impetus for my colleagues and me in considering this alternative conceptualization of scene perception has been *boundary extension*, a constructive error in memory for views of the world that was first reported in Intraub and Richardson (1989).

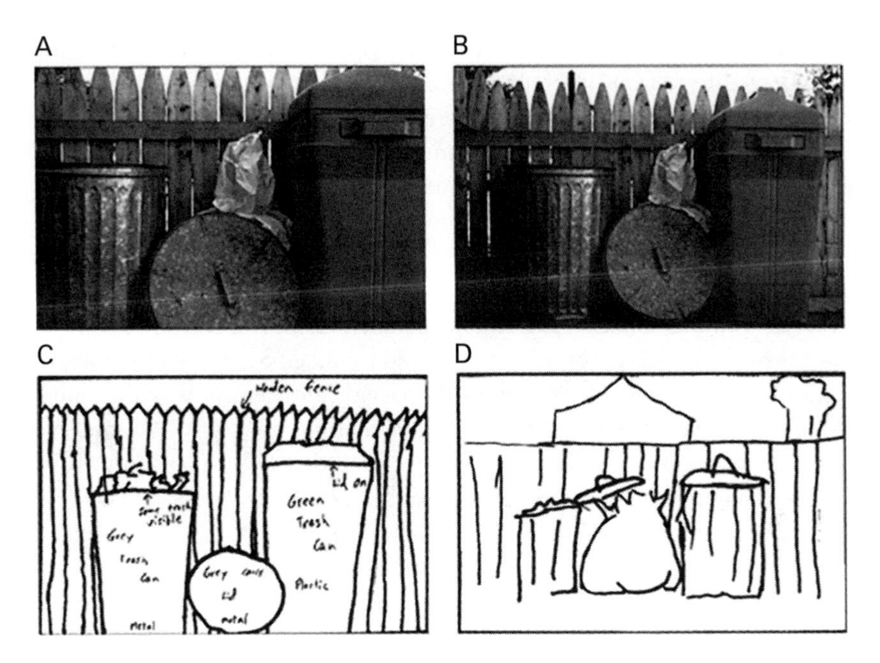

Figure 1.1
A pair of photographs (close-up and more wide-angle view) and a drawing of each one made by different participants. Note the content at the edges of the photographs and the edges of the drawings; participants remembered seeing more of the world than was shown (boundary extension). Based on figure 1.1 in Intraub and Richardson (1989), *Journal of Experimental Psychology: Learning, Memory & Cognition.*

Boundary Extension

Boundary extension (Intraub, 2010, 2012) is an error in which the observer remembers having seen beyond the physical boundaries of a view. Figure 1.1 shows an example of boundary extension in participants' drawings of photographs from memory (Intraub & Richardson, 1989). As shown in the figure, their drawings included more of the scene than was shown in the photograph. When the close-up view (panel A) was drawn from memory, not only did participants remember seeing complete garbage cans, but they remembered seeing a continuation of the fence beyond each one as well as more of the world beyond the upper and lower boundaries (as shown in the example in panel C). Although this overinclusive memory was an error with respect to the stimulus view, a comparison of this drawing with the wider-angle photograph of the same scene (panel B) shows that this error was also a good prediction of the world just beyond the boundaries of the original view. This effect was first discovered in the context of long-term memory (retention intervals of minutes to days: Intraub, Bender, & Mangels, 1992; Intraub & Richardson, 1989), but subsequent research has shown that boundary extension can occur across a masked retention interval as brief as 42 ms (commensurate with a saccade) as well as across an actual saccade when the stimulus and test picture fall on different sides of the screen (Dickinson & Intraub, 2008; Intraub & Dickinson, 2008). The rapid presence of extrastimulus scene layout is challenging to explain in terms of visual memory alone.

Boundary extension may not occur for all types visual stimuli (e.g., an object on a blank background) or at least not to the same degree (Gottesman & Intraub, 2002; Intraub, Gottesman, & Bills, 1998). It appears to be strongly associated with perceiving or thinking about scenes. For example, Intraub et al. (1998) presented line drawings of single objects on blank backgrounds. Participants were instructed to remember the objects and their sizes, but one group was induced to imagine a specific real-world location in the blank background. Although the visual information was the same, boundary extension occurred only in the scene imagination condition. An imagery control condition revealed that it was not imagery per se that had caused boundary extension in the "imagine-background" condition because when participants were instructed to imagine the colors of the objects during presentation, again boundary extension did not occur. When, in another condition, a simple background was added to each line drawing (consistent with the imagery inducement descriptions), again boundary extension occurred. In fact, performance was virtually identical when the background was imagined as when it was perceived. This, and related observations (Gottesman & Intraub, 2002) suggested that boundary extension is associated with the recruitment of processes associated with perceiving or thinking about spatial layout and meaningful locations (i.e., scenes).

Different aspects of spatial layout have been shown to affect boundary extension. The scope of the view is one such factor. Boundary extension is greatest in the case of tight close-ups and decreases, ultimately disappearing, as the view widens (Intraub et al., 1992). Close-ups yield the largest error even when there is a clear marker for boundary placement such as when a boundary slightly crops the main object in a picture. In an eye-tracking study, Gagnier, Dickinson, and Intraub (2013) presented participants with close-up photographs of single-object scenes either with or without a cropped object. Participants fixated the picture boundaries and, in the case of the cropped objects, fixated the place where the boundary cut across the object, and they refixated those areas, minutes later, at test. Yet, when participants adjusted the boundaries using the mouse to reveal more or less of the background at each picture boundary, they moved the boundaries outward to reveal unseen surrounding space in spite of their well-placed eye fixations.

To determine if knowing what would be tested in advance would influence oculomotor activity and eliminate boundary extension, another group of participants was forewarned about the nature of the test at the start of the experiment. Fixations to the boundary region and to the cropped region increased for this group, indicating that participants were attending to these critical areas. However, at test, although the size of the boundary error was reduced, boundary extension again occurred. In spite of numerous fixations to the location where the boundary had cropped the object, participants moved the cropping boundary outward such that they not only completed the object, but showed additional background beyond. Thus, knowing what would be tested in advance, increasing eye fixations to the most informative areas during encoding (including a clear marker of boundary placement) and then fixating those regions again at test were not sufficient to overcome boundary extension. Memory was not constrained to the high-acuity visual information available during multiple fixations but also included additional nonvisual information consistent with representation of the view within a larger spatial context.

Spatial factors such as the distance of an object to a boundary (Bertamini, Jones, Spooner, & Hecht, 2005; Intraub et al., 1992) impact boundary extension. But in the context of the current discussion, it is important to note that this effect is not tied solely to distance *within* the picture space (e.g., "the object is 1 cm from the picture's left boundary") but reflects how much of the real world scene is depicted within that distance in the picture (e.g., how much of the background can be seen in that 1-cm space). Gagnier, Intraub, Oliva, and Wolfe (2011) kept the distance between the main object and the edges of the picture constant, but across conditions they changed the camera's viewpoint (0° degrees, straight ahead; or 45° angle, at an angle) so that the picture would include more or less of the scene's background within the space between the object and the picture's boundaries. Results showed that in this case, although the distance between the object and the picture's boundaries was the same, boundary extension was affected by how much of the world could be seen within that space. They found that at a 45° angle, with more of the background space visible than at the 0° angle (straight ahead), boundary extension was attenuated (similar to the reduction in boundary extension between more wide-angle and more close-up views).

The observations described to this point demonstrate that scene representation is not simply a visual representation of the photographs presented in these experiments but instead draws upon other sources of information about the likely surrounding scene that the photograph only partially revealed. Further support for this point is the observation that boundary extension occurs even in the absence of vision, when blindfolded participants use haptic exploration to perceive and remember objects that are meaningfully arranged on a natural surface (e.g., a place setting on a table: Intraub, 2004; also see Intraub, 2010). In Intraub (2004) each stimulus region was bounded by a wooden frame that limited the haptic participants' exploration. The frame was removed, and minutes later participants reconstructed the boundaries. Results revealed that these participants remembered having felt the "unfelt" world just beyond the edges of the original stimulus. To determine if this "haptic" boundary extension may have been mediated by visual imagery, in the same study a "haptic expert," a woman who had been deaf and blind since early life, explored the same scenes. She too increased the area of the regions to include more of the background than she had actually touched.

In closing this section I should point out that although most research on boundary extension has been conducted with young adults, research thus far indicates that it occurs throughout the life span. Boundary extension has been reported in children's memory (4–10 years of age; Candel, Merckelbach, Houben, & Vandyck, 2004; Kreindel & Intraub, 2012; Seamon, Schlegel, Hiester, Landau, & Blumenthal, 2002), in older adults (Seamon et al., 2002), and in infants as young as 3 months of age (Quinn & Intraub, 2007). In sum, participants remembered having seen, felt, and imagined more of a view than was physically presented, even when they fixated the region near the boundary (visual exploration) or touched it (haptic exploration). In vision, they failed to recognize the identical view across a retention interval lasting less than 1/20th of a second. Intraub and Dickinson (2008) proposed a framework they referred to as the *multisource model* of scene perception that offers an explanation for these observations.

A Multisource Model of Scene Perception

As discussed earlier, Intraub and Dickinson (2008; Intraub, 2010, 2012) proposed that visual scene representation draws on multiple sources of top-down information in addition to the visual input. A depiction of the model is presented in figure 1.2. The construction of a multisource scene representation is depicted in the top panel of figure 1.2. The visual input is organized within the observer's spatial framework along with amodal perception beyond the boundaries and expectations and constraints based on rapidly available scene knowledge. In the case of a photograph, the observer takes the viewpoint of the camera (e.g., "in front of me" in typical photographs; "below me" in the case of a bird's-eye view). In fact, in photography, viewpoints such as low-angle, eye-level, and high-angle have been shown to influence the observer's interpretation of characters and events (Kraft, 1987). This organized multisource representation can be thought of as a mental *simulation* of the world that the visual information only partially reveals (Barsalou et al., 2003).

The top panel of figure 1.2 shows that this occurs while the sensory information is available. The presentation of a view elicits top-down sources of information about the larger scene (the likely surrounding world associated with the current view). All sources of information are available. While the sensory information is present, the dividing line between the sensory input and the top-down continuation of the scene is very clear. The observer can readily *see* the boundary of the view, even while thinking about the surrounding context. However, once the sensory input is gone, as in the lower panel of figure 1.2, what is now available is a remembered scene representation in which different parts were originally derived from different sources. Following the

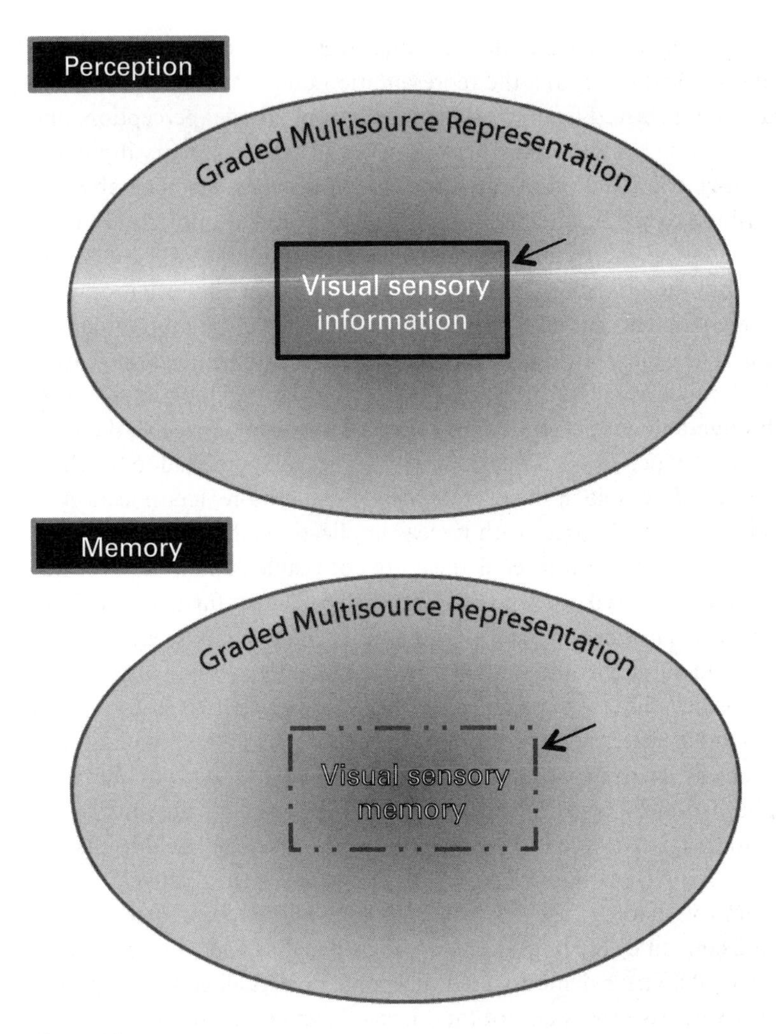

Figure 1.2
An illustration of how the multisource model accounts for boundary extension. In stage 1 (top panel), the sensory input, along with multiple top-down sources of information, creates a simulation of the likely surrounding world; the dividing line (designated by the arrow) between visible information in a photograph and the top-down continuation of the scene is easy to discern. In stage 2 (bottom panel), after the sensory input is gone, there are no tags to specify source; the dividing line (designated by the arrow) between visual memory for the once-visible information and memory for top-down generated information just outside the view is no longer clear. Information from just beyond the original boundaries was so well constrained by the visual input that in memory it is misattributed to a visual source, resulting in boundary extension.

key insight raised by Marcia Johnson and her colleagues (Johnson, Hashtroudi, & Lindsay, 1993; Johnson & Raye, 1981), the representation does not contain "tags" to indicate which parts were derived from which source (vision, amodal perception, and so forth). Now the dividing line between what was originally visual sensory input and originally the top-down continuation of the scene is no longer distinct. The observer may falsely misattribute to vision, the highly constrained expected information from just beyond the boundaries, and boundary extension is the result.

The idea that during the first stage several sources of top-down information contribute to scene perception can be thought of as follows. In the close-up photograph in figure 1.1A, note that while viewing the picture one can see that the garbage cans are cropped by the boundaries of the view on the left and right of the picture yet at the same time (through amodal perception and object knowledge) perceive the pails as whole. The pails are not perceived as "broken" pails. They are perceived to be whole pails that are just not fully visible in the current view. Furthermore, identification of the view as an "outdoor scene" carries with it clear implications that there must be a sky above and a ground plane below (even if these are not visible), and if the observer lives in the United States, the type of fence and type of pails may further specify an "outside area" that is "a suburban neighborhood" rather than "a city scene."

The simulation includes not only the studied view but an understanding of the likely surrounding scene, which differs across observers based on experience and world knowledge. The boundary extension error does not include the entire simulation; it involves only a relatively small expanse of space just beyond the edges of the view. Only the region that is tightly constrained by the visual information just inside the boundary is likely to be misattributed to vision. Again, like the visual field itself, this scene representation should be thought of as being graded, with relatively highly detailed visual memory shading into increasingly less well-specified expectations about the layout and content of the larger surrounding scene. Bar and colleagues (Bar, Aminoff, & Schacter, 2008) have demonstrated that some objects elicit a very specific surrounding context whereas others do not (and instead can be tied to multiple possible locations). Thus, different views of the world may yield very specific or very vague expectations about the surrounding world (including expectations about what one is most likely to see if one *could* turn one's head to see it).

In this account, boundary extension occurs very rapidly because the simulation is not generated *after* the stimulus is gone. What will become the boundary-extended region in memory is part of the scene simulation that becomes active within the first fixation. Once the stimulus is gone, even for a fraction of a second (Dickinson & Intraub, 2008; Intraub & Dickinson, 2008), the observer may misattribute to vision a small swath of surrounding space just beyond the original boundaries. The mistaken region is misattributed only because it so closely resembles the remembered visual information just inside the view. In the example in figure 1.1, completion of the garbage pails and continuation of the fence are so highly constrained by the visual information that they are readily misattributed to having been seen. This, of course, is a theoretical hypothesis, but there are some observations in the literature that are consistent with this possibility..

In the source-monitoring framework (Johnson et al., 1993), the decision about the source of a memory is affected by the qualities of the remembered information. In the case of boundary extension, factors that increase the similarity between memory for the visually presented information just inside the boundary and the imagined continuation of that information just outside the boundary should therefore affect how much imagined space will be misattributed to vision. Consistent with this idea, Intraub, Daniels, Horowitz, and Wolfe (2008) found that when participants viewed 750-ms photographs under conditions of divided attention, which would be expected to compromise the quality of the visual input, they experienced *greater* boundary extension than when attention was not divided. Gagnier and Intraub (2012) found that memory for line drawings of complex scenes led to greater boundary extension than memory of color photographs of the same scenes. They suggested that the mental representation was more similar across the boundary for the simple lines in the line drawing than for the more complex visual information in the photograph. It is difficult to argue that divided attention would cause greater computation of surrounding space or that line drawings would evoke a greater sense of a specific surrounding world than would naturalistic photographs. Instead, the authors proposed that in all cases, the view rapidly activated a representation of the likely surrounding spatial layout, but that the attribution of source (seen vs. imagined) differed. Divided attention and simple line stimuli in the two examples just described may have helped to increase that similarity and thus led to more of the imagined surrounding space being misattributed to vision (i.e., a greater boundary extension error).

The boundary extension error itself has been described as an adaptive error in that predicting upcoming layout might facilitate view integration as we sample the world (Intraub, 1997). Evidence for its presence across a saccade (Dickinson & Intraub, 2008; Intraub & Dickinson, 2008) suggests that it is at least available to support integration of successive views. More direct evidence of its potential impact has been provided by Gottesman (2011) in a priming task. Using a modified version of Sanocki and Epstein's (1997) layout-priming paradigm, she demonstrated that, when boundary extension occurs, the falsely remembered region beyond the boundary can prime visual perception of that region when it is visually presented later. Perhaps similar priming occurs in the haptic modality, but as yet this has not been tested.

In evaluating the first stage of the model (generation of the multisource scene representation), is there any evidence to support the idea that a view elicits a representation of surrounding space? In the next sections I describe behavioral evidence (descriptions of remembered views) and neuropsychological evidence for this proposition, and then I describe neuroimaging studies that suggest the neural architecture that may underlie both scene representation and boundary extension.

Scene Simulation: Evidence from Scene Descriptions

In day-to-day scene perception we are embedded within the scenes we perceive. What is suggested here is that the mental representation of a scene may reflect this physical reality (see Shepard, 1984, for a discussion of internalized constraints in perception). In Intraub (2010) I described an anecdotal illustration of scene simulation in which observers' interpretation of a scene had clearly drawn on expectations that went well beyond the visual information in the picture. The picture that elicited this simulation was the mundane photograph of garbage cans from Intraub and Richardson (1989) shown in figure 1.1. I had always interpreted this photograph as depicting garbage awaiting pickup on the street. I thought that my co-author, Mike Richardson, had set the tripod in a suburban street with another neighbor's house behind him. In fact, for years, I admonished students not to stand in the street when taking pictures, for safety's sake. When more recently I asked a colleague if from memory he had a sense of the camera's location and what was behind the photographer, he quickly reported that the garbage was in a backyard and that the photographer was standing with his back to the owner's house. To my surprise, another colleague, when asked, immediately said that the photographer was in an alley with the other side of the alley behind him and added, "Where else would he be?" After some confusion (for me, "alley" brought to mind a dark New York City–style alley between large buildings), she explained that in the southwestern United States where she had spent most of her life, suburban streets are aligned such that backyards on adjacent streets abut a street that serves as an alleyway behind the houses for garbage pickup.

Recently I contacted six researchers[1] in different regions of the United States and one from the United Kingdom who I thought would have long-term memory for this photograph, and asked them the same questions. Their responses are shown in table 1.1. Respondents 1–5 are from the United States (respondent 2 is from the Southwest). Of interest is the very different response offered by respondent 6, who remembered the picture but clearly had no sense of a locale. He offered only that the camera was positioned in front of the garbage cans. Further inquiry revealed that where he lives, these types of receptacles (particularly the metal can) are atypical and that he associates them mostly with old U.S. cartoons. Thus, the context for him was *weak* without a strong sense of locale. This may be an example of the observation by Bar et al. (2008) that different objects can elicit either strong or weak contexts, in this case specifically tied to this respondent's experience with the objects.

Table 1.1
Answers to where the camera was and what was behind it based on six researchers' long-term memory for the photograph of garbage cans by a fence (see figure 1.1)

Respondent	Response to location of the camera and what was behind the photographer.
1	I've always thought of it as a scene at the side fence of a house accessible to the front (so that the garbage collectors can get to the cans) but not in plain view when you're out in the backyard.
2	Photographer was standing in an alley (extending to the left and right). Behind him/her was another fence, and beyond that, another house.
3	[The camera is] in front of the fence, as if the photographer was standing at the back of a house looking into the backyard.
4	The cans were against a wooden picket fence, so I assumed that the photographer was on the far side of a driveway or possibly a small parking area.
5	The photographer was probably standing on the street.
6	[After providing description of objects in the picture] ... I guess the photographer must have been straight in front of the fence. No idea what was behind him.

Note: Respondents 1–5 are based in the United States; respondent 6 is based in the United Kingdom (see text).

These reports are anecdotal, but they suggest that even a view as mundane as in the photos in figure 1.1 can evoke a representation of a coherent surrounding world—especially when familiar objects are presented. Unbidden, specific surroundings came to mind and appear to have been part of the interpretation of the view (e.g., garbage awaiting pickup on the street or garbage in a backyard or garbage in an alleyway). An interesting aspect of the reports is the commitment to a particular locale in the participant's mind that the view evoked. Those who were queried claimed to have always "thought of the picture this way." There is no reason to think that any of these different mental scenarios would impact boundary extension (how much extrastimulus information they later attributed to vision) as, in all scenarios, immediately beyond the edges of the view there is likely to be more of the fence and a continuation of the background above and below the given view. However, the commitment to a locale and an imagined surrounding world raise the question of whether observers who have a deficit in imagining a surrounding world might also be prone to little or no boundary extension.

A Neuropsychological Approach to Scene Simulation and Boundary Extension

Hassabis and colleagues (Hassabis, Kumaran, Vann, & Maguire, 2007) reported an interesting, previously unknown deficit associated with bilateral hippocampal lesions. In addition to the expected memory deficit of anterograde amnesia, all but one of the patients tested (a patient with some spared hippocampal tissue) also exhibited a markedly impaired ability to imagine a coherent surrounding scene that they were asked to create in response to a verbal cue (e.g., "Imagine you are lying on a white sandy beach in a beautiful tropical bay"). Ten such scenarios (referring to common locales, such as visiting an art museum or market, or self-referential future scenarios, such as imagining a possible event over the next weekend) were presented, and both the patients and the matched controls were encouraged to "give free reign to their imaginations and create something new." They were encouraged to think of themselves as physically present in the surroundings and were asked to describe as many sensory details and feelings as they could.

Patients' descriptions seemed to lack spatial coherence. Their imagined worlds were fragmented and lacking in detail. It is important to note that the patients could report appropriate objects that matched the semantic context of the specified scenario, but spatial references that were apparent in the control participants' descriptions (e.g., "behind me is a row of palm trees….") were lacking in the patients' descriptions. The content of their descriptions and their subjective reports about the problems they encountered in trying to imagine a coherent world differed markedly from that of the matched control participants. Hassabis, Kumaran, Vann, et al. (2007; see also Hassabis, Kumaran, & Maguire, 2007) suggested that underlying the ability either to reconstruct a scenario from one's past or to imagine a new one (in one's future, or simply a new event based on one's general knowledge) relies on the ability to maintain a coherent, multimodal spatial representation of the event.

If we consider the multisource model depicted in figure 1.2, how might a lack of a spatially coherent scene simulation in the first stage impact boundary memory later? Mullally, Intraub, and Maguire (2012) sought to determine if patients with bilateral hippocampal lesions would be more resistant to boundary extension than their matched control participants. In terms of the multisource model, if the surrounding context lacks spatial coherence and detail, then very little if any of that imagined representation will be misattributed to vision after the stimulus is gone. Paradoxically, this hypothesis predicts that patients who suffer from severe memory deficits would actually have a more veridical memory for views of the world than would their matched controls. To test this hypothesis, Mullally, Intraub, and Maguire (2012) chose three different protocols for assessing boundary extension that would fall within the patients' memory span. The first was a brief presentation paradigm (Intraub & Dickinson, 2008), the second was an immediate drawing task (Kreindel & Intraub, 2012; Seamon et al., 2002), and the third was a haptic border reconstruction task (Intraub, 2004).

To ascertain if this group of patients showed the same scene construction deficits as in Hassabis, Kumaran, Vann, et al. (2007), a similar set of scene construction tasks was also administered. As in that study, the assessment revealed a deficient ability to construct a spatially coherent imagined world. Patients offered such comments as the imagined space being "squashed," and they provided fragmented descriptions. A new scene probe task was developed in which the patients looked at photographs and were asked to describe what they thought would be likely to exist just beyond the boundaries of the view. They did not differ from the controls in naming semantically appropriate objects, sensory description, or thoughts, emotions, or actions. However, they produced significantly fewer spatial references in describing the content. Given these problems, how did they fare on the boundary extension tasks?

In the brief presentation task, on each trial they were presented with a photograph for 250 ms. The view was interrupted by a 250-ms mask and then reappeared and stayed on the screen. The participant then rated the test view as being the same, closer up (bigger object, less surrounding space) or farther away (smaller object, more surrounding space) than before on a five-point scale. In all cases, the picture following the 250-ms masked retention interval was identical to the stimulus view. Boundary extension occurred in both groups, but was greater in the control group. Figure 1.3 shows the number of times participants in each group classified the same view as "more close up," "the same," and "farther away." As the figure shows, control participants were more likely to erroneously rate the identical test view as looking "too close up" (indicating that they remembered the view before the mask as having shown more of the scene). Patients were more accurate in recognizing that the views were actually the same. Classifying the test views as "farther away" (smaller object, more surrounding space) was relatively rare and did not differ between groups. Thus, patients did not appear to be randomly selecting responses. They appeared to be more accurate in recognizing identical views after a 250-ms masked retention interval. This better accuracy was mirrored in the pattern of confidence ratings. Control participants were

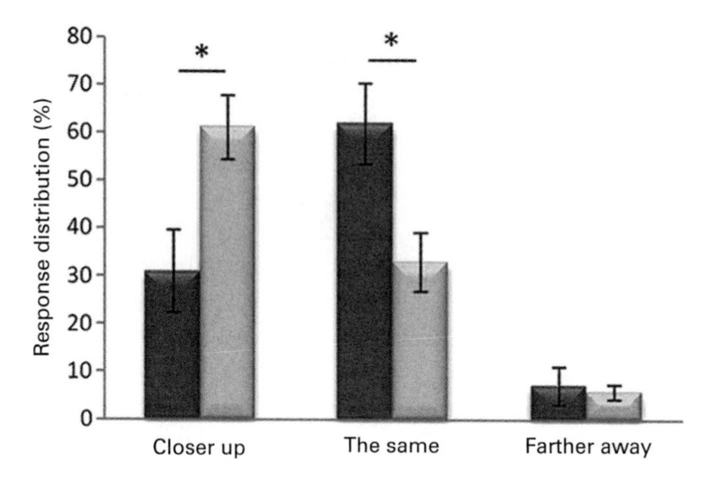

Figure 1.3
The proportion of trials classified as either "closer up," "the same" (correct answer), or "farther away" was calculated and represented as a percentage response distribution score for the patients (bilateral hippocampal lesions) and their matched control participants. Reproduction of a panel presented in figure 1.2 of Mullally, Intraub, and Maguire (2012), *Current Biology*.

more confident in their erroneous boundary extension responses than in their correct "same" responses, whereas patients were more confident of their same responses than their erroneous boundary extension responses.

In the drawing task participants viewed a photograph for 15 seconds and then immediately drew it from memory. Both the patients and their matched control participants drew boundary-extended pictures. They reduced the size of the main object and included more surrounding background in their drawings than was shown in the photograph they had just studied. However, again, patients exhibited less boundary extension than did their matched control participants. Their drawings more accurately captured the view. The three photographs that served as stimuli and the drawings made by one patient and her matched control participants are shown in the upper panel of figure 1.4. In the lower panel the graph shows the reduction in size of the main object in the drawing (as compared with the object in the photograph) for the patient and her control participants. The patients' objects and the amount of visible background drawn were more similar to the view in the original photograph than were those of their matched control participants. A group of independent judges rated all the drawings in the study and found no difference in detail or quality between the pictures drawn by patients and those drawn by control participants. A separate group of independent judges could not discriminate which pictures were drawn by patients versus control participants. The patients simply appeared to be more resistant to the boundary extension error.

In the third task haptic exploration of objects in settings similar to those in Intraub (2004) was undertaken by both groups. Participants were blindfolded and felt the objects and backgrounds of small scenarios bounded by a wooden frame. After they had explored each scenario, the frame was removed, and the participants immediately reexplored the region, indicating where each boundary had been located. In this case the control participants showed significant boundary extension, setting the boundaries outward; on average they increased the total area by about 12%. No reliable change in position was observed for the patients, the *N* was small, but the direction of the mean area remembered did not suggest boundary extension (it was reduced by about 5%). In sum, across all three tasks, patients who showed poor spatial coherence in scene construction and imagination tasks were also more resistant to boundary extension than were their matched controls.

Neuroimaging and Boundary Extension

The first report of the parahippocampal place area (PPA) (Epstein & Kanwisher, 1998) was published the same year as Intraub et al. (1998). What was striking to us was that Epstein and Kanwisher had found that pictures of locations (e.g., the corner of a room) caused heightened activation in PPA, whereas objects without scene

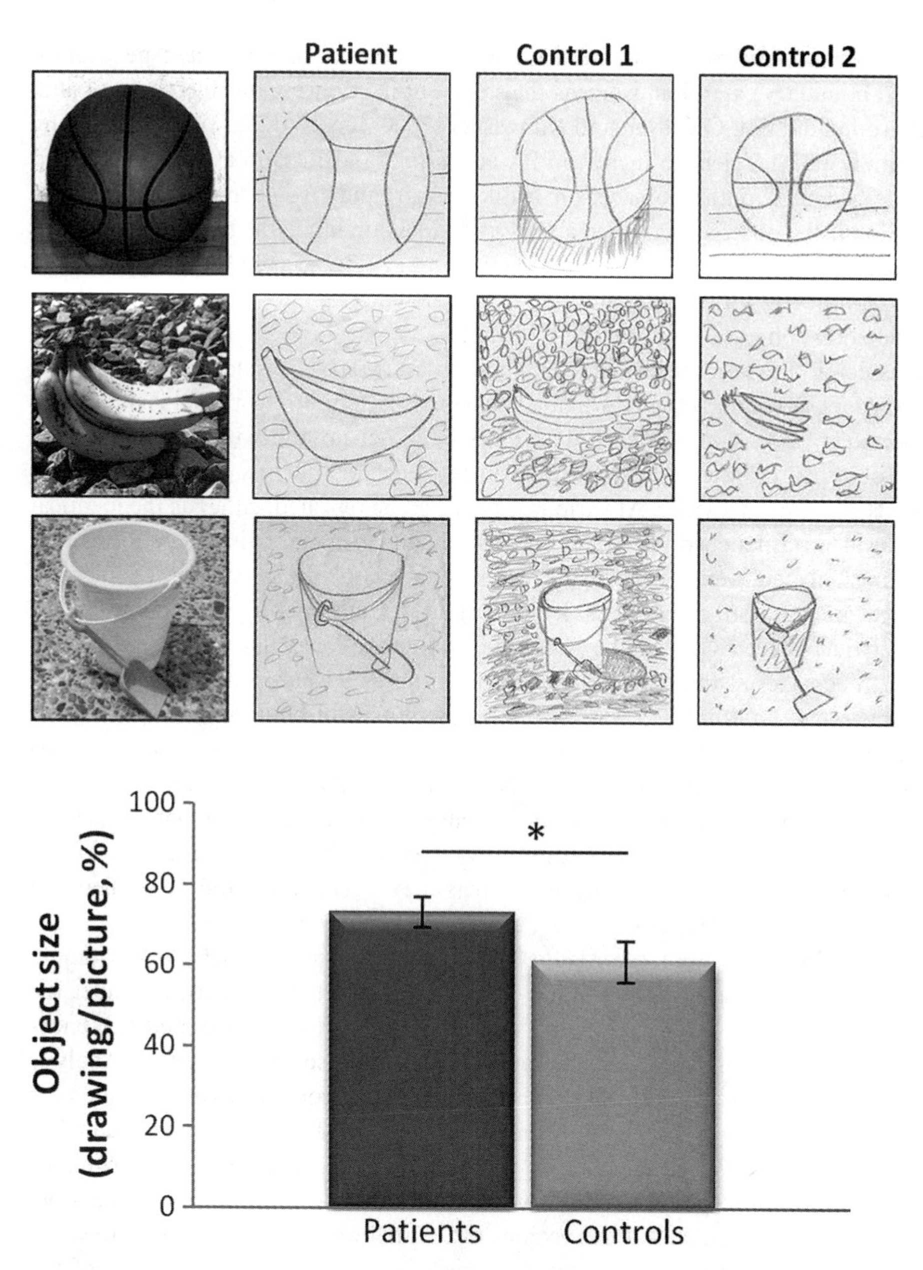

Figure 1.4
The left panel displays the three scene stimuli. Sample drawings by a patient and her two matched control participants are displayed in the middle and left panels. The graph shows that the control subjects reduced the size of the main object, incorporating more background in their drawings than did the patients. Data are presented as means +1 SEM; $*p < 0.05$. Reproduction of figure 1.3 of Mullally, Intraub, and Maguire (2012), *Current Biology*.

context did not, and we had found that outline drawings of objects in a scene location elicited boundary extension, whereas the same outline drawing of objects on a blank background did not. O'Craven and Kanwisher (2000) reported that simply imagining a location was sufficient to increase PPA activation, and Intraub et al. (1998) found that drawings of outline objects on blank backgrounds would result in boundary extension if the observer imagined a real-world location filling the blank background. Because both boundary extension and activation in PPA seemed to be tied in some way to layout and scene representation, we wondered if boundary extension might be associated in some way with the PPA.

This possibility was subsequently explored in an fMRI experiment (Park, Intraub, Yi, Widders, & Chun, 2007) in which a behavioral pattern of errors that is diagnostic of boundary extension was exploited. Pairs of closer-up and wider-angle views of single object scenes were used in the experiment. Each picture was presented for a few seconds at a time in a series. At various lags, the scene repeated, either as the identical view or as a mismatched view (i.e., the other member of the pair). Repetitions that were mismatched were of particular interest because the same pair of pictures was presented in both cases, just in a different order (the closer-up view followed later by the wider-angled view or the wider-angled view followed later by the closer-up view). In behavioral studies (beginning with Intraub & Richardson, 1989), when participants rated a mismatched test picture on the five-point boundary scale described earlier, a marked asymmetry was observed. When the order was *close-then-wide*, participants rated the mismatched view as being more similar to the original view than when the order was *wide-then-close*. Presumably this is because boundary extension in memory for the first picture caused it to more closely match perception of the second picture in the *close-wide* case, whereas it exaggerated the difference between pictures in the *wide-close* case.

When fMRI data were recorded, the participants simply watched the stimuli, making no behavioral responses at all. They were instructed to try to remember the pictures in as much detail as possible (focusing on both the object and the background) and were informed that the same scenes would sometimes repeat. Analysis of adaption responses in both PPA and the retrosplenial complex (RSC) revealed the diagnostic asymmetry. The neural response to the second picture was attenuated in the *close-wide* case (suggesting that to these regions the stimuli were very similar), whereas the neural response to the second picture showed no attenuation in the *wide-close* case (suggesting that to these regions the stimuli were quite different). However, in lateral occipital cortex (associated with object recognition, but not the size of the object) attenuation occurred in both cases (to this region the stimuli were the same regardless of the view). The pattern of neural attenuation in PPA and RSC suggested that both areas were sensitive to the boundary-extended representations of the pictures rather than to the physical views that were presented. Following the fMRI study, the same participants took part in a behavioral boundary extension experiment, and their explicit boundary ratings also revealed the typical asymmetry.

Epstein (2008, 2011; Epstein & Higgins, 2007) has proposed that PPA and RSC are part of the neural architecture that underlies navigation and integration of views within larger contexts (although Bar et al., 2008, have suggested that these regions may also be involved in conceptual processing of contextual associations; please see chapter 6 by Epstein and chapter 7 by Aminoff in this volume). Most research on scene representation that explores these ROIs has made use of visual stimuli, but research contrasting neural responses in PPA and RSC to haptic exploration of Lego scenes in contrast to Lego objects (no scene context) by blindfolded participants, as well as by congenitally blind participants, supports the idea that these areas may be responding to the spatial structure that underlies both visual and haptic exploration (Wolbers, Klatzky, Loomis, Wutte, & Giudice, 2011). Epstein (2011), in a discussion of the implications of the Wolbers et al. finding, drew the comparison to boundary extension, which, as described earlier, occurs whether exploration is visual or haptic (when the latter condition included either sighted participants who were blindfolded or a woman who had been deaf and blind since early life; Intraub, 2004). These results provide support for the value of moving from a modality-centric view of scene perception to a spatio-centric conceptualization.

The Park et al. (2007) study addressed attenuation of responses in PPA and RSC, thereby focusing on memory (the second stage of the multisource model). It cannot provide insight into brain activity associated with the first stage of the model—generation of a scene simulation. The Mullally et al. (2012) study of boundary extension in patients with bilateral hippocampal lesions suggested a possible role of hippocampus in developing a coherent spatial representation of a scene. Chadwick, Mullally, and Maguire (2013) conducted an fMRI experiment to test this possibility. They used a modified version of Intraub and Dickinson's (2008) brief presentation task but on all trials presented a close-up tested by the identical close-up (timing was similar to that in the brief presentation Experiment in Mullally et al., 2012). In brief-presentation boundary extension experiments (unlike those using longer multisecond presentation), although boundary extension occurs overall, on many trials ratings suggest no boundary extension; for example, in Intraub and Dickinson (2008) boundary extension occurred for the majority of pictures (64%) but not for all pictures. Chadwick et al. (2013) cleverly exploited this and compared the neural response in hippocampus in trials in which boundary extension occurred and trials in which it did not (based on the participants' behavioral rating in each trial).

In line with Chadwick et al.'s hypothesis, greater hippocampal activation was associated with stimuli in trials in which the behavioral response to the test picture was consistent with boundary extension than in those in which it was not. They reported that this neural response occurred before the onset of the test stimulus (the retention interval ranged from 1.2 to 3.2 seconds), so this response could not be attributed to the memory error at test.

Chadwick et al. also analyzed habituation responses in PPA (here referring to specific regions of PHC), RSC, and visual cortex. PHC and RSC were sensitive to the boundary extension error (as in the Park et al., 2007, study). Greater adaptation occurred in trials during which participants had correctly identified the views as being the same, and less adaptation occurred in trials during which they did not, and a similar adaptation effect was observed in visual cortex. What is important to remember is that the stimulus and test pictures in the experiment were always identical, so the differences observed were not mediated by any visual differences between stimulus and test. Chadwick et al. conducted DCM connectivity analyses that suggested that the hippocampus was driving the responses in the other regions of interest (no habituation effects were observed in the hippocampus). They concluded by suggesting that the neural responses observed are consistent with the two-stage model of boundary extension in which the first stage involves computation of a spatially coherent scene representation and the second stage involves the boundary error (PHC and RSC are sensitive to that *boundary-extended* representation).

The early hippocampal response (putatively tied to the first stage) in conjunction with the adaptation responses (in PHC and RSC) led Chadwick et al. to suggest that the hippocampus plays a fundamental role in supporting construction of a spatially coherent scene representation that is "channeled backwards through the processing hierarchy via PHC and as far as early visual cortex" to provide predictions about the likely surrounding world. This parallels the first stage of processing described earlier in figure 1.2. Subsequently, adaptation responses arise in response to a representation that now includes extended boundaries. The neuroimaging data in combination with the neuropsychological data suggest that the hippocampus might be involved in scene construction when the observer is presented with a view of the world. It is interesting that other research has suggested a role for the hippocampus not only in supporting episodic memory (and reconstruction of past events) but in supporting simulation of future events (e.g., future planning: Addis, Cheng, Roberts, & Schacter, 2011; Hassabis & Maguire, 2007). It is suggested here that scene perception itself (within the present moment as we sample the world around us) may involve many of the same simulation processes. The behavioral and neuroimaging research on boundary extension reviewed here suggests that the traditional modality-centric approach to scene perception does not capture the complexity of what it means to understand a scene and that a spatiocentric approach provides a viable alternative that would incorporate the visual input within a multisource cognitive representation of surrounding space.

Conclusion

Boundary extension provides an unusual means for exploring scene representation because people remember having seen beyond the boundaries of the physical view. They remember experiencing a region of space in the absence of any corresponding sensory input. Participants include this unseen region in their drawings, they move the boundaries outward to reveal this space in interactive border adjustment tasks, and they rate the scope of the view incorrectly, indicating that they remember seeing more of the scene than was actually presented (Intraub, 2010). Neural responses in PPA and RSC reflect this same overinclusive memory for recently presented views (Chadwick et al., 2013; Park et al., 2007) in tasks that elicit boundary extension. The multisource model (Intraub, 2010, 2012; Intraub & Dickinson, 2008) can account for these observations by replacing a modality-specific framework of scene representation (e.g., a visual representation) with a multisource representation organized around the observer's sense of surrounding space. Recent neuropsychological and neuroimaging evidence have suggested that the hippocampus may play a role in the mental construction of this surrounding space (Chadwick et al., 2013; Mullally et al., 2012). This provides a potential bridge between research on scene perception and research on mental constructions (and their associated neural structures) that are thought to be involved in remembering past scenarios and in generating representations of future scenarios (Addis et al., 2011; Addis, Wong, & Schacter, 2007; Hassabis & Maguire, 2007; Johnson & Sherman, 1990). In the case of scene perception, the brain's ongoing constructive activity is focused not only on long-term memory or on distant future projections, but on the present, as we perceive and interact with our immediate surrounding world.

Note

1. Many thanks to Marco Bertamini, Tim Hubbard, Geoff Loftus, Greta Munger, Dan Reisberg, and Dan Simons for their descriptions.

References

Addis, D. R., Cheng, T., Roberts, R. P., & Schacter, D. L. (2011). Hippocampal contributions to the episodic simulation of specific and general future events. *Hippocampus*, *21*(10), 1045–1052.

Addis, D. R., Wong, A. T., & Schacter, D. L. (2007). Remembering the past and imagining the future: Common and distinct neural substrates during event construction and elaboration. *Neuropsychologia*, *45*(7), 1363–1377.

Bar, M. (2004). Visual objects in context. *Nature Reviews Neuroscience*, *5*(8), 617–629.

Bar, M., Aminoff, E. M., & Schacter, D. L. (2008). Scenes unseen: The parahippocampal cortex intrinsically subserves contextual associations, not scenes or places per se. *Journal of Neuroscience*, *28*(34), 8539–8544.

Barsalou, L. W., Kyle Simmons, W., Barbey, A. K., & Wilson, C. D. (2003). Grounding conceptual knowledge in modality-specific systems. *Trends in Cognitive Sciences*, *7*(2), 84–91.

Bertamini, M., Jones, L. A., Spooner, A., & Hecht, H. (2005). Boundary extension: The role of magnification, object size, context, and binocular information. *Journal of Experimental Psychology. Human Perception and Performance*, *31*(6), 1288–1307.

Candel, I., Merckelbach, H., Houben, K., & Vandyck, I. (2004). How children remember neutral and emotional pictures: Boundary extension in children's scene memories. *American Journal of Psychology*, *117*, 249–257.

Chadwick, M. J., Mullally, S. L., & Maguire, E. A. (2013). The hippocampus extrapolates beyond the view in scenes: An fMRI study of boundary extension. *Cortex*, *49*(8), 2067–2079.

Dickinson, C. A., & Intraub, H. (2008). Transsaccadic representation of layout: What is the time course of boundary extension? *Journal of Experimental Psychology. Human Perception and Performance*, *34*(3), 543–555.

Epstein, R. A. (2008). Parahippocampal and retrosplenial contributions to human spatial navigation. *Trends in Cognitive Sciences*, *12*(10), 388–396.

Epstein, R. A. (2011). Cognitive neuroscience: Scene layout from vision and touch. *Current Biology*, *21*(11), R437–R438.

Epstein, R. A., & Higgins, J. S. (2007). Differential parahippocampal and retrosplenial involvement in three types of visual scene recognition. *Cerebral Cortex*, *17*(7), 1680–1693.

Epstein, R. A., & Kanwisher, N. (1998). A cortical representation of the local visual environment. *Nature*, *392*(6676), 598–601.

Fantoni, C., Hilger, J. D., Gerbino, W., & Kellman, P. J. (2008). Surface interpolation and 3D relatability. *Journal of Vision*, *8*(7), 1–19.

Gagnier, K. M., Dickinson, C. A., & Intraub, H. (2013). Fixating picture boundaries does not eliminate boundary extension: Implications for scene representation. *Quarterly Journal of Experimental Psychology*, *66*, 2161–2186.

Gagnier, K. M., & Intraub, H. (2012). When less is more: Line-drawings lead to greater boundary extension than color photographs. *Visual Cognition*, *20*(7), 815–824.

Gagnier, K. M., Intraub, H., Oliva, A., & Wolfe, J. M. (2011). Why does vantage point affect boundary extension? *Visual Cognition*, *19*, 234–257.

Gottesman, C. V. (2011). Mental layout extrapolations prime spatial processing of scenes. *Journal of Experimental Psychology. Human Perception and Performance*, *37*(2), 382–395.

Gottesman, C. V., & Intraub, H. (2002). Surface construal and the mental representation of scenes. *Journal of Experimental Psychology. Human Perception and Performance*, *28*(3), 589–599.

Greene, M. R., & Oliva, A. (2009). The briefest of glances: The time course of natural scene understanding. *Psychological Science*, *20*(4), 464–472.

Hassabis, D., Kumaran, D., & Maguire, E. A. (2007). Using imagination to understand the neural basis of episodic memory. *Journal of Neuroscience*, *27*(52), 14365–14374.

Hassabis, D., Kumaran, D., Vann, S. D., & Maguire, E. A. (2007). Patients with hippocampal amnesia cannot imagine new experiences. *Proceedings of the National Academy of Sciences of the United States of America*, *104*(5), 1726–1731.

Hassabis, D., & Maguire, E. A. (2007). Deconstructing episodic memory with construction. *Trends in Cognitive Sciences*, *11*(7), 299–306.

Hochberg, J. (1986). Representation of motion and space in video and cinematic displays. In K. J. Boff, L. Kaufman, & J. P. Thomas (Eds.), *Handbook of perception and human performance* (Vol. 1, pp. 22.21–22.64). New York: Wiley.

Intraub, H. (1997). The representation of visual scenes. *Trends in Cognitive Sciences*, *1*(6), 217–222.

Intraub, H. (2004). Anticipatory spatial representation of 3D regions explored by sighted observers and a deaf-and-blind observer. *Cognition*, *94*(1), 19–37.

Intraub, H. (2010). Rethinking scene perception: A multisource model. In B. Ross (Ed.), *Psychology of learning and motivation: Advances in research and theory* (Vol. 52, pp. 231–264). San Diego: Elsevier Academic Press.

Intraub, H. (2012). Rethinking visual scene perception. *Wiley Interdisciplinary Reviews: Cognitive Science, 3*(1), 117–127.

Intraub, H., Bender, R. S., & Mangels, J. A. (1992). Looking at pictures but remembering scenes. *Journal of Experimental Psychology. Learning, Memory, and Cognition, 18*(1), 180–191.

Intraub, H., Daniels, K. K., Horowitz, T. S., & Wolfe, J. M. (2008). Looking at scenes while searching for numbers: Dividing attention multiplies space. *Perception & Psychophysics, 70*(7), 1337–1349.

Intraub, H., & Dickinson, C. A. (2008). False memory 1/20th of a second later: What the early onset of boundary extension reveals about perception. *Psychological Science, 19*(10), 1007–1014.

Intraub, H., Gottesman, C. V., & Bills, A. J. (1998). Effects of perceiving and imagining scenes on memory for pictures. *Journal of Experimental Psychology. Learning, Memory, and Cognition, 24*(1), 186–201.

Intraub, H., & Richardson, M. (1989). Wide-angle memories of close-up scenes. *Journal of Experimental Psychology. Learning, Memory, and Cognition, 15*(2), 179–187.

Irwin, D. E. (1991). Information integration across saccadic eye movements. *Cognitive Psychology, 23*(3), 420–456.

Johnson, M. K., Hashtroudi, S., & Lindsay, D. S. (1993). Source monitoring. *Psychological Bulletin, 114*(1), 3–28.

Johnson, M. K., & Raye, C. L. (1981). Reality monitoring. *Psychological Review, 88*, 67–85.

Johnson, M. K., & Sherman, S. J. (1990). Constructing and reconstructing the past and the future in the present. In E. T. Higgins & R. M. Sorrentino (Eds.), *Handbook of motivation and social cognition: Foundations of social behavior* (pp. 482–526). New York: Guilford Press.

Kanizsa, G. (1979). *Organization in vision*. New York: Praeger.

Kraft, R. N. (1987). The influence of camera angle on comprehension and retention of pictorial events. *Memory & Cognition, 15*(4), 291–307.

Kreindel, E., & Intraub, H. (2012). *Boundary extension in children vs. adults: What developmental differences may tell us about scene representation.* Paper presented at the Vision Sciences Society, Naples, FL.

Mullally, S. L., Intraub, H., & Maguire, E. A. (2012). Attenuated boundary extension produces a paradoxical memory advantage in amnesic patients. *Current Biology, 22*(4), 261–268.

O'Craven, K. M., & Kanwisher, N. (2000). Mental imagery of faces and places activates corresponding stiimulus-specific brain regions. *Journal of Cognitive Neuroscience, 12*(6), 1013–1023.

O'Regan, J. K. (1992). Solving the "real" mysteries of visual perception: The world as an outside memory. *Canadian Journal of Experimental Psychology, 46*(3), 461–488.

Park, S., Intraub, H., Yi, D. J., Widders, D., & Chun, M. M. (2007). Beyond the edges of a view: Boundary extension in human scene-selective visual cortex. *Neuron, 54*(2), 335–342.

Potter, M. C. (1976). Short-term conceptual memory for pictures. *Journal of Experimental Psychology. Human Learning and Memory, 2*(5), 509–522.

Quinn, P. C., & Intraub, H. (2007). Perceiving "outside the box" occurs early in development: Evidence for boundary extension in three- to seven-month-old infants. *Child Development, 78*(1), 324–334.

Rayner, K. (2009). Eye movements and attention in reading, scene perception, and visual search. *Quarterly Journal of Experimental Psychology (Hove), 62*(8), 1457–1506.

Rayner, K., & Pollatsek, A. (1992). Eye movements and scene perception. *Canadian Journal of Psychology, 46*(3), 342–376.

Sanocki, T., & Epstein, W. (1997). Priming spatial layout of scenes. *Psychological Science, 8*(5), 374–378.

Seamon, J. G., Schlegel, S. E., Hiester, P. M., Landau, S. M., & Blumenthal, B. F. (2002). Misremembering pictured objects: People of all ages demonstrate the boundary extension illusion. *American Journal of Psychology, 115*(2), 151–167.

Shepard, R. N. (1984). Ecological constraints on internal representation: Resonant kinematics of perceiving, imagining, thinking, and dreaming. *Psychological Review, 91*(4), 417–447.

Thorpe, S., Fize, D., & Marlot, C. (1996). Speed of processing in the human visual system. *Nature, 381*(6582), 520–522.

Tversky, B. (2009). Spatial cognition: Embodied and situated. In P. Robbins & M. Aydede (Eds.), *The Cambridge handbook of situated cognition* (pp. 201–216). Cambridge: Cambridge University Press.

Volkmann, F. C. (1986). Human visual suppression. *Vision Research, 26*(9), 1401–1416.

Wolbers, T., Klatzky, R. L., Loomis, J. M., Wutte, M. G., & Giudice, N. A. (2011). Modality-independent coding of spatial layout in the human brain. *Current Biology, 21*(11), 984–989.

Yin, C., Kellman, P. J., & Shipley, T. F. (2000). Surface integration influences depth discrimination. *Vision Research, 40*(15), 1969–1978.

2

More Than Meets the Eye: The Active Selection of Diagnostic Information across Spatial Locations and Scales during Scene Categorization

George L. Malcolm and Philippe G. Schyns

Imagine flipping through a friend's holiday photos. In one you see a number of buildings and streets and realize that she had been in a city. But if you knew she was visiting several cities during her trip and wanted to know which particular one this photo was taken in, you might recognize the castle on the hilltop in the background and conclude that your friend went to Edinburgh. Or maybe you are less concerned with the city she visited and more by the weather during the trip. On the same photo you notice that the sky is dark and that there are umbrellas out, and you infer that your friend had the misfortune to suffer a rainy day.

Complex images such as scenes contain a wealth of information that can be selected and processed in different ways to satisfy a viewer's ongoing task demands (e.g., identify the city, determine weather conditions). The visual information that allows these conceptual judgments is processed in less than a second from the time when light hits the retina, allowing viewers to interact with their environment in a timely and appropriate manner. Scene processing can thus be seen as a gateway process that informs subsequent behaviors and, in many ways, represents an ultimate goal in visual cognition research. However, despite its importance, investigations into the topic have been surprisingly sparse.

Mary Potter's seminal research in the 1960s and 1970s introduced the idea that the general meaning of scenes can be recognized within the duration of a single glance. Her participants identified a cued scene (e.g., a boat on a beach) from a series of rapidly presented scene images, even when presented for as little as 125 ms each. Critically, participants could recognize target scenes when cued with only a semantic description. Thus, even when none of the exact visual features were available, participants were able to rapidly process briefly flashed scenes to an abstract level of meaning (Potter, 1975).

In these studies, object-centric terms (e.g., a woman on the phone) were used to describe target scenes, suggesting that very fast object processing was taking place. Numerous studies have since corroborated the incredible speed with which objects embedded in natural environments can be categorized (Kirchner & Thorpe, 2006;

Rousselet, Macé, & Fabre-Thorpe, 2003; Thorpe, Fize, & Marlot, 1996; VanRullen & Thorpe, 2001b). The ability to quickly process scenes based on object-centric terms harmonized with research suggesting that recognizing objects (Friedman, 1979) and spatial relationships between objects (De Graef, Christiaens, & d'Ydewalle, 1990) predicates scene categorization (see also Biederman, 1987). Likewise this object-centric approach fit with serial, bottom-up integration models in which scene categorization was the endpoint, preceded by object recognition and before that by edge detection processes that indicated boundaries and surfaces (Büllthoff & Mallot, 1988; Hildreth & Ullman, 1990; Marr, 1982).

However, these theories are inconsistent with various studies demonstrating that scenes can be categorized accurately even when displayed too briefly to allow exhaustive object processing. For instance, Biederman (1972) found that the ability to recognize an object embedded in a scene declined when the surrounding scene was jumbled. Furthermore, Rousselet et al. (2005) found that when viewers had to categorize briefly presented complex scenes (sea, mountain, urban, and indoor), early response latencies were around 260–300 ms. When compared with the 210 ms it takes to detect the appearance of a scene (VanRullen & Thorpe, 2001a), this suggests that the extra time needed to make a complex scene judgment can be as low as 30–70 ms. Although the exact nature of what is categorized about a stimulus over such very brief durations must still be established, the strict bottom-up, objects-before-scenes hypothesis is now challenged. Within these time frames, both spatial and object information could be accessed to inform the scene schema (Biederman, Mezzanotte, & Rabinowitz, 1982). The question has thus evolved into how is perceptual information accrued to form a semantically informative representation of the external environment.

Schyns and Oliva (1994; Oliva & Schyns, 1997) investigated this problem from a spatial-scale approach. In reference to Fourier's theorem, the authors noted that sinusoids of varying amplitudes, phases, and angles could characterize two-dimensional signals across multiple spatial scales. In vision these signals are referred to as spatial frequencies (SF) and are expressed by the number of cycles per degree of visual angle (or sometimes cycles per image). A SF channel is tuned to filter incoming visual signals, so it allows a subset of the full SF spectrum of information to pass through (De Valois & De Valois, 1990). Three types of SF channels exist: low-pass, which selects all visual information below a set spatial frequency threshold; high-pass, which selects all visual information above a set spatial frequency threshold; and band-pass, which selects spatial frequency information between an outlying pair of thresholds.

Humans have evolved to utilize these spatial scales with the visual system consisting of four to six overlapping, quasilinear band-pass filters (Campbell & Robson, 1968; Ginsburg, 1986; Marr & Hildreth, 1980; Wilson & Bergen, 1979; Wilson & Wilkinson, 1997), each tuned to a specific SF band. These channels act early in the visual system, transmitting filtered signals prior to many critical visual processes such as motion

(Morgan, 1992), stereopsis (Legge & Gu, 1989; Schor, Wood, & Ogawa, 1994), depth perception (Marshall, Burbeck, Ariely, Rolland, & Martin, 1996), as well as saccade programming (Findlay, Brogan, & Wenban-Smith, 1993), meaning that these filters act as a conduit to information that becomes part of our conscious experience.

Critically, from a scene perception standpoint, different channels provide the viewer with different properties of a scene image: low-spatial-frequency (LSF) information indicates size, lightness, and spatial layout of blobs, revealing the general spatial relationship within the image; high-spatial-frequency information (HSF) provides detailed information about edges and boundaries, often relating to object and texture processing (Biederman & Ju, 1988). The construction of an internal representation of a scene could potentially be built from a multiscaled integration of information passing through these channels, with the selection of scales depending on task demands—e.g., categorizing a scene as a city from the overall layout of buildings at a coarse scale or as New York City from the distinctive fine-scale tiles on the roof of the Chrysler building.

To begin with, Schyns and Oliva (1994) noted that scenes can often be characterized by their spatial properties. For instance, the spatial properties of a mountainous terrain are very different from those of a forest, which are very different from those of a city, and so forth. The authors therefore hypothesized that the regularities of these spatial properties within basic scene categories are exploited by the visual system early on to inform the viewer of a scene's schema. The fact that spatial properties can be established through coarse, LSF information suggests that there is a coarse-to-fine (CtF) preference for extracting features. Relying initially on the coarse spatial properties would offer three potential benefits for activating an early scene schema: (1) utilizing LSF information bypasses a potentially time-consuming object recognition bottleneck (particularly if a diagnostic object is not present near the fovea); (2) psychophysical and physiological evidence indicates that LSF is processed more quickly than HSF information (De Valois & De Valois, 1990; Navon, 1977); and (3) LSF is less prone to noise and can coerce HSF processing (Marr, 1982).

It should be noted that there is a distinct difference between CtF processing being discussed here and global-to-local processing made popular by the now famous Navon stimuli (e.g., a large F made up of small Ls; Navon, 1977). Viewers prefer processing the global structure (the F in figure 2.1) before its respective local components (the Ls). The explanation for this global precedence is based on CtF order for extracting image information; however the global precedence operates in a two-dimensional plane (the viewed image) while CtF processing operates in *n*-dimensional scale space, where an orthogonal axis contains two-dimensional images from each scale (or *n* scales) (Oliva & Schyns, 1997). Oliva and Schyns (1997) elegantly demonstrated this by creating an image with two Navon stimuli, one high-pass filtered and the other low-pass filtered. Despite the different filtering, both the global structure and local

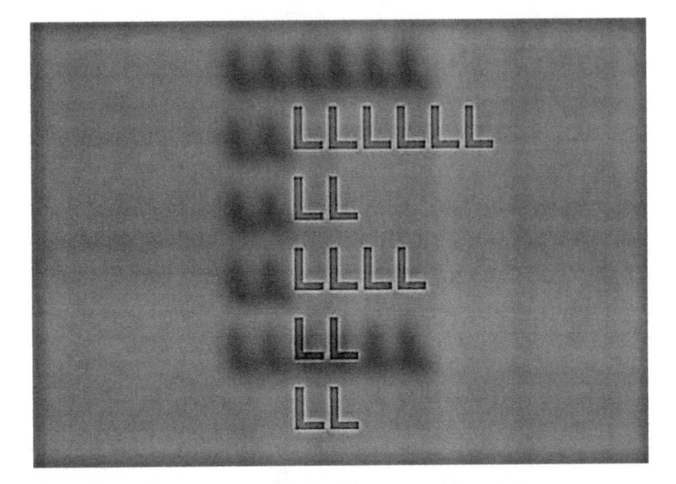

Figure 2.1
Two Navon stimuli are present within the image, the global F (for Fine) made up of local Ls and the global C (for Coarse) also made up of locals Ls. Critically, the first stimulus was high-pass filtered, whereas the second was low-pass filtered. The ability to read both the C and F at the global level as well as both sets of Ls at the local level demonstrates the orthogonal relationship between coarse-to-fine and global-to-local processing. Image taken from Oliva and Schyns (1997).

components for each Navon letter were evident (figure 2.1), thus demonstrating that CtF processing can occur across various spatial scales.

Schyns and Oliva (1994) investigated this CtF hypothesis by providing viewers with scenes in either full resolution, HSF, LSF, or, critically, hybrid images (see figure 2.2) that contained information pertaining to one scene in HSF (e.g., a highway, figure 2.2, top image) and information pertaining to a completely different scene in LSF (e.g., a city, figure 2.2, top image), followed by a test scene in full resolution for which participants had to make a yes/no matching decision. The filtered images were presented for either 30 or 150 ms. Participants could accurately match scene images from either the HSF or LSF stimuli at both durations, meaning that both sets of visual information were available to viewers. Therefore, any bias to either SF band in a hybrid image would be a result of a preference inherent within the visual system. When scenes were presented for only 30 ms, participants performed better at matching the test scene to the LSF scene within the hybrid image, but at 150 ms, participants responded more accurately to the HSF scene.

The results suggest that during fast scene analysis the visual system teases apart coarse and fine information, relying initially on spatial relationships of blobs while downplaying distinct boundary edges when beginning to construct a conceptual representation. However, over extended viewing, boundary edges become more diagnostic. Collectively, this pattern of results supported the CtF processing hypothesis

Figure 2.2
Examples of the hybrid stimuli used. Hybrids combined LSF components from one scene (a city in the top image, a highway in the bottom one) and HSF components of another scene (a highway in the top image, a city in the bottom one). Image taken from Schyns and Oliva (1994).

and, furthermore, demonstrated that coarse LSF features carry enough information for the initial scene schema activation.

In order to determine whether this CtF result generalizes from a simple matching task to a categorization process, the authors expanded the methodology to include two hybrid images with reciprocating information. For example, the first hybrid image would show one scene in LSF (e.g., a city) and another in HSF (e.g., a highway), while the second image would contain the same scenes but with interchanged SF bands (now the city information is in HSF and the highway information in LSF). This meant that a CtF and FtC approach would lead to two different scenes. Each hybrid was presented for 45 ms in succession, and participants responded after the presentation by naming the scene category they saw.

If there is a CtF preference, viewers would encode the LSF blobs in the first hybrid, then add the reciprocal HSF boundary edges in the second image, and report the respective scene; if there is an FtC preference, however, viewers would first encode HSF edges in the initial hybrid and then the reciprocal LSF blobs in the second image. If no bias exists in the visual system, then viewers should name either scene comprising the respective hybrid image with equal probability. However, in support of the previous experiment, viewers were found to prefer a CtF presentation, selecting the CtF possibility 67% of the time compared to the FtC 29% of the time.

Cumulatively, these results suggest that the scene recognition process is time and spatial-scale dependent: the initial stage utilizes coarser spatial information, whereas later stages utilize finer detail. This interpretation is consistent with subsequent research demonstrating that spatial organization information (Castelhano & Heaven, 2011; Greene & Oliva, 2009) is activated initially, allowing for initial superordinate categorizations (Kadar & Ben-Shahar, 2012; Loschky & Larson, 2010). From such LSF-engendered schemas, feedback processes facilitating object recognition can be activated (Bar et al., 2006).

However, it is important to bear in mind that these experiments did not test for any potential relationship between the categorization task and the information available within disparate spatial scales. Thus, the study could not comment on whether the selection of spatial scales was perceptually or diagnostically driven. Although the first experiment indicated that both LSF and HSF information were perceptually available at short durations, suggesting that any difference in processing was the result of a top-down bias within the visual system, the resulting CtF preference in experiment 2 could still have been perceptually driven. LSF, with its high contrast, could have interfered with HSF registration and created a perceptual bias that coerced a favored coarse-to-fine recognition scheme (the fixed-usage scenario).

On the other hand, the low-level bias could act independently of the observed CtF preference. Converging evidence suggests that there might be a diagnostically driven process for selecting task-relevant information that includes specific spatial scales.

Schyns and Rodet (1997) demonstrated that the nature of a categorization task affects low-level perceptual processes. By varying the order of category learning between participant groups, the authors induced orthogonal perceptions of identical stimuli. And although that particular study did not directly investigate the use of varying spatial scales, other studies have demonstrated that the visual system can selectively attend different SF bands within an image (Miellet, Caldara, & Schyns, 2011; Rothstein, Schofield, Funes, & Humphreys, 2010; Schyns & Oliva, 1999; Shulman & Wilson, 1987; Wong & Weisstein, 1983).

The collective evidence thus suggests the possibility that viewers could potentially select diagnostically driven spatial scales to facilitate timely scene recognition that befits the viewer's immediate task demands. Coarse-scale information might be sufficient for an express categorization (e.g., a room, a field), but more precise categorization (e.g., a restaurant, a cricket pitch) might require the active selection of comparatively finer SF information (Archambault, Gosselin, & Schyns, 2000; Collin & McMullen, 2005; Oliva, 2005; Schyns, 1998; Malcolm, Nuthmann, & Schyns, 2014).

In order to test the hypothesis that scale selection in complex scene images is flexible and determined by diagnostic informational needs, Oliva and Schyns (1997) ran another study with hybrid images and assigned diagnostic information to a single spatial scale. In an initial sensitization phase, two groups of participants were shown hybrids that were only meaningful at one scale. Either they saw scenes that contained LSF information of a scene along with HSF noise, or else HSF information of a scene along with LSF noise, depending on their assigned group. Participants were thus trained to extract diagnostic information at a set spatial scale. Participants viewed six of these hybrid images, reporting the category to which the scene image belonged as quickly and accurately as possible. Then, without pause, the hybrids changed from showing category/noise hybrids to showing hybrids from two different scene categories (e.g., LSF city and HSF of a highway). If scene recognition involves the strategic selection of a diagnostic spatial scale, then the participants should categorize the test phase hybrids according to which sensitization group they were assigned to. However, if CtF processing is fixed, participants should revert to the LSF information when it portrays scene information.

The results found mutually exclusive categorization that aligned with the participant's sensitizing group. That is, if they were sensitized to the HSF/noise scenes, participants kept reporting the HSF scenes in the test phase; on the other hand, if they were sensitized to the LSF/noise scenes, they kept reporting the LSF scenes in the test phase. Furthermore, in a debriefing stage, all but one participant reported that they did not notice a second scene in the test hybrids, ruling out the possibility that the results were due to strategic reporting. The results therefore suggest that responses were motivated by a continued reliance on information that was diagnostic in the sensitization phase, supporting a flexible processing explanation.

The results suggest that despite participants in each condition being shown the exact same bottom-up input, viewers extract the spatial scale they determine to be diagnostic. This, however, then begs the question as to what happened to the unreported spatial scale that participants claimed not to notice. The diagnostic selection process could have perceptually inhibited the unreported scale, or it might have registered but simply ignored this information. If the unreported scale were indeed registered at some level, could its properties affect the categorization process?

To test this, Oliva and Schyns (1997, experiment 3) again presented hybrid images whose category participants had to report in their response. Hybrid scenes were shown in groups of three: scene $n-1$ was a LSF/noise hybrid, scene n was a LSF/HSF hybrid whose respective spatial scales came from different scene categories; and scene $n+1$ again showed a LSF/noise hybrid, although, critically, the LSF information came from the same scene as the HSF in scene n. These triplets were interleaved into a string of LSF/noise stimuli.

After screening out participants who became aware of the HSF information in the hybrids during testing as well as trial sequences where the HSF scene was inadvertently reported, the authors found that LSF/noise hybrids on trials $n+1$ were reported faster when the previous scene contained priming HSF information. Thus, while subjects were unaware of HSF information while categorizing scenes at LSF, this information facilitated categorization of the subsequent scene. So although participants selected spatial scales that were diagnostic for the categorization task, this did not result in the perceptual inhibition of the nondiagnostic information at other scales.

A follow-up experiment attempted to characterize this priming process. The priming could be perceptual or semantic in nature. If the priming of scene $n+1$ by scene n's HSF information was perceptual, this would suggest that the LSF information was simply being mapped onto the stored perceptual HSF information from the previous trial. However, if the priming of scene $n+1$ was semantic in nature, this would suggest that both scenes in the LSF/HSF hybrids were categorized at some level (although only one explicitly) before the unused scene was used to facilitate the subsequent trial.

The previously described priming experiment was repeated, except now the primes and targets were taken from different scenes that were still in the same category. For example, if scene n was a LSF valley/HSF highway, scene $n+1$ would contain the LSF information from a different highway along with HSF noise. Critically, the scenes belonging to the same category were organized along two orthogonal measures: (1) the subjective similarity of the two scenes as judged by humans and (2) the objective similarity as judged by a perceptually based, translation-invariant metric that compared energies at different angles across multiple scales. In this way the prime and target scene could be subjectively and objectively correlated, subjectively similar but objectively uncorrelated, subjectively dissimilar and objectively correlated, or, finally, semantically dissimilar and objectively uncorrelated. If priming is conceptual in nature, then it should prime regardless of the subjective and objective similarity between the two scenes; even when they are subjectively and objectively dissimilar, the fact that they come from the same recognized category would promote faster RTs. However, if priming is restricted to when the two scenes are similar but does not occur when they are dissimilar, this would suggest that the covert processing that led to priming of the following trial was perceptual in nature only.

Oliva and Schyns (1997) found evidence for this latter interpretation. When scenes were similar subjectively and correlated objectively, priming was observed to occur with significantly shorter RTs; however, when scenes were subjectively dissimilar and objectively uncorrelated, no priming was observed to occur. Thus, while viewers can flexibly select a diagnostic scale that facilitates an ongoing task, the unattended scales are processed at a perceptual, but not a conceptual, level.

These collective findings have a broad impact on our understanding of scene processing. Research has tended to assume that coarse blobs that result from LSF information should be recognized and processed before fine boundary edges from HSF information. Empirical research has supported this hypothesis with studies demonstrating that the nature of semantic properties that can be gleaned from very short durations unfolds in a superordinate-to-basic level progression (Fei-Fei, Iyer, Koch, & Perona, 2007; Kadar & Ben-Shahar, 2012; Loschky & Larson, 2010) and then, presumably, from basic to subordinate level (although to the authors' knowledge, this has not explicitly been tested; however, see Malcolm, Nuthmann, & Schyns, 2014).

However, these studies do not consider the information demands of the recognition task. When multiple spatial scales are perceptually available, the visual system can select which scale(s) to continue processing from a perceptual to a conceptual level of understanding based on the needs of the observer. This suggests that rather than this analysis being a fixed process, there exists a dynamic bidirectional interplay between available features and the ongoing task. When scenes are viewed under normal conditions with a full resolution of scales available, the task can bias information demands to process the most informative scale(s) available, affecting the recognition process—and thereby conscious experience—accordingly.

Sampling Diagnostic Spatial Regions

The research summarized above has demonstrated that the human visual system does not passively accrue spatial frequency information in order to form a categorical scene judgment but actively selects band-pass information that facilitates task-related categorical discrimination.

However, even when categorical processing actively samples particular SF bands, there is still potentially too much information within the scene image for the visual system to process and integrate into a semantic representation. In order to work around the potential restriction of information, an efficient visual system would prioritize not only sampling of diagnostic SF bands but also information from discrete spatial locations within each SF band. Recall figure 2.1, which demonstrated the orthogonal relationship between the *n* spatial scales that comprises an image and the "flattened" two-dimensional global and local spatial layout of that image. In other words, the diagnostic value of information may vary not just across SF bands but also orthogonally within these two-dimensional SF bands.

The sampling of information across both two-dimensional spaces and SF scales within an image has been previously demonstrated during facial expression processing. For instance, Smith, Cottrell, Sowden, and Schyns (2005) presented observers with face stimuli wearing different expressions (the six universally recognized fear, happiness, sadness, disgust, anger, and surprise as well as neutral; Ekman & Friesen, 1975; Izard, 1971). Each image was segmented into five SF bands, wherein scale-adjusted Gaussian windows revealed a subset of the image from each band. Based on the information revealed across all five SF bands, participants categorized the facial expression. The amount of information shown in each band was adjusted over the experiment to maintain 75% accuracy. For each image, each band was analyzed separately, with pixels that occurred at a greater than chance rate on correctly identified face images being considered diagnostic for that particular SF. If the human brain evolved to decode facial expressions efficiently, then its decoding routines should seek to minimize ambiguities by selecting subspaces of the input across SF bands that yield the greatest task-relevant information. The authors found supportive evidence (see figure 2.3); as an example, fearful and happy relied on diagnostic information from completely different regions of the face as well as across SF bands. Furthermore, evidence of coding of these diagnostic regions was found as early as the peak of the occipitotemporal evoked potential N170 (i.e., about 170 ms following stimulus onset)

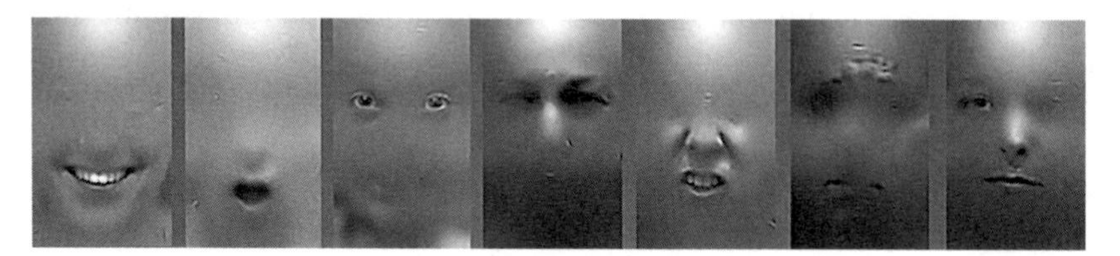

Figure 2.3
Diagnostic information for expressions (from left to right): happy, surprised, fearful, angry disgusted, sad and neutral. The diagnostic information for each expression represents a localized subspace of the image with information coming from a range of independent SF bands. Image taken from Smith, Cottrell, Gosselin, and Schyns (2005).

(Schyns, Petro, & Smith, 2007; Schyns, Thut, & Gross, 2011), with a trimming of the spatial frequency content of the representations, keeping only high scale details, between the N170 endpoint and the parietal P300 evoked potential subtending perceptual decisions (van Rijsbergen & Schyns, 2009).

These findings can be extended to ask the question whether viewers strategically extract diagnostic scene features from discrete locations within separable SF bands. Unlike faces, scenes do not dynamically adjust their visage in order to facilitate categorical decoding by a viewer (no matter how solipsistic that viewer may be!). However, as suggested above, an efficient visual system would have evolved to quickly locate potentially diagnostic information before integrating the localized information into a category representation.

Recall the example in the beginning of this chapter in which you flipped through your friend's photos. If you categorized your friend's photo as a city, information from the buildings and cars will be more diagnostic than those belonging to trees, clouds, people, and so forth, even if all these spatially distinct objects contain information within diagnostic SF bands. When the same image is categorized more specifically as Edinburgh, the spatial location of diagnostic information should change again: the location in the visual field correlating with the castle on the hill for instance or the Georgian windows on the buildings in the foreground may be critical in recognizing the Scottish capital but provide comparatively less information that would indicate that this is a city.

An efficient visual system would have evolved to minimize sampled information to a few select locations within a scene that would still yield enough diagnostic information to construct a semantic representation. Moreover, the research described above demonstrated that the human visual system could select specific SF bands of a scene image to process in a top-down manner. It should therefore follow that the visual system should also be able to sample two-dimensional locations within the different SF bands of a scene image in a top-down manner, reinforcing a bidirectional interplay between the incoming perceptual information and the viewer's goals.

Recent research by Malcolm, Nuthmann, and Schyns (2014) addressed both of these points. The authors wanted to investigate whether diagnostic information varied as a function of how specifically a scene was categorized and, if so, whether there was evidence of a top-down extraction that complemented bottom-up visual information. In an initial experiment participants were presented with scene images and asked to categorize them at either the basic (e.g., restaurant) or subordinate level (e.g., cafeteria) with the bottom-up information held constant while the top-down task varied. Critically, scenes were low-pass filtered to varying degrees. Participants could accurately categorize scenes at the basic level when presented with information low-pass filtered to 25×18.75 cycles/image[1] (viewing distance 90 cm), while accurate subordinate judgments required viewers being given more information with a less-low-pass-filtered image (50 × 37.5 cycles/image). Thus, more specific scene category judgments require the availability of comparatively finer SF information, extending previous object categorization findings from Collin and McMullen (2005) to scene images.

The authors then presented all the scenes again, low-pass filtered at the 25 × 18.75 cycles/image level, which was found to reduce categorization accuracy at the basic level to 78% and the subordinate level to 23% (in other words, chance level, as it was a four-alternative forced choice task), and a new set of participants had to again categorize the images. However, in addition to the low-pass-filtered scene images, participants were given a gaze-contingent window, 2.5° in diameter, which provided full resolution information to the fovea (figure 2.4, plate 1, top left). This foveal window allowed participants to supplement their view of the ever-present scene spatial layout (that is, the low-passed scene information) with HSF visual details to make a

Figure 2.4 (plate 1)
Counterclockwise from top left. Top left, an example of what a participant might see during fixation with full resolution information at the fovea and low-passed information in the periphery. Bottom left, the low-pass-filtered scene with diagnostic objects for a subordinate categorization in red coloring and full resolution (cafeteria). Bottom right, the low-pass-filtered scene with diagnostic objects for a basic categorization in blue coloring and full resolution (restaurant). Top right, the low-pass-filtered scene with fixated regions from all participants shown in full resolution. Regions in red represent objects fixated at a greater-than-chance rate during subordinate categorization, blue during basic categorization, and purple (red + blue) during both. All potential objects are outlined.

categorical judgment. By recording fixation locations, the authors could track which objects were fixated at a greater than chance rate when a scene was correctly categorized at either the basic or subordinate level of specificity. These objects were considered to be diagnostic at that particular level of category specificity.

Viewers took longer to make subordinate than basic-level judgments and made more fixations in the process. However, further analysis revealed that eye movements were not made randomly but showed evidence of strategic deployment: participants in the subordinate condition directed their gaze progressively further from the center of the image and more into the periphery as the trial went on, whereas in the basic condition viewers stayed closer to the middle of the image over the same epoch. This differing gaze strategy suggests that participants were deploying gaze actively to sample information from discrete spatial locations within the image as a function of task.

Corroborating this strategic search, each scene was found to contain diagnostic objects (objects fixated at a greater than chance rate) for both the basic and subordinate conditions as well as at least one object (and usually many more) that was diagnostic at one categorization level but not the other (see figure 2.4 (plate 1), top right). The results suggest that objects diagnostic to the needed categorization level were sought, located, and integrated with the low-passed gist representation in order to facilitate an accurate basic or subordinate judgment. Within this framework, objects provide easily recognizable, semantically informative distal units of information whose semantic properties can facilitate (Malcolm, Nuthmann, & Schyns, 2014) or hinder (Joubert, Rousselet, Fize, & Fabre-Thorpe, 2007) categorization depending on their relation to the rest of the image.

However, although this result suggests that category judgment uses active sampling to locate diagnostic objects, it still leaves open the question as to why there was a diverging gaze pattern. Participants could not guide gaze by the semantic properties of objects before they were fixated as the low-pass filtering masked their identities. Instead a likely situation was that the set size of diagnostic objects in an image varied as a function of categorization specificity. For example, a table and chair and menu might be diagnostic in suggesting that a scene is, at the basic level, a restaurant, but it is likely that it is not those specific tables, chairs, and menus in the image that were diagnostic, simply that those fixated provided enough information. On the contrary, to make a subordinate judgment that the image is a cafeteria, the buffet might need to be recognized. In other words, the range of objects needed to indicate that an image was a restaurant could be comparatively larger than the range of objects needed to identify what specific type of restaurant it is. The progressive exploration stretching into the periphery of the scene image during subordinate judgments suggests a search for a smaller set of less-prevalent diagnostic objects; the comparatively moderate exploration during basic judgments suggests a search for a larger, readily available set of diagnostic objects.

To test this hypothesis Malcolm, Nuthmann, and Schyns (2014) had participants again categorize scenes at either the basic or subordinate levels, but instead of providing a gaze-contingent window to allow participants to sample information anywhere in the image, scenes appeared for only 150 ms before being masked. In addition, each scene image was low-pass filtered apart from the objects previously identified as diagnostic at either the basic or subordinate level, and these were shown in full resolution (meaning there were two separate versions of each scene: LSF with basic diagnostic objects in full resolution and LSF with subordinate diagnostic objects in full resolution; see the bottom two images of figure 2.4, plate 1). If categorizing a scene involves identifying one of a potentially large set of objects, then manipulating which objects are revealed at full resolution should have a minimal effect on accuracy. However, if the range of objects that reveal a scene's category is relatively small, then manipulating which objects are identifiable should have a significantly larger effect on categorization accuracy.

The results confirmed this hypothesis. Subordinate judgments were affected by which diagnostic objects were shown (with better accuracy when subordinate rather than basic level objects were available). However, basic-level judgments were not affected by which set of diagnostic objects was shown, suggesting that the objects that had been identified as diagnostic for basic judgments made up only a subset of potential diagnostic objects available in the images and presenting an explanation as to why participants did not need to spread their search so widely over the scenes.

Conclusions

Scene categorization plays an integral role in our ability to interact with the world around us in an appropriate manner. Yet studies examining how scenes are categorized are relatively few and tend to employ an implicit framework that categorization is the end result of a passive, structured accrual of information—the "one scale fits all" approach.

Here we have summarized evidence from several studies that suggest that although scene categorization involves the accrual of information, this process involves a task-driven selection of visual information by the viewer that complements the incoming feedforward information to create a bidirectional cycle. The results revealed that participants have the ability to sample high-spatial-frequency information, providing detailed object information just as quickly as low-spatial-frequency scene layout information. Furthermore, observers can sample the specific spatial frequency band to process depending on its diagnosticity, contradicting a fixed process. Finally, these spatial frequency bands can be orthogonally accessed with viewers actively seeking information depending on the category task needs. However, although the presented research highlights the existence of the feedback component to this bidirectional cycle, further research is needed to identify the properties of this function.

We close this chapter with a few questions that we believe could articulate future research in the field. The first concerns the future of the "one scale fits all gist processing" as defined by the LSF of an image. Although the results from these studies have greatly added to our knowledge of scene representation, it is our view that focusing on gist limits research to investigating specific image information (e.g., the blobs that are revealed when a scene is low-pass filtered) even when we know that other image properties are routinely used (e.g., the HSF of an image). Even if we focus on how SFs from any band combine to form a scene representation, this focus still limits our ability to understand the scene categorization process. Collective results suggest that scene categorization should be considered even beyond the limits of SF. For example, it is easy to demonstrate (cf. figure 2.3, with faces) that a more suitable visual information framework would embrace multiscale, localized information afforded by wavelets (in figure 2.3, it is clear that different components of the face require full resolution, such as the eyes in the "fear" condition, whereas others do not, such as the cheeks). Scenes are similarly likely to require simultaneous accrual of local and global information from the same image at different levels of spatial resolution. And looking into the future, real scenes are three-dimensional events, not two-dimensional images. Multiscale generative models of 3D scenes will need to be developed to test for more complex representations that are not restricted by the 2D image space.

The second research question stems from the observation that "one scale fits all gist processing" has so far restricted the conceptual framework of scene recognition to the type of scene categorizations that can be performed with this limited information. It remains an empirical question precisely what categorizations LSF information affords. Our view is that the answer might not be the expected one. Undoubtedly, some categories can be inferred from sparse LSF cues, but for those categories that can be recognized (e.g., mountains vs. forests), it is not clear whether processing of stripped-down spatial information represents the actual scene categorization mechanism or whether participants have simply been able to deduce the correct scene from the available experimental choices based on limited information provided to the visual system. The real questions in psychology (as opposed to, say, pattern classification in engineering or AI) are (1) what scene categories can be inferred from these cues and which ones cannot, and (2) whether those categories that can be inferred correspond to a psychological reality—for example, are they basic level categories, superordinate categories, or do they cut across all levels? Considerable research has so far embraced gist processing without a validation of its psychological reality. To reiterate, the point is not whether some categories can be inferred from LSF because this is undoubtedly true. The point is whether the interaction between scene categories in memory and

LSF information represents a psychologically meaningful interaction, such as the sort of interaction posited by Biederman between a geometric description of objects and the basic level of their categorization (Biederman, 1987).

It is clear to us that it is now critical to go beyond "one size fits all gist processing" because the conceptual framework is too restrictive to account for the variety of hierarchical scene categorizations humans perform. Research should focus on formalizing the visual information that is diagnostic for hierarchically organized visual categorization tasks and accept that the hierarchy of spatial frequencies, although providing a tempting visual analogue to the hierarchy of scene categories, limits more than it informs the process.

Note

1. The smallest unit for which a light-dark transition can be represented on a screen is two pixels (e.g., one black, the other white). This means that in an 800 × 600 screen, the greatest number of transitions is 400 × 300 cycles/image in the width and height directions, respectively. These two unit transitions contain the finest details of an image (boundaries, textures, etc.). If we peel away these fine transitions, we are left with light-dark transitions represented every four pixels and greater (or 200 × 150 cycles/image). When we say something is low-passed to 25 × 18.75 cycles/image, this refers to the greatest number of transitions (or overall level of detail) that can be displayed in a screen image in the width and height directions.

References

Archambault, A., Gosselin, F., & Schyns, P. G. (2000). *A natural bias for the basic level?* Paper presented at the Proceedings of the Twenty-Second Annual Conference of the Cognitive Science Society.

Bar, M., Kassam, K. S., Ghuman, A. S., Boshyan, J., Schmidt, A. M., Dale, A. M., et al. (2006). Top-down facilitation of visual recognition. *Proceedings of the National Academy of Sciences of the United States of America*, *103*(2), 449–454.

Biederman, I. (1972). Perceiving real-world scenes. *Science*, *177*(4043), 77–80.

Biederman, I. (1987). Recognition-by-components: A theory of human image understanding. *Psychological Review*, *94*(2), 115–147.

Biederman, I., & Ju, G. (1988). Surface versus edge-based determinants of visual recognition. *Cognitive Psychology*, *20*, 38–64.

Biederman, I., Mezzanotte, R. J., & Rabinowitz, J. C. (1982). Scene perception: Detecting and judging objects undergoing relational violations. *Cognitive Psychology*, *14*(2), 143–177.

Büllthoff, H. H., & Mallot, H. (1988). Integration of depth modules: Stereo and shading. *Journal of the Optical Society of America*, *5A*, 1749–1758.

Campbell, F. W., & Robson, J. G. (1968). Application of the Fourier analysis to the visibility of gratings. *Journal of Physiology*, *197*(3), 551–566.

Castelhano, M. S., & Heaven, C. (2011). Scene context influences without scene gist: Eye movements guided by spatial associations in visual search. *Psychonomic Bulletin & Review*, *18*(5), 890–896.

Collin, C. A., & McMullen, P. A. (2005). Subordinate-level categorization relies on high spatial frequencies to a greater degree than basic-level categorization. *Perception & Psychophysics*, *67*(2), 354–364.

De Graef, P., Christiaens, D., & d'Ydewalle, G. (1990). Perceptual effects of scene context on object identification. *Psychological Research*, *52*, 317–329.

De Valois, R. L., & De Valois, K. K. (1990). *Spatial vision*. New York: Oxford University Press.

Ekman, P., & Friesen, W. V. (1975). *Unmasking the face*. Englewood Cliffs, NJ: Prentice Hall.

Fei-Fei, L., Iyer, A., Koch, C., & Perona, P. (2007). What do we perceive in a glance of a real-world scene? *Journal of Vision*, *7*(1), 1–29.

Findlay, J. M., Brogan, D., & Wenban-Smith, M. (1993). The visual signal for saccadic eye movements emphasizes visual boundaries. *Perception & Psychophysics*, *53*, 633–641.

Friedman, A. (1979). Framing pictures: The role of knowledge in automatized encoding and memory for gist. *Journal of Experimental Psychology. General*, *108*, 316–355.

Ginsburg, A. P. (1986). Spatial filtering and visual form perception. In K. R. Boff, L. Kaufman, & J. P Thomas (Eds.), *Handbook of perception and human performance* (Vol. II, pp. 1–41). New York: John Wiley & Sons.

Greene, M. R., & Oliva, A. (2009). The briefest of glances: The time course of natural scene understanding. *Psychological Science*, *20*(4), 464–472.

Hildreth, E. C., & Ullman, S. (1990). The computational study of vision. In M. Posner (Ed.), *Foundations of cognitive science* (pp. 581–630). Cambridge, MA: MIT Press.

Izard, C. (1971). *The face of emotion*. New York: Appleton-Century-Crofts.

Joubert, O. R., Rousselet, G., Fize, D., & Fabre-Thorpe, M. (2007). Processing scene context: Fast categorization and object interference. *Vision Research*, *47*, 3286–3297.

Kadar, I., & Ben-Shahar, O. (2012). A perceptual paradigm and psychophysical evidence for hierarchy in scene gist processing. *Journal of Vision*, *12*(13), 1–17.

Kirchner, H., & Thorpe, S. J. (2006). Ultra-rapid object detection with saccadic eye movements: Visual processing speed revisited. *Vision Research*, *46*(11), 1762–1776.

Legge, G. E., & Gu, Y. (1989). Stereopsis and contrast. *Vision Research*, *29*, 989–1004.

Loschky, L. C., & Larson, A. M. (2010). The natural/man-made distinction is made prior to basic-level distinctions in scene gist processing. *Visual Cognition*, *18*(4), 513–536.

Malcolm, G. L., Nuthmann, A., & Schyns, P. G. (2014). Beyond gist: Strategic and incremental information accumulation for scene categorization. *Psychological Science*, *25*, 1087–1097.

Marr, D. (1982). *Vision*. San Francisco, CA: Freeman.

Marr, D., & Hildreth, E. C. (1980). Theory of edge detection. *Proceedings of the Royal Society of London*, *2078*, 187–217.

Marshall, J. A., Burbeck, C. A., Ariely, D., Rolland, J. P., & Martin, K. E. (1996). Occlusion edge blur: A cue to relative visual depth. *Journal of the Optical Society of America. A, Optics, Image Science, and Vision*, *13*, 681–688.

Miellet, S., Caldara, R., & Schyns, P. G. (2011). Local Jekyll and global Hyde: The dual identity of face identification. *Psychological Science*, *22*(12), 1518–1526.

Morgan, M. J. (1992). Spatial filtering precedes motion detection. *Nature*, *355*, 344–346.

Navon, D. (1977). Forest before trees: The precedence of global features in visual perception. *Cognitive Psychology*, *9*(3), 353–383.

Oliva, A. (2005). Gist of the scene. In L. Itti, G. Rees, & J. K. Tsotsos (Eds.), *Encyclopedia of neurobiology of attention* (pp. 251–256). San Diego, CA: Elsevier.

Oliva, A., & Schyns, P. G. (1997). Coarse blobs or fine edges? Evidence that information diagnosticity changes the perception of complex visual stimuli. *Cognitive Psychology*, *34*, 72–107.

Potter, M. C. (1975). Meaning in visual scenes. *Science*, *187*, 965–966.

Rothstein, P., Schofield, A., Funes, M. J., & Humphreys, G. W. (2010). Effects of spatial frequency bands on perceptual decision: It is not the stimuli but the comparison. *Journal of Vision*, *10*(25), 1–20.

Rousselet, G. A., Joubert, O. R., & Fabre-Thorpe, M. (2005). How long to get to the "gist" of real-world natural scenes? *Visual Cognition*, *12*(6), 852–877.

Rousselet, G. A., Macé, M. J.-M., & Fabre-Thorpe, M. (2003). Is it an animal? Is it a human face? Fast processing in upright and inverted natural scenes. *Journal of Vision*, *3*(6), 440–455.

Schor, C. M., Wood, I. C., & Ogawa, J. (1994). Spatial tuning of static and dynamic local steropsis. *Vision Research*, *24*, 573–578.

Schyns, P. G. (1998). Diagnostic recognition: Task constraints, object information, and their interactions. *Cognition*, *67*, 147–179.

Schyns, P. G., & Oliva, A. (1994). From blobs to boundary edges: Evidence for time- and spatial-scale-dependent scene recognition. *Psychological Science*, *5*(4), 195–200.

Schyns, P. G., & Oliva, A. (1999). Dr. Angry and Mr. Smile: When categorization flexibly modifies the perception of faces in rapid visual presentations. *Cognition*, *69*, 243–265.

Schyns, P. G., Petro, L. S., & Smith, M. L. (2007). Dynamics of visual information integration in the brain for categorizing facial expressions. *Current Biology*, *17*(18), 1580–1585.

Schyns, P. G., & Rodet, L. (1997). Categorization creates functional features. *Journal of Experimental Psychology. Learning, Memory, and Cognition*, *23*(3), 681–696.

Schyns, P. G., Thut, G., & Gross, J. (2011). Cracking the code of oscillatory activity. *PLoS Biology*, *9*(5), 1–8.

Shulman, G. L., & Wilson, J. (1987). Spatial frequency and selective attention to local and global information. *Perception*, *16*, 89–101.

Smith, M., Cottrell, G., Sowden, P., & Schyns, P. G. (2005). Transmitting and decoding facial expressions of emotions. *Psychological Science*, *16*, 184–189.

Thorpe, S., Fize, D., & Marlot, C. (1996). Speed of processing in the human visual system. *Nature*, *381*(6582), 520–522.

van Rijsbergen, N. J., & Schyns, P. G. (2009). Dynamics of trimming the content of face representations for categorization in the brain. *PLoS Computational Biology*, *5*(11), 1–9.

VanRullen, R., & Thorpe, S. J. (2001a). Is it a bird? Is it a plane? Ultra-rapid visual categorization of natural and artifactual objects. *Perception*, *30*(6), 655–668.

VanRullen, R., & Thorpe, S. J. (2001b). The time course of visual processing: From early perception to decision making. *Journal of Cognitive Neuroscience*, *13*(4), 454–461.

Wilson, H. R., & Bergen, J. R. (1979). A four mechanism model for spatial vision. *Vision Research*, *19*, 1177–1190.

Wilson, H. R., & Wilkinson, F. (1997). Evolving concepts of spatial channels in vision: From independence to nonlinear interactions. *Perception*, *26*, 939–960.

Wong, E., & Weisstein, N. (1983). Sharp targets are detected better against a figure and blurred targets are detected better against a background. *Journal of Experimental Psychology. Human Perception and Performance*, *9*, 194–202.

3

The Constructive Nature of Scene Perception

Soojin Park and Marvin M. Chun

Humans have the remarkable ability to recognize complex, real-world scenes in a single, brief glance. The *gist*, the essential meaning of a scene, can be recognized in a fraction of a second. Such recognition is sophisticated, in that people can accurately detect whether an animal is present in a scene or not, what kind of event is occurring in a scene, as well as the scene category, all in as little as 150 ms (Potter, 1976; Schyns & Oliva, 1994; Thorpe, Fize, & Marlot, 1996; VanRullen & Thorpe, 2001). With this remarkable ability, the experience of scene perception feels effortless. It is ubiquitous as it is fundamental—after all, every image that comes into our brain is a scene. Scene perception directly impacts our actions in a 3D world by providing information about where we are as well as where we should navigate. This requires the integration of views across eye movements and across time to connect the present view with past memory. Thus, as effortless as it seems, scene perception involves many different levels of computation that integrate space, time, and memory. In this chapter we demonstrate the constructive nature of scene perception involving different brain regions to achieve a meaningful experience of the visual world.

The human visual system has three overarching goals in processing the visual environment. First, at the moment of physical input, the visual system must rapidly compute diagnostic properties of space and objects contained in the scene. Aside from recognizing faces and communicating with people, our daily activities require comprehension of the environment's spatial layout for the purposes of navigation as well as recognition of objects contained within that environment. As you view a particular scene, you are rapidly computing its spatial structure: determining where buildings are located and identifying paths through which you might navigate. At the same time, you can recognize a scene as a part of the broader environment and as a familiar scene in your memory. Visual scene understanding thus involves integrating a series of computations to enable coherent and meaningful scene perception.

Spatial structure, landmarks, and navigational paths are the major structural properties that define a scene. Recognizing these different structural properties is central to scene perception. Scenes with similar sets of structural properties will be grouped into similar scene categories. For example, if two scenes both have an open spatial layout, natural content, and strong navigability, they will both be categorized as fields. On the other hand, if two scenes have some overlapping structural properties but differ largely in other properties, they will be categorized differently. For example, if both scenes have an open spatial layout, but one has urban content and the other has natural content, they will be categorized differently (e.g., a highway vs. a field). Thus, a scene category is defined by combinations of different structural features (e.g., spatial layout and objects), and these structural features dictate how the viewer will recognize the space and function within it. In the first part of the chapter we examine how the brain represents structural property dimensions of scenes.

If the initial problem of scene recognition involves integrating multiple structural properties into a representation of a single view of a scene, then the second major challenge for the visual system is the problem of perceptual integration. To describe this problem, we should define the following terms—*view, scene,* and *place*—which depend on the observer's interactions with the environment (Oliva, Park, & Konkle, 2011). When an observer navigates in the real world, the observer is embedded in a space of a given "place," which is a location or landmark in the environment and often carries semantic meaning (e.g., the Yale campus, my kitchen). A "view" refers to a particular viewpoint that the observer adopts at a particular moment in one fixation (e.g., a view of the kitchen island counter when standing in front of the refrigerator), and a "scene" refers to the broader extension of space that encompasses multiple viewpoints. For example, a scene can be composed of multiple viewpoints taken by an observer's head or eye movements (e.g., looking around your kitchen will reveal many views of one scene). Visual input is often dynamic, as the viewer moves through space and time in the real environment. In addition, our visual field is spatially limited, causing the viewer to sample the world through constant eye and head movement. Yet, in spite of this succession of discrete sensory inputs, we perceive a continuous and stable perceptual representation of our surroundings. Thus, the second challenge for scene recognition is to establish coherent perceptual scene representations from discrete sensory inputs. Specifically, this involves the balancing of two opposing needs: each view of a scene should be distinguished separately to infer the viewer's precise position and direction in a given space, but these disparate views must be linked to surmise that these scenes are part of the same broader environment or "place." In the second part of this chapter we discuss how the human visual system represents an integrated visual world from multiple discrete views that change over time. In particular, we focus on different functions of the parahippocampal place area (PPA) and retrosplenial complex (RSC) in representing and integrating multiple views of the same place.

A third challenge for the visual system is to mentally represent a scene in memory after the viewer moves away from a scene and the perceptual view of the scene has disappeared. We often bring back to our mind what we just saw seconds ago, or need to match the current view with those in memory that reflect past experience. Such memory representations can closely reflect the original visual input, or they may be systematically distorted in some way. In the last part of the chapter we describe studies that test the precise nature of scene memory. In particular, we show that the scene memory is systematically distorted to reflect a greater expanse than the original retinal input, a phenomenon called *boundary extension*.

These complex visual and memory functions are accomplished by a network of specialized cortical regions devoted to processing visual scene information (figure 3.1, plate 2). Neuroimaging studies of scene recognition have provided insight about the functioning of these specialized cortical regions. Among them, the most well-known region is the parahippocampal place area (PPA) near the medial temporal region, which responds preferentially to pictures of scenes, landmarks, and spatial layouts depicting 3D space (Aguirre, Zarahn, & D'Esposito, 1998; Epstein, Harris, Stanley, & Kanwisher, 1999; Epstein & Kanwisher, 1998; Janzen & Van Turennout, 2004). The PPA is most sensitive to the spatial layout or 3D structure of an individual scene, although some recent work suggests that the PPA also responds to object information such as the presence of objects in a scene (Harel, Kravitz, & Baker, 2013), large real-world objects (Konkle & Oliva, 2012), and objects with strong context (Aminoff, Kveraga, & Bar, 2013). The complexity and richness of the PPA representation are discussed further under Representing Structural Properties of a Scene.

The PPA has been one of the most studied regions to represent "scene category-specific" information; however, more recent findings suggest that there is a family of regions that respond to scenes beyond the PPA, including the retrosplenial cortex and the transverse occipital sulcus. The retrosplenial complex (RSC), a region superior to the PPA and near the posterior cingulate, responds strongly to scenes compared to other objects (just as the PPA does). Yet, the RSC shows unique properties that may be important for spatial navigation rather than visual analysis of individual scenes (Epstein, 2008; Park & Chun, 2009; Vann, Aggleton, & Maguire, 2009). For example, the RSC shows relatively greater activations than the PPA for route learning in a virtual environment, mentally navigating in a familiar space, and recognizing whether a scene is a familiar one in memory (Epstein, 2008; Ino et al., 2002; Maguire, 2001). The section on Integrating a View to a Scene focuses on comparing the different functions of the PPA and RSC. The transverse occipital sulcus (TOS) also responds selectively to scenes compared to other visual stimuli. Recent findings suggest that the TOS is causally involved in scene recognition and is sensitive to mirror-reversal changes in scene orientation, whereas the PPA is not (Dilks, Julian, Kubilius, Spelke, & Kanwisher, 2011; Dilks, Julian, Paunov, & Kanwisher, 2013). Finally, in contrast to the regions above that prefer scenes over objects, the lateral occipital complex (LOC) represents object shape and category (Eger, Ashburner, Haynes, Dolan, & Rees, 2008; Grill-Spector, Kushnir, Edelman, Itzchak, & Malach, 1998; Kourtzi &

Kanwisher, 2000; Malach et al., 1995; Vinberg & Grill-Spector, 2008). Because scenes contain objects, we also consider the role of the LOC in representing the object contents and object interactions in a scene.

The goal of this chapter is to review studies that characterize the nature of scene representation within each of these scene-sensitive regions. In addition, we address how the functions of scene-specific cortical regions are linked at different stages of scene integration: structural construction, perceptual integration, and memory construction.

We propose a theoretical framework showing distinct but complementary levels of scene representation across scene-selective regions (Park & Chun, 2009; Park, Chun, & Johnson, 2010; Park, Intraub, Yi, Widders, & Chun, 2007), illustrated in figure 3.1 (plate 2). During navigation and visual exploration different physical views are perceived, and the PPA represents the visuostructural property of each view separately (Epstein & Higgins, 2007; Epstein & Kanwisher, 1998; Goh et al., 2004; Park, Brady, Greene, & Oliva, 2011; Park & Chun, 2009), encoding the geometric properties of

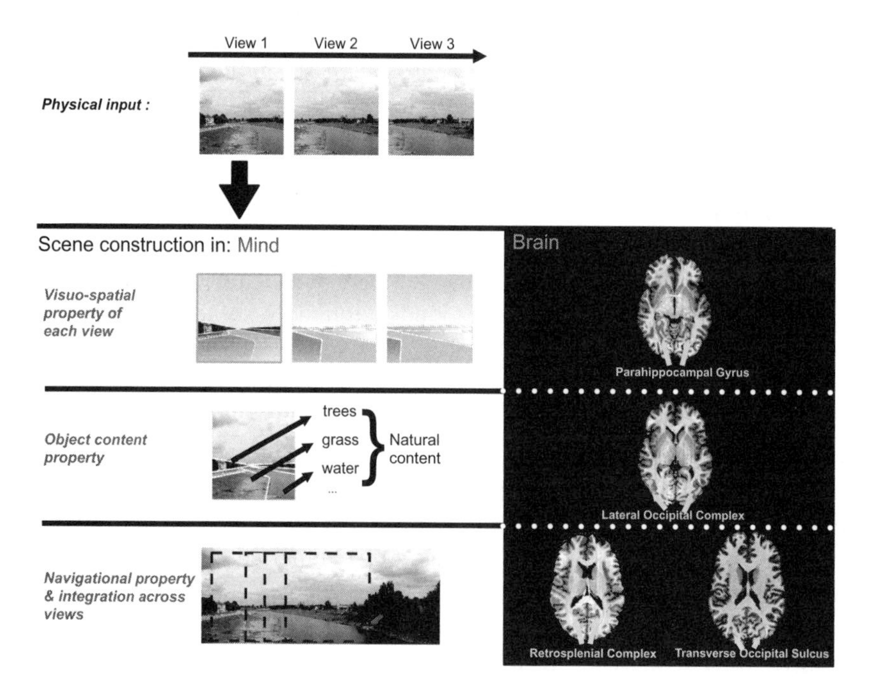

Figure 3.1 (plate 2)
A schematic illustration of three levels of scene processing. As the viewer navigates in the world, different views of scenes enter the visual system (view 1, view 2, view 3). The PPA treats each view of scenes as different from the others and is involved in analyzing the spatial properties of each specific view, such as the spatial layout and structure. The LOC processes object content properties in a scene, such as whether scenes have natural or urban content. The RSC and TOS analyze the navigationally relevant functional properties of a scene, creating an integrated representation of a scene across views.

scenes such as perspective, volume, and open/closed spatial layout, regardless of what types of objects fill in the space (Kravitz, Peng, & Baker, 2011; Park, Brady, et al., 2011; Park, Konkle, & Oliva, 2014). In parallel, the LOC represents the object properties in a scene, such as whether the scene has natural content (e.g., trees and vegetation) or whether the scene has urban content (e.g., buildings and cars; Park, Brady, et al., 2011). None of these regions represents scenes solely based on semantic category; for example, a city street and a forest will be represented similarly in the PPA as long as they have similar spatial layout, regardless of their differing semantic categories (Kravitz et al., 2011; Park, Brady, et al., 2011). The RSC represents scenes in an integrated/view-independent manner, treating different views that are spatiotemporally related as the same scene (Epstein & Higgins, 2007; Park & Chun, 2009). Given its involvement in spatial navigation in humans and rodents (Kumaran & Maguire, 2006), the RSC may also represent a scene's functional properties, such as how navigable a scene is, how many possible paths there are, or what actions the observer should take within the environment. The TOS may also represent the navigability of a scene, given that this region is sensitive to mirror-reversal changes of scenes, which alter the direction of a path (e.g., a path originally going to the left now will become a path going to right; Dilks et al., 2011). This pattern of response is similar to that of the RSC but different from that of the PPA, which does not show any sensitivity to mirror-reversal changes.

In the current chapter we present evidence that demonstrates how the distinct regions illustrated in figure 3.1 (plate 2) play a complementary role in representing the scene at the visuostructural level, perceptual integration level, and memory level.

Representing Structural Properties of a Scene

People are good at recognizing scenes, even when these scenes are presented very rapidly (Potter, 1975; also see chapter 9 by Potter in this volume). For example, when a stream of images is presented at a rapid serial visual presentation rate of around 100 ms per item, people can readily distinguish if a natural forest scene appeared among a stream of urban street images (Potter, 1975; Potter, Staub, & O'Connor, 2004). Even though people are able to recognize objects in rapidly presented scenes such as "trees," what subjects often report is in the basic-level category of a scene, such as a forest, beach, or a field (Rosch, 1978). Thus, one might assume that scenes are organized in the brain according to basic-level categories, with groups of neurons representing forest scenes, field scenes, and so on. However, recent computational models and neuroimaging studies suggest that the visual system does not classify scenes as belonging to a specific category per se but rather according to their global properties, that is, their spatial structure (Hoiem, Efros, & Hebert, 2006; Torralba & Oliva, 2003; Torralba, Oliva, Castelhano, & Henderson, 2006).

Object information and the spatial structure of a scene are extracted separately but in parallel (Oliva & Torralba, 2001) and then are later integrated to arrive at a decision about the identity of the scene or where to search for a particular object. In other words, when the visual system confronts a scene, it first decomposes the input into multiple layers of information, such as naturalness of object contents, density of texture, and spatial layout. This information is later combined to give rise to a meaningful scene category (in this example, a forest). Behavioral studies also suggest that object and scene recognition take place in an integrated manner (Davenport & Potter, 2004; Joubert, Rousselet, Fize, & Fabre-Thorpe, 2007). Target objects embedded in scenes are more accurately identified in a consistent than an inconsistent background, and scene backgrounds are identified more accurately when they contain a consistent rather than inconsistent object (Davenport & Potter, 2004; Loftus & Mackworth, 1978; Palmer, 1975). We also almost never see objects devoid of background context, and many scenes are in fact defined by the kinds of objects they contain—a pool table is what makes a room a pool hall, and recognizing a pool hall thus involves the recognition of the pool table in it, in addition to the indoor space around it. Taken together, these facts indicate that objects and scenes usefully constrain one another and that any complete representation of a visual scene must integrate multiple levels of these separable properties of spatial layout and object content.

Natural scenes can be well described on the basis of global properties such as different degrees of openness, expansion, mean depth, navigability, and others (Greene & Oliva, 2009b; Oliva & Torralba, 2006). For example, a typical "field" scene has an open spatial layout with little wall structure, whereas a typical "forest" scene has an enclosed spatial layout with strong perspective of depth (figure 3.2, plate 3). In addition, a field has natural objects or textures such as grass and trees, and a forest scene typically has natural objects such as trees, rocks, and grass. Similarly, urban scenes such as a street or highway can also be decomposed according to whether the scene's horizon line is open and visible (e.g., highway) or enclosed (e.g., street), in addition to its manmade contents (e.g., cars, buildings). We recognize a field as belonging to field category and a street as belonging to a street category because the visual system immediately computes the combination of structural scene properties (e.g., spatial layout and object content). The combination of such scene properties thus constrains how we interact with scenes or navigate within them.

In the example above we mentioned the spatial and object dimensions of a scene, but it is worth noting that real-world scenes have much higher degrees of complexity and dimensionality of structural information (Greene & Oliva, 2009a, 2009b; Oliva & Torralba, 2006). In a complex real-world scene these numerous properties are often entangled and are difficult to examine separately. Indeed, most investigations concerning the neural coding of scenes have focused on whether brain regions respond to one type of category-specific stimulus compared to others (e.g., whether the PPA

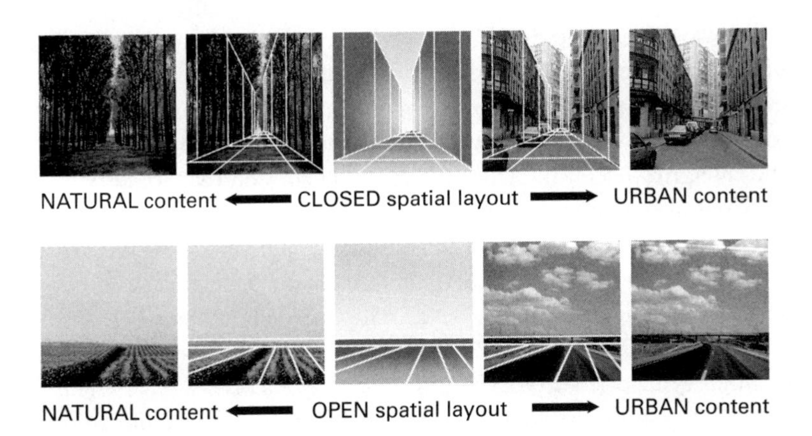

Figure 3.2 (plate 3)
A schematic illustration of spatial layout and content properties of scenes. Note that the spatial layout can correspond between natural and urban scenes. If we keep the closed spatial layout and fill in the space with natural contents, the scene becomes a forest, whereas if we fill in the space with urban contents, the scene becomes an urban street scene. Likewise, if we keep the open spatial layout and fill in the space with natural contents, the scene becomes a field; if we fill in the space with urban contents, the scene becomes a highway scene. Figure adapted from Park et al. (2011).

responds to a field vs. forest or whether LOC responds to a cut-out tree on a blank background). However, such category-specific representation may be a product of how the visual system reduces the complex dimensionality of a visual scene into a tractable set of scene categories. Thus, it is important to identify the precise dimensions in which neurons in scene-selective visual areas encode scene information.

An initial step to study scene processing in the brain should involve examining if scene categories are even represented to start with. After all, scenes in the same category (e.g., two scenes in the field category) are the scenes that share the most similar spatial and object properties (e.g., both scenes have open spatial layout, similar expansion, and natural contents and textures). Research has demonstrated that scene-responsive cortical regions such as the PPA and RSC represent the semantic category of scenes. Walther, Caddigan, Fei-Fei, and Beck (2009) used multivoxel analysis to test if patterns of fMRI activity in scene-selective cortices could classify six different natural scene categories (beach, buildings, forests, highways, industry, and mountains). Analysis of patterns of neural activity can offer more precise information about representation in a particular brain region compared to conventional methods, which average activity across voxels (Cox & Savoy, 2003; Kamitani & Tong, 2005). Machine learning methods, such as support-vector machine (SVM) classification, enable classification of different patterns of activity associated with different categories of scenes. Walther et al. (2009) found high classification performance in the PPA and RSC for distinguishing scene categories. Interestingly, they ran a separate behavioral study to measure errors in categorizing these scenes when presented very briefly (e.g., miscategorizing a highway scene as a beach). These behavioral error patterns were then compared to fMRI multivoxel classification error patterns, and a strong correlation was found between the two. In other words, scenes that had similar patterns of brain activity (e.g., beaches and highways) were scenes that were often confused in the behavioral scene categorization task. This elegant study showed that scene representations in the PPA reflect semantic categories and that scenes that are behaviorally confusable have similar patterns of voxel activity in this region.

What are the similarities across scene categories that made particular scenes highly confusable both behaviorally and at the neural level? The confusability between scene categories may be due to similarity in their spatial layouts (e.g., open spaces with a horizontal plane), similarity among the types of objects contained in these scenes (e.g., trees, cars, etc.), or similarity in the everyday function of scenes (e.g., spaces for transportation, spaces for social gatherings). Determining what types of scenes are systematically confused with one other can reveal whether a brain region represents spatial properties or object properties. Park et al. (2011) directly tested for such confusion errors using multivoxel pattern analysis. They asked whether two different properties of a scene, such as its spatial layout and its object content, could be dissociated within a single set of images. Instead of asking whether the PPA and LOC could accurately represent different categories of scenes, they focused on the confusion errors of a multivoxel classifier to examine whether scenes were confused based on similarity in spatial layout or object contents. There were four types of scene groups defined by spatial layout and object content (figure 3.3, plate 4: open natural scenes, open urban scenes, closed natural scenes, and closed urban scenes). Open versus closed defined whether the scene had an open spatial layout or a closed spatial layout. The natural versus urban distinction defined whether the scene had natural or urban object contents. Although both the PPA and LOC had similar levels of accurate classification performance, the patterns of confusion errors were strikingly different. The PPA made more confusion errors across images that shared the same spatial layout, regardless of object contents, whereas the LOC made more confusion errors across images that shared similar objects, regardless of spatial layout. Thus, we may conclude that a street and a forest will be represented similarly in the PPA as long as they have similar spatial layout, even though a street is an urban scene and a forest is a natural scene. On the other hand, a forest and field scene will be represented similarly in the LOC because they have similar natural contents.

Another study computed a similarity matrix of 96 scenes and also found that PPA representations are primarily based on spatial properties (whether scenes have open spatial layout vs. closed spatial layout), whereas representations in early visual cortex (EVC) are primarily based on the relative distance to the central object in a scene (near vs. far; Kravitz et al., 2011). Using a data-driven approach, the authors measured multivoxel patterns for each of 96 individual scenes. They then cross-correlated these response patterns to establish a similarity matrix between each pair of scenes.

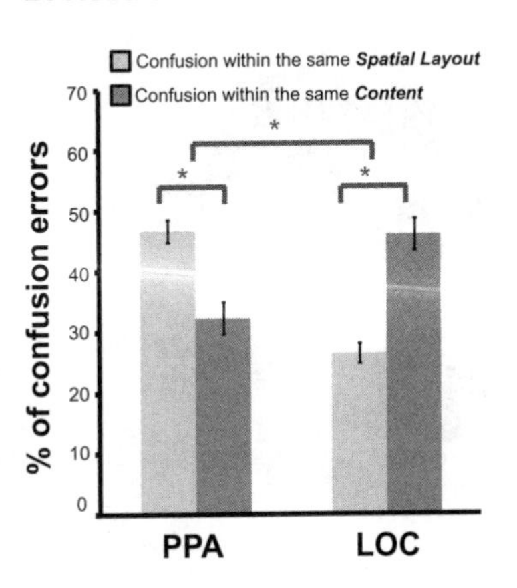

Figure 3.3 (plate 4)
(A) Hypothetical patterns of confusion errors based on the spatial layout or object content similarity. The rows represent the scene image conditions as presented to the participants, and the columns represent the scene condition that the classifier predicted from the fMRI patterns of activity. If spatial layout properties of scenes are represented in a particular brain area, we expect confusion within scenes that share the same spatial layout (marked in light gray). If content properties of scenes are important for classification, we expect confusion within scenes that share the same content (dark gray cells). (B) Confusion errors (percentage) are shown for the PPA and the LOC. Figure adapted from Park et al. (2011).

When the matrix was reorganized according to dimensions of space (open vs. closed), objects (natural vs. urban) and distance (near vs. far), there was a high correlation in the PPA for scenes that shared dimensions of space (figure 3.4A, plate 5), and high correlation in EVC for scenes that shared the dimension of distance. These results highly converge with those of Park et al. (2011), together suggesting that scene representations in the PPA and RSC are primarily based on spatial layout information and not scene category per se.

Park et al. (2011) and Kravitz et al. (2011) indicate that the PPA and LOC have relatively specialized involvement in representing spatial or object information. However, one should be careful in drawing conclusions about orthogonal or categorical scene representations across the PPA and LOC. The PPA does not exclusively represent spatial information, and the LOC does not solely represent object information. For example, Park, Brady et al. (2011) found above-chance levels of classification accuracy for four groups of scene types (open natural, open urban, closed natural, and closed urban) in both the PPA and LOC. To accurately classify these four groups of scenes, the PPA and LOC must encode both spatial layout (open vs. closed) and object information (natural vs. urban). Thus, even though the confusion error patterns suggest a preference for information concerning spatial layout in the PPA and a preference for object content information in the LOC, these functions are not exclusively specialized. In fact, scene information spans a gradient across ventral visual regions.

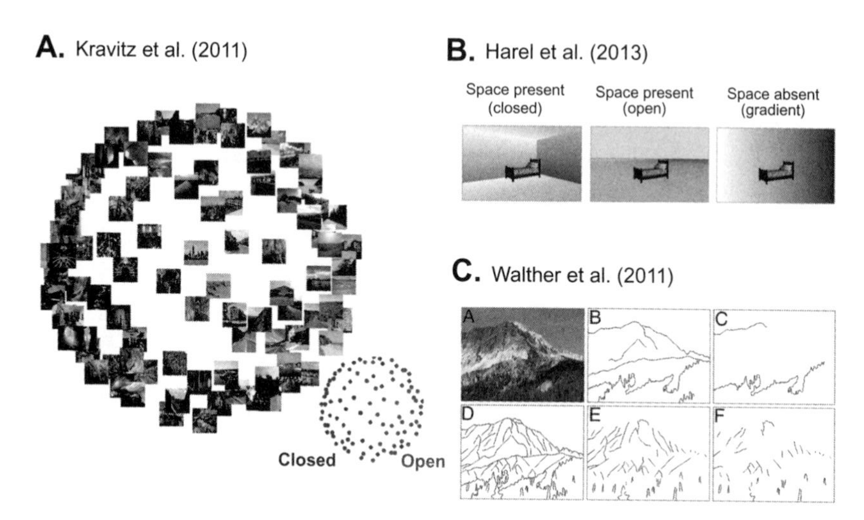

Figure 3.4 (plate 5)
(A) Multidimensional scaling plot for the PPA. Scenes are shown in a two-dimensional plane with the distance between pairs of scenes reflecting the correlation between their response patterns. Pairs of images that had higher correlations are shown closer together. Here, you can see that scenes that had similar spatial layout (closed or open) are clustered closely together (Kravitz, Peng, & Baker, 2011). (B) Stimuli used in Harel et al. (2013). Participants saw minimal scene stimuli that are composed of objects (e.g., furniture) combined with three different types of background stimuli (closed-space present, open-space present, and gradient-space absent). (C) Examples of line drawing images used in Walther et al. (2011). A corresponding line drawing is shown (D) for a photograph of a scene (A). This line drawing scene was degraded by either removing 50% of its pixels by removing local contours (short lines) (B) or global contours (long lines) (E); or by removing 75% of pixels by removing short (C) or long contours (F). The category identification performance was significantly impaired when global contours were removed (E and F) compared to when local contours were removed (B and C), suggesting that global spatial layout information is important. Figures adopted from Kravitz, Peng, and Baker (2011), Harel, Kravitz, and Balker (2013), and Walther et al. (2011).

Harel, Kravitz, and Baker (2013) manipulated spatial layout (e.g., open spatial layout, closed spatial layout, or no spatial layout) and object content (furniture present or absent; figure 3.4B, plate 5). They tested if the PPA, LOC, and RSC could correctly decode whether a scene background contained spatial layout information (space absence decoding) and whether a scene contained an object (object absence decoding). Multivoxel pattern analysis showed that RSC was able to decode whether a scene included spatial layout information but not whether a scene contained objects. In contrast, the LOC was able to decode whether a scene contained objects but not whether a scene's background included spatial layout information. The PPA was able to decode both whether the scene contained spatial layout information or object information. These results suggest that there is a gradient of representation: strong spatial representation with little object representation in the RSC; some spatial and some object representation in the PPA; and strong object representation with little spatial representation in the LOC.

Other studies have tested whether different cues for defining spatial layout information matter for scene categorization. Walther et al. (2011) suggests that scene categorization is largely based on the global structure of a scene, such as its global contours. To test whether global or local contours have different degrees of impact, Walther et al. (2011) selectively removed equal pixel amounts of global (long) or local (short) contours from a line drawing scene (figure 3.4C, plate 5). Participants performed significantly worse in identifying the categories of scenes that had global contours removed compared to scenes that had local contours removed, suggesting that global spatial layout information is more important for scene identification.

Although the studies described above have investigated spatial representation, other studies have focused on the representation of object properties in scenes. MacEvoy and Epstein (2011) were able to predict a scene category from multiple objects in the lateral occipital cortex (LO) but not in the PPA. That is, multivoxel patterns in the LO for a scene category (e.g., kitchen) were highly correlated with the average of the patterns elicited by signature objects (e.g., stove or refrigerator). These results support earlier views of scene perception, which held that real-world scene identification emerges by identifying a set of objects in it (Biederman, 1981; Friedman, 1979). However, a scene is not just a bag of objects or linear combinations of them but reflects the semantic co-occurrence or spatial composition between these objects. Objects often appear in meaningful spatial arrangements based on the functional or semantic relationship between them (e.g., a cup on a table; a pot pouring water into a cup). This interacting relationship enhances the identification of individual objects (Green & Hummel, 2006) and scenes (Biederman, 1981). Kim and Biederman (2011) tested how a collection of objects may be processed together as a scene. They asked whether a particular brain region encodes meaningful relationships among multiple objects. They showed that the LOC responds strongly to a pair of objects presented in an interacting position (e.g., a bird in front of a bird house) compared to a pair of objects presented side by side and not interacting in a semantically meaningful way. They did not find any preference for interacting objects in the PPA, consistent with the idea that the PPA does not care about object information (MacEvoy & Epstein, 2011). These studies suggest that the LO represents more than simple object shape and should be considered a scene-processing region, representing multiple objects and their relationships to one another. On the other hand, the PPA seems to represent geometric space beyond an object or multiple objects consistent with recent computational findings that suggest parallel processing of objects and spatial information (Fei-Fei & Perona, 2005; Lazebnik, Schmid, & Ponce, 2006; Oliva & Torralba, 2001).

Thus, both the PPA and LO contribute to scene processing: the PPA represents geometric aspects of space, and the LO represents multiple objects and their relationships. Although this suggests that the spatial and object properties of scenes are represented differently in distinctive brain regions, defining what constitutes an object property or spatial property can be ambiguous. Objects sometimes define a scene's spatial layout—for example, a fountain can be categorized as an object or as a landmark. Size, permanence, and prior experience with a particular object modulate whether it is treated as an object or a scene. When objects gain navigational (landmark) significance, they may be treated as scenes and activate the PPA. Janzen and Van Turennout (2004) had subjects view a route through a virtual museum while target objects were placed at either an intersection of a route (decision point relevant for navigation) or at simple turns (nondecision point). High PPA activity was found for objects at intersections, which were critical for navigation, compared to objects at simple turns, which were equally familiar but did not have navigational value. This finding suggests that prior experience with an object in a navigationally relevant situation transforms these objects to relevant landmarks, which activates the PPA. Konkle and Oliva (2012) also showed that large objects such as houses activate the PPA region more than small objects.

One can also ask whether the PPA and RSC differentiate landmark properties. Auger, Mullally, and Maguire (2012) characterized individual landmarks by multiple properties such as size, visual salience, navigational utility, and permanence. They found that the RSC responded specifically to the landmarks that were consistently rated as permanent, whereas the PPA responded equally to all types of landmarks. In addition, they showed that poor navigators, compared to good navigators, were less reliable and less consistent in their ratings of a landmark's permanence. Thus, the primary function of the RSC may be processing the most stable or permanent feature of landmarks, which is critical for navigation. Altogether, the above studies suggest that object and scene representations are flexible and largely modulated by object properties or prior interactions with an object, especially when the objects may serve a navigational function, which we discuss further in the next section.

Integrating a View to a Scene

Once an immediate view is perceived and a viewer moves through the environment, the visual system must now confront the problem of integration. There are two seemingly contradictory computational problems that characterize this process. First, the visual system has to represent each individual view of a scene as unique in order to maintain a record of the viewer's precise position and heading direction. At the same time, however, the visual system must recognize that the current view is a part of a broader scene that extends beyond the narrow aperture of the current view. Constructing such an integrated representation of the environment guides navigation, action, and recognition from different views. How does the brain construct such stable percepts of the world? In this section, we discuss how the human visual system perceives an integrated visual world from multiple specific views that change over time.

For this purpose, we focus on two scene-specific areas in the brain, the PPA and the RSC. Both of these regions may be located by using a scene localizer, exhibiting strong preference to scenes over other visual stimuli. However neurological studies with patients suggest that the PPA and the RSC may play different roles in scene perception and navigation. Patients who have damage to the parahippocampal area cannot identify scenes such as streets or intersections and often rely on identification of small details in a scene such as street signs (Landis, Cummings, Benson, & Palmer, 1986; Mendez & Cherrier, 2003). However, these patients are able to draw a map or a route that they would take in order to navigate around these landmarks (Takahashi & Kawamura, 2002). Another patient with PPA damage showed difficulty learning the structure of new environments but had spared spatial knowledge of familiar environments (Epstein, DeYoe, Press, Rosen, & Kanwisher, 2001). This contrasts with patients with RSC damage, who were able to identify scenes or landmarks but had lost the ability to use these landmarks to orient themselves or to navigate through a larger environment (Aguirre & D'Esposito, 1999; Maguire, 2001; Valenstein et al., 1987). For example, when patients with RSC damage saw a picture of a distinctive landmark near their own home, they would recognize the landmark but could not use this landmark to find their way to their house. These neurological cases suggest that the parahippocampal and retrosplenial areas encode different kinds of scene representations: the parahippocampal area may represent physical details of the view of a scene, and the retrosplenial area may represent navigationally relevant properties such as the association of the current view to other views of the same scene in memory. These functional differences in the PPA and RSC may account for two different approaches taken to explain visual integration across views. The PPA, with higher sensitivity to perceptual details of a scene, may encode specific features of each view individually. On the other hand, the RSC, with its involvement in navigationally relevant analysis of a scene, may encode spatial regularities that are common across views, representing the scene in a view-invariant way.

Park and Chun (2009) directly tested viewpoint specificity and invariance across the PPA and RSC. When the same stimulus is repeated over time, the amount of neural activation for the repeated stimulus is significantly suppressed in comparison to the activity elicited when it was first shown. This robust phenomenon, called repetition suppression, may be used as a tool to measure whether a particular brain region represents two slightly different views of scenes as the same or different (see Grill-Spector, Henson, & Martin, 2006). Park and Chun (2009) presented three different views from a single panoramic scene to mimic the viewpoint change that may occur during natural scanning (for example, when you move your eyes from the left to the right corner of a room; figure 3.5, plate 6). If scene representations in the brain are view specific, then physically different views of the same room will be treated differently, so that no repetition suppression will be observed. Conversely, if scene

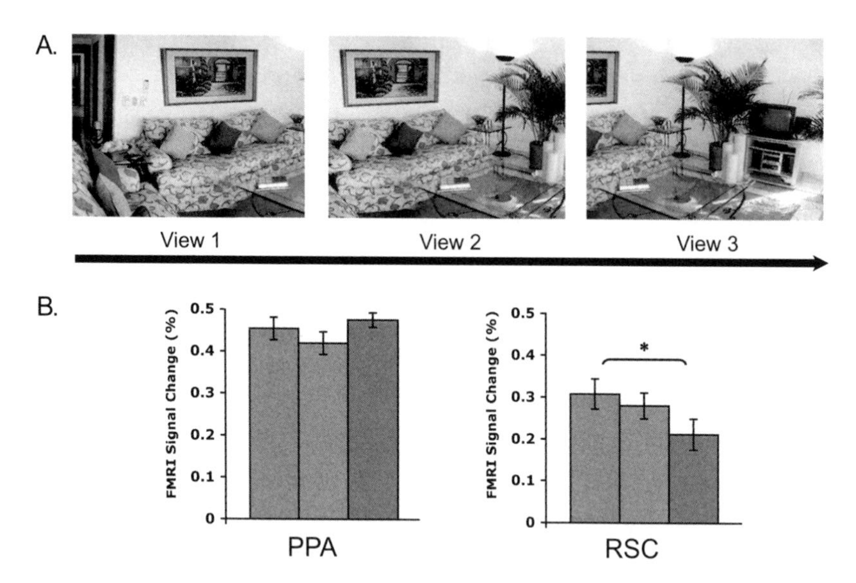

Figure 3.5 (plate 6)
(A) Example of panoramic first, second, and third images. These views were taken from a single panoramic scene. These panoramic scenes were presented in order at fixation. The PPA panoramic third image was taken from a single panoramic view. Panoramic first, second, and third images were sequentially presented one at a time at fixation. (B) Mean peak hemodynamic responses for panoramic first, second, and third in the PPA and RSC. The PPA showed no repetition suppression from the first to the third panoramic image, suggesting view specificity, whereas the RSC showed a significant repetition suppression, suggesting scene integration. Figure adapted from Park and Chun (2009).

representations in the brain are view invariant, then these views will be integrated into the representation of a single continuous room, yielding repetition suppression for different views from the same scene. The results revealed that the PPA exhibits view specificity, suggesting that this area focuses on selective discrimination of different views, whereas the RSC shows view-invariance, suggesting that RSC focuses on the integration of scenes under the same visual continuity. Viewpoint specificity in the PPA is supported by previous literature (Epstein, Graham, & Downing, 2003; Epstein & Higgins, 2007), and viewpoint integration in RSC fits with its characterization as an area that is important in navigation and route learning in humans and rodents (Burgess, Becker, King, & O'Keefe, 2001; Aguirre & D'Esposito, 1999; see also Vann et al., 2009 for review). This finding of two distinct but complementary regions in scene perception suggests that the brain develops ways to construct our perception with both specificity and stability from fragmented visual input. In addition, the experiment showed that spatiotemporal continuity across multiple views is critical to build an integrated scene representation RSC. When different views of panoramic scenes were presented with a long lag and intervening items, the RSC no longer showed patterns of neural attenuation consistent with scene integration. Thus, the continuous percept of time and space across changing views provides important cues for building a coherent visual world.

Other researchers have also found that the PPA and RSC distinctively represent individual scenes as components of broader unseen spaces. Epstein, Parker, and Feiler (2007) tested whether a specific view of a scene (e.g., a view of school library) is represented neurally as part of a broader real-world environment beyond the viewer's current location (e.g., the whole campus). In their study they presented participants from the University of Pennsylvania community with views of familiar places around the campus or views from a different, unfamiliar campus. Participants judged either the location of the view (e.g., whether the view of a scene is on the west or east of a central artery road through campus) or its orientation (e.g., whether the view is facing west or east of the campus). The PPA responded equally to all conditions regardless of the task, but the RSC showed stronger activation to location judgments compared to orientation judgments. The location judgment required information about the viewer's current location as well as the location of the current scene within the larger environment. The RSC also showed much higher activity for familiar scenes than for unfamiliar scenes. Thus, the RSC is involved in the retrieval of specific location information of a view and how this view is situated relative to the surrounding familiar environment.

In a related vein, researchers found different levels of specificity and invariance across other scene selective areas including the transverse occipital sulcus (TOS). The TOS specifically responds to scenes compared to objects and often shows up along with the PPA and RSC in scene localizers. It is more posterior and lateral and is also often referred to as an occipital place area. Dilks et al. (2011) tested mirror-viewpoint change sensitivity in object- and scene-specific brain areas. When a scene image is mirror-reversed, the navigability of the depicted scene changes fundamentally as a path in the scene will reverse direction (e.g., a path originally going to the left now will become a path going to the right). Using repetition suppression they found that the RSC and the TOS were sensitive to mirror-reversals of scenes, treating two mirror-reversed scenes as different from each other. On the other hand, they found that the PPA was invariant to mirror-reversal manipulations, which challenges the idea that the PPA is involved in navigation and reorientation. Although these results seemingly contradict other findings showing viewpoint specificity in the PPA, they fit with the idea that the PPA represents the overall spatial layout of a given view, which is unchanged by mirror-reversal, as an image that has a closed spatial layout will remain as a closed scene; an open spatial layout will remain the same. What the mirror reversal changes is the functional navigability or affordance within a scene, such as in which direction the viewer should navigate. Thus, it makes sense that mirror-reversals did not affect the PPA, which represents visuospatial layout, but they affected the RSC, which represents the navigational properties of a scene. The function of the TOS is still not well known, although recent research with transcranial magnetic stimulation (TMS) over the TOS suggests that the TOS is causally involved in scene perception (Dilks et al., 2013). Dilks et al. delivered TMS to the TOS and to the nearby face-selective occipital face area (OFA) while participants performed discrimination tasks involving either scenes or faces. Dilks et al. found a double dissociation, in that TMS to the TOS impaired discrimination of scenes but not faces, whereas TMS to the OFA impaired discrimination of faces but not scenes. This finding suggests that the TOS is causally involved in scene processing, although the precise involvement of TOS is still under investigation.

Another related question is whether scene representations in the PPA, RSC, and TOS are specific to retinal input. MacEvoy and Epstein (2007) presented scenes to either the left or right visual hemifields. Using repetition suppression, they tested if identical scenes repeated across different hemifields are treated as the same or differently in the brain. They found position invariance in the PPA, RSC, and TOS, suggesting that these scene-selective regions contain large-scale features of the scene that are insensitive to changes of retinal position. In addition, Ward et al. (2010) found that when stimuli are presented at different screen positions while fixation of the eyes is permitted to vary, the PPA and TOS respond equally to scenes that are presented at the same position relative to the point of fixation but not to scenes that are presented at the same position relative to the screen. This suggests an eye-centered frame of reference in these regions. In another study that controlled fixations within a scene, Golomb et al. (2011) showed that active eye movements by the viewer play an important role in scene integration. Stimuli similar to those depicted in figure 3.5A (plate 6) were used. The PPA showed repetition suppression to successive views when participants actively made saccades across a stationary scene (e.g., moving their eyes from left, middle, and right fixation points embedded in a scene) but not when the eyes remained fixed and a scene scrolled in the background across fixation, controlling for local retinal input between the two manipulations. These results suggest that active saccades may play an important role in scene integration, perhaps providing cues for retinotopic overlap across different views of the same scene.

So far in this chapter, we have focused on the parahippocampal and retrosplenial cortices. However, it is important to mention the role of the hippocampus in scene and space perception. Functional connectivity analysis suggests that parahippocampal and retrosplenial regions have strong functional connectivity with the hippocampus and other medial temporal regions such as the entorhinal and perirhinal cortices (Rauchs et al., 2008; Summerfield, Hassabis, & Maguire, 2010). A long history of rodent work has demonstrated hippocampal involvement in spatial cognition, such as maze learning and construction of a "cognitive map," a mental representation of one's spatial environment, in the hippocampus (Knierim & Hamilton, 2010; O'Keefe & Nadel, 1979). In particular, hippocampal neurons provide information about both the rat's external and internal coordinate systems. Place cells are one type of hippocampal neuron that fires when a rat is at a specific location defined by an external coordinate system. Head-direction cells fire when the rat's head is oriented at a certain direction in the rat's internal coordinate system. The hippocampus also contains boundary cells, which respond to the rat's relative distance to an environmental boundary (Bird & Burgess, 2008; O'Keefe & Burgess, 1996). These findings suggest that the hippocampus is a critical region that represents the viewer's position in the external world. In addition, the division of labor described above for the PPA and RSC is interesting to think about in relation to computational models of hippocampal function in rats. Recent studies suggest that pattern separation, which amplifies differences in input, and pattern completion, which reconstructs stored patterns to match with current input, occur in different parts of the hippocampus: CA3/DG is involved in pattern separation, whereas CA1 is involved in pattern completion (see Yassa & Stark, 2011, for review). Even though it is difficult to make a direct comparison between hippocampal subregions and outer cortical regions such as the parahippocampal and retrosplenial regions, these complementary functions found in the rodent hippocampus seem to correspond to the complementary functions found in the PPA and RSC. For example, the PPA may rely on pattern separation to achieve view specificity, and the RSC may perform pattern completion to enable view integration.

Recently, fMRI studies have probed for cognitive map-like representations in human hippocampus. Morgan et al. (2011) scanned participants while viewing photographs of familiar campus landmarks. They measured the real-world (absolute) distance between pairs of landmarks and tested whether responses in the hippocampus, the PPA, and the RSC were modulated by the real-world distance between landmarks. They found a significantly attenuated response in the left hippocampus for a pair of landmarks that are closer in the real world compared to a pair of landmarks that are farther from one another in the real world. In contrast, the PPA and RSC encoded the landmark identity but not the real-world distance relationship between landmarks (Morgan et al., 2011). These results suggest that the hippocampus encodes landmarks in a map-like representation, reflecting relative location and distance between landmarks. Another study using multivoxel pattern analysis with high-spatial-resolution fMRI found that the position of an individual within an environment was predictable based on the patterns of multivoxel activity in the hippocampus (Hassabis et al., 2009). In this experiment participants navigated in an artificially created room that had four different corners (corners A–D). In each trial the participants navigated to an instructed target position (e.g., go to the corner A). When they reached the corner, they pressed a button to adopt a viewpoint looking down, which revealed a rug on the floor. This rug view visually looked the same across four corners; thus, the multivoxel activity collected during this period was based on the viewer's position in a room and not on any visual differences between the four corners. Multivoxel patterns in the hippocampus enabled classification of which of the four corners the participant was positioned. These results are similar to the rat's place cells, which fire in response to the specific location of the rat within an environment. With high-resolution fMRI techniques and computational models that enable segmentation of hippocampal subregions, future research should be aimed at identifying whether the human hippocampus, like that of the rat, also contains both external and internal coordinate systems facilitated by place cells and head direction cells as well as boundary cells that encode the viewer's distance to environmental boundaries.

Representing Scenes in Memory

We sample the world through a narrow aperture that is further constrained by limited peripheral acuity, but we can easily extrapolate beyond this confined window to perceive a continuous world. The previous section reviewed evidence that coherent scene perception is constructed by integrating multiple continuous views. Such construction can occur online but can also occur as we represent scene information in memory that is no longer in view. In this section we discuss how single views are remembered and how the visual system constructs information beyond the current view. Traditionally, the constructive nature of vision has been tested in low-level physiological studies, such as filling in of the retinal blind spot or contour completion. However, less is known about what type of transformations or computations are performed in higher-level scene-processing regions. Yet, expectations about the visual world beyond the aperture-like input can systematically distort visual perception and memory of scenes. Specifically, when people are asked to reconstruct a scene from memory, they often include additional information beyond the initial boundaries of the scene, in a phenomenon called *boundary extension* (Intraub, 1997, 2002; also see chapter 1 by Intraub in this volume). The boundary-extension effect is robust across various testing conditions and various populations, such as recognition, free recall, or directly adjusting borders of the boundary both visually and haptically (Intraub, 2004, 2012). Boundary extension occurs in children and infants as well (Candel, Merckelbach, Houben, & Vandyck, 2004; Quinn & Intraub, 2007; Seamon, Schlegel, Hiester, Landau, & Blumenthal, 2002). Interestingly, boundary extension occurs for scenes with background information but not for scenes comprising cutout objects on a black screen (Gottesman & Intraub, 2002). This systemic boundary extension error suggests that our visual system is constantly extrapolating the boundary of a view beyond the original sensory input (figure 3.6).

Boundary extension is a memory illusion, but this phenomenon has adaptive value in our everyday visual experience. It provides an anticipatory representation of the upcoming layout that may be fundamental to the integration of successive views. Using boundary extension, we can test whether a scene is represented in the brain as it is presented in the physical input or as an extended view that observers spatially extrapolated in memory. Are there neural processes in memory that signal the spatial

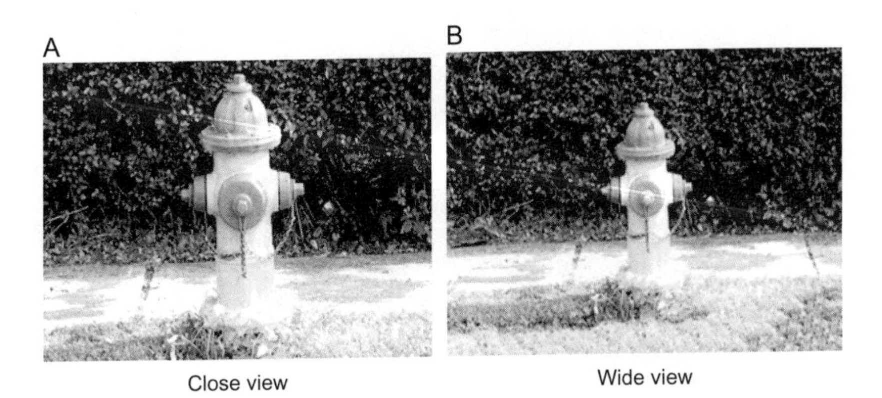

Figure 3.6
Example of boundary extension. After viewing a close-up view of a scene (A), observers tend to report an extended representation (B).

extrapolation of physically absent but mentally represented regions of a scene? If so, this would demonstrate that higher-level scene-processing areas such as the PPA and RSC facilitate the perception of a broader continuous world through the construction of visual scene information beyond the limits of the aperture-like input.

Park et al. (2007) tested for such effects of boundary extension. They used fMRI repetition suppression for close views and wide views of scenes to reveal which scene pairs were treated as similar in scene regions of the brain. When the original view of a scene is a close-up view, boundary extension predicts that this scene will be extended in memory and represented as a wider view than the original. Thus, if the same scene is presented with a slightly wider view than the original, this should match the boundary-extended scene representation in scene-selective areas and should result in repetition suppression. On the other hand, if a wide view of a scene is presented first, followed by a close view (wide-close condition), there should be no repetition suppression even though the perceptual similarity between close-wide and wide-close repetitions is identical. This asymmetry in neural suppression for close-wide and wide-close repetition was exactly what Park et al. (2007) observed (figure 3.7). Scene-processing regions such as the PPA and RSC showed boundary extension in the form of repetition suppression for close-wide scene pairs but not for wide-close scene pairs. In contrast, there were no such asymmetries in the LOC. This reveals that the brain's scene-processing regions reflect a distorted memory representation, and such boundary extension is specific to background scene information and not to foreground objects. Such extended scene representations may reflect an adaptive mechanism that allows the visual system to perceive a broader world beyond the sensory input.

Another fMRI study on boundary extension points to further involvement of the hippocampus (Chadwick, Mullally, & Maguire, 2013). Online behavioral boundary

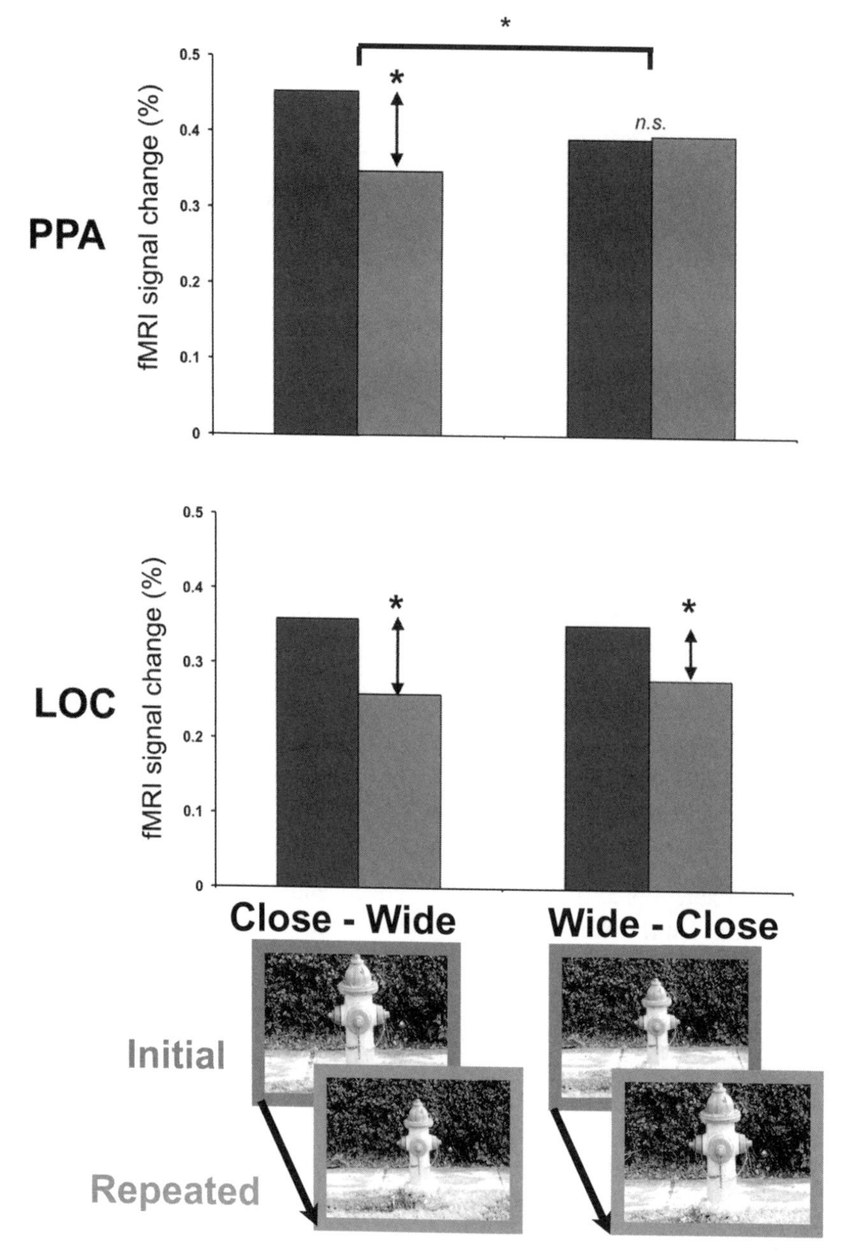

Figure 3.7
Peaks of the hemodynamic responses for close-wide and wide-close conditions are shown for the PPA and LOC. Examples of close-wide and wide-close condition are presented at the bottom. An interaction between the close-wide and wide-close condition activation, representing boundary extension asymmetry, was observed in the PPA but not in the LOC. Figure adapted from Park et al. (2007).

effects were measured for individual scenes as participants viewed scenes in the scanner. When scenes that showed the boundary extension effect were compared to scenes that did not show it, there was a significant difference in activity in the hippocampus. Moreover, functional connectivity analysis showed that scenes with boundary extension had high connectivity between the hippocampus and the parahippocampal cortex, whereas scenes without boundary extension effect did not. A neurological study with patients who have hippocampal damage also found that these patients had less or no boundary extension compared to a control group (Mullally, Intraub, & Maguire, 2012). For example, when the same close view of a scene was repeated following a close view, normal controls would respond that the second close view was different from the original, showing the usual boundary extension distortion. However, patients with hippocampal damage were more accurate at rating the second close view as identical to the original, showing no distortion from boundary extension. These are intriguing results because patients with hippocampal damage actually showed more accurate scene memory than controls, immune from the boundary extension error. These results suggest that the hippocampus may play a central role in boundary extension, and hence, the boundary extension effect found in the parahippocampal cortex in Park et al. (2007) may reflect such feedback input from the hippocampus. Because boundary extension is an example of constructive representations of scenes in memory, these results further support the role of the hippocampus in the anticipation and construction of memory (Addis, Wong, & Schacter, 2007; Buckner & Carroll, 2007; Turk-Browne, Scholl, Johnson, & Chun, 2010). In addition, the boundary extension distortion should not simply be viewed as a memory error but rather as a successful adaptive mechanism that enables anticipation of a broader perceptual world from limited input.

Because the amount of visual information a human can see at one time is limited, we have also evolved mechanisms for bringing to mind recent visual information that is no longer present in the current environment. Such acts are called *refreshing* and occur when one briefly thinks back to a stimulus one just saw. The act of refreshing may facilitate scene integration by foregrounding the information to facilitate the binding of the previous and the current views. Given the potential role of refreshing in scene integration, Park, Chun, and Johnson (2010) asked whether discrete views of scenes are integrated during refreshing of these views. Similar to results found with physical scenes, when participants refreshed different views of scenes, the PPA showed view-specific representations, and the RSC showed view-invariant representations. Research directly comparing cortical activity for perception and refreshing showed that activity observed in the RSC and precuneus for refreshing closely mirrored the activity for perceiving in these regions (Johnson, Mitchell, Raye, D'Esposito, & Johnson, 2007). Thus, the act of refreshing in these high-level regions might play an important role during perceptual integration by mirroring the activity of perceiving panoramic views of scenes in continuation.

Conclusion

Constructing a rich and coherent percept of our surroundings is essential for navigating and interacting with our environment. How do we recognize scenes? In this chapter we reviewed key studies in cognitive neuroscience that investigated how we construct a meaningful scene representation at the structural, perceptual, and memory levels. Multiple brain regions play distinctive functions in representing different properties of scenes, and the PPA, RSC, and LOC areas represent a continuum of specialized processing for spatial properties, from navigational features (RSC and PPA) to that of diagnostic objects (LOC).

However, even the useful distinction between spatial and object representation may be an oversimplification. Real-world objects or scenes have enormous complexity and vary along an exceptionally high number of dimensions such as layout, texture, color, depth, density, and so on. The next major goal in the field will be to understand the precise neural processing mechanisms in the PPA, RSC, and LOC areas. An essential first step is to identify the dimensions in which neurons encode scene information. The enormous, megapixel dimensionality of a visual scene must be reduced by the ventral pathway to a tractable set of dimensions for encoding scene information. These coding dimensions must be flexible enough to support robust categorization but also sensitive to parametric variations necessary for discriminating different exemplars and specific views.

Scene research lags behind that for faces and objects. Electrophysiological recordings reveal neurons in the face-selective cortex that encode specific and parametric components of face parts, geometry, and configuration (Freiwald & Tsao, 2010; Freiwald, Tsao, & Livingstone, 2009) and object-selective regions that encode parametric dimensions of 2D contour, 3D surface orientation, and curvature of objects (Hung, Carlson, & Connor, 2012). Research in human IT cortex has also begun to show not only categorical but continuous representations of individual objects (Kriegeskorte et al., 2008). Similarly detailed representations at the neuronal level have yet to be discovered for scene perception. Hence, one of the next goals in the field of scene perception should be to test precise coding dimensions of scene-selective neurons reflecting continuous and parametric changes in the coding dimension (e.g., varying degrees of the size of space; varying degrees of openness in spatial layout). For example, do the PPA and RSC discriminate the size of space independent of the clutter or density of objects within a given space? Estimating the size of space and the level of clutter in a scene is central to our interactions with scenes—for example, when deciding whether or not to take a crowded elevator or when driving through downtown traffic. Park et al. (2011) varied both the size of space and levels of object clutter depicted within scenes and discovered that the anterior portions of the PPA and RSC responded parametrically to different sizes of space in a way that generalized across scene categories.

Another major goal in the field of scene understanding is to describe how information from multiple scene-selective regions, representing different properties of scenes, is synthesized. Scene categorization and scene gist recognition are so rapid and efficient, some of this information must be combined at very early stages of visual processing. How and where are these properties weighted and combined to enable scene categorization that rapidly occurs within 200 ms? Future research should aim to reveal the interaction across the family of scene-selective regions within the rapid time course of scene recognition.

How do we represent a coherent scene from constantly changing visual input? Converging evidence throughout this chapter suggests that the brain overcomes multiple constraints of our visual input by constructing an anticipatory representation beyond the frame of the current view. The visual system assumes what may exist just beyond the boundaries of a scene or what may exist when our eyes are successively moved to the next visual frame. Such assumptions are represented in high-level visual areas and produce a rich and coherent perceptual experience of the world. Is there a functional architecture in the brain that enables such extrapolation of scene information? The PPA represents scenes not just based on the current visual stimulus but within the temporal context in which these scenes were presented (Turk-Browne, Simon, & Sederberg, 2012). A scene that was embedded in a predictable temporal context had greater PPA repetition suppression than a scene that did not have any predictable temporal context preceding it. These results show that the PPA not only represents the present input but integrates predictable contexts created from the past. Such predictive coding found in high-level visual cortex may support navigation by integrating past and present input. Related to this, an important future direction for scene perception research would be to show how real-world scene perception unfolds over time. In many real-world circumstances, information at the current moment becomes meaningful only in the context of a past event. For example, if you present frames of a movie trailer in a randomized order, the whole trailer will be incomprehensible. Indeed, our brain is sensitive to sequences of visual information across different time scales (Hasson, Yang, Vallines, Heeger, & Rubin, 2008; Honey et al., 2012). When a meaningful visual event is presented over time (e.g., a movie clip), early visual areas such as V1 are involved in frame-by-frame analysis of single snapshots; midlevel areas such as the FFA and the PPA are involved in integration over a short time scale (e.g., a few seconds); and higher-order areas such as the temporoparietal junction (TPJ) are involved in integration and reasoning of an event over a longer time scale, creating a hierarchy of temporal receptive fields in the brain. Although these studies tested higher-level understanding of the meaning of complex event sequences, one can imagine a similar hierarchy of temporal receptive fields in daily navigation. For example, recognizing that the current view is continuous from the previous view (e.g., integrating panoramic views over time) might require integration over a short time scale, whereas recognizing where you are in a city may require integration of the route you took over a longer time scale. More research on scene and spatial navigation should integrate how our brain combines scene information presented over different temporal contexts and scales.

Altogether, the rich and meaningful visual experience that we take for granted relies on the brain's elegant functional architecture of multiple brain regions with complementary functions for scene perception. Research in the field of scene understanding has grown rapidly over the past few years, and the field has just begun to distinguish which structural and conceptual properties of visual scenes are processed at different stages of the visual pathway. In combination with fMRI multivoxel pattern analysis and computational models of low-level visual systems, we are at the stage of being able to roughly reconstruct what the viewer is currently seeing (akin to "mind reading") (Kay, Naselaris, Prenger, & Gallant, 2008). The next major goal in the field is to discover the precise dimensions of scenes that are encoded in multiple scene-selective regions, to figure out how these dimensions are synthesized to give rise to the perception of a complete scene, and to understand how these representations change or integrate over time as the viewer navigates in the world.

References

Addis, D. R., Wong, A. T., & Schacter, D. L. (2007). Remembering the past and imagining the future: Common and distinct neural substrates during event construction and elaboration. *Neuropsychologia*, *45*(7), 1363–1377.

Aguirre, G. K., & D'Esposito, M. (1999). Topographical disorientation: A synthesis and taxonomy. *Brain*, *122*, 1613–1628.

Aguirre, G. K., Zarahn, E., & D'Esposito, M. (1998). An area within human ventral cortex sensitive to "building" stimuli: Evidence and implications. *Neuron*, *21*, 373–383.

Aminoff, E. M., Kveraga, K., & Bar, M. (2013). The role of parahippocampal cortex in cognition. *Trends in Cognitive Sciences*, *17*(8), 379–390.

Auger, S. D., Mullally, S. L., & Maguire, E. A. (2012). Retrosplenial cortex codes for permanent landmarks. *PLoS ONE*, *7*(8), e43620.

Biederman, I. (1981). On the semantics of a glance at a scene. In M. Kubovy & J. R. Pomerantz (Eds.), *Perceptual organization* (pp. 213–253). Hillsdale, NJ: Lawrence Erlbaum Associates.

Bird, C. M., & Burgess, N. (2008). The hippocampus and memory: Insights from spatial processing. *Nature Reviews Neuroscience*, *9*, 182–194.

Buckner, R. L., & Carroll, D. C. (2007). Self-projection and the brain. *Trends in Cognitive Sciences*, *11*(2), 49–57.

Burgess, N., Becker, S., King, J. A., & O'Keefe, J. (2001). Memory for events and their spatial context: Models and experiments. *Philosophical Transactions of the Royal Society of London. Series B, Biological Sciences*, *356*, 1493–1503.

Candel, I., Merckelbach, H., Houben, K., & Vandyck, I. (2004). How children remember neutral and emotional pictures: Boundary extension in children's scene memories. *American Journal of Psychology*, *117*, 249–257.

Chadwick, M. J., Mullally, S. L., & Maguire, E. A. (2013). The hippocampus extrapolates beyond the view in scenes: An fMRI study of boundary extension. *Cortex*, *49*(8), 2067–2079.

Cox, D. D., & Savoy, R. L. (2003). Functional magnetic resonance imaging (fMRI) "brain reading": Detecting and classifying distributed patterns of fMRI activity in human visual cortex. *NeuroImage*, *19*, 261–270.

Davenport, J. L., & Potter, M. C. (2004). Scene consistency in object and background perception. *Psychological Science*, *15*(8), 559–564.

Dilks, D., Julian, J. B., Kubilius, J., Spelke, E. S., & Kanwisher, N. (2011). Mirror-image sensitivity and invariance in object and scene processing pathways. *Journal of Neuroscience*, *33*(31), 11305–11312.

Dilks, D. D., Julian, J. B., Paunov, A. M., & Kanwisher, N. (2013). The occipital place area (OPA) is causally and selectively involved in scene perception. *Journal of Neuroscience*, *33*(4), 1331–1336.

Eger, E., Ashburner, J., Haynes, J., Dolan, R. J., & Rees, G. (2008). fMRI activity patterns in human LOC carry information about object exemplars within category. *Journal of Cognitive Neuroscience*, *20*, 356–370.

Epstein, R. A. (2008). Parahippocampal and retrosplenial contributions to human spatial navigation. *Trends in Cognitive Sciences*, *12*(10), 388–396.

Epstein, R., DeYoe, E. A., Press, D. Z., Rosen, A. C., & Kanwisher, N. (2001). Neuropsychological evidence for a topographical learning mechanism in parahippocampal cortex. *Cognitive Neuropsychology*, *18*(6), 481–508.

Epstein, R., Graham, K. S., & Downing, P. E. (2003). Viewpoint specific scene representations in human parahippocampal cortex. *Neuron*, *37*, 865–876.

Epstein, R., Harris, A., Stanley, D., & Kanwisher, N. (1999). The parahippocampal place area: Recognition, navigation, or encoding? *Neuron*, *23*(1), 115–125.

Epstein, R. A., & Higgins, J. S. (2007). Differential parahippocampal and retrosplenial involvement in three types of visual scene recognition. *Cerebral Cortex*, *17*(7), 1680–1693.

Epstein, R. A., & Kanwisher, N. (1998). A cortical representation of the local visual environment. *Nature*, *392*(6676), 598–601.

Epstein, R. A., Parker, W. E., & Feiler, A. M. (2007). Where am I now? Distinct roles for parahippocampal and retrosplenial cortices in place recognition. *Journal of Neuroscience*, *27*(23), 6141–6149.

Fei-Fei, L., & Perona, P. (2005). A Bayesian hierarchical model for learning natural scene categories. *Computer Vision and Pattern Recognition*, *2*, 524–531.

Freiwald, W. A., & Tsao, D. Y. (2010). Functional compartmentalization and viewpoint generalization within the macaque face-processing system. *Science*, *330*(6005), 845–851.

Freiwald, W. A., Tsao, D. Y., & Livingstone, M. S. (2009). A face feature space in the macaque temporal lobe. *Nature Neuroscience*, *12*(9), 1187–1196.

Friedman, A. (1979). Framing pictures: The role of knowledge in automatized encoding and memory for gist. *Journal of Experimental Psychology. General*, *108*, 316–355.

Goh, J. O. S., Siong, S. C., Park, D., Gutchess, A., Hebrank, A., & Chee, M. W. L. (2004). Cortical areas involved in object, background and object-background processing revealed with functional magnetic resonance adaptation. *Journal of Neuroscience*, *24*(45), 10223–10228.

Golomb, J. D., Albrecht, A., Park, S., & Chun, M. M. (2011). Eye movements help link different views in scene-selective cortex. *Cerebral Cortex*, *21*(9), 2094–2102.

Gottesman, C. V., & Intraub, H. (2002). Surface construal and the mental representation of scenes. *Journal of Experimental Psychology. Human Perception and Performance*, *28*(3), 589–599.

Green, C., & Hummel, J. E. (2006). Familiar interacting object pairs are perceptually grouped. *Journal of Experimental Psychology. Human Perception and Performance*, *32*, 1107–1119.

Greene, M. R., & Oliva, A. (2009a). The briefest of glances: The time course of natural scene understanding. *Psychological Science*, *20*(4), 464–472.

Greene, M. R., & Oliva, A. (2009b). Recognition of natural scenes from global properties: Seeing the forest without representing the trees. *Cognitive Psychology*, *58*(2), 137–176.

Grill-Spector, K., Henson, R., & Martin, A. (2006). Repetition and the brain: Neural models of stimulus-specific effects. *Trends in Cognitive Sciences*, *10*(1), 14–23.

Grill-Spector, K., Kushnir, T., Edelman, S., Itzchak, Y., & Malach, R. (1998). Cue-invariant activation in object-related areas of the human occipital lobe. *Neuron*, *21*, 191–202.

Harel, A., Kravitz, D. J., & Baker, C. I. (2013). Deconstructing visual scenes in cortex: Gradients of object and spatial layout information. *Cerebral Cortex*, *23*(4), 947–957.

Hassabis, D., Chu, C., Rees, G., Weiskopf, N., Molyneux, P. D., & Maguire, E. A. (2009). Decoding neural ensembles in the human hippocampus. *Current Biology*, *19*, 546–554.

Hasson, U., Yang, E., Vallines, I., Heeger, D. J., & Rubin, N. (2008). A hierarchy of temporal receptive windows in human cortex. *Journal of Neuroscience*, *28*, 2539–2550.

Hoiem, D. H., Efros, A. A., & Hebert, M. (2006). Putting objects in perspective. *International Journal of Computer Vision*, *80*, 3–15.

Honey, C. J., Thesen, T., Donner, T. H., Silbert, L. J., Carlson, C. E., Devinsky, O., et al. (2012). Slow dynamics in human cerebral cortex and the accumulation of information over long timescales. *Neuron*, *76*(2), 423–434.

Hung, C.-C., Carlson, E. T., & Connor, C. E. (2012). Medial axis shape coding in macaque inferotemporal cortex. *Neuron*, *74*(6), 1099–1113.

Ino, T., Inoue, Y., Kage, M., Hirose, S., Kimura, T., & Fukuyama, H. (2002). Mental navigation in humans is processed in the anterior bank of the parieto-occipital sulcus. *Neuroscience Letters*, *322*, 182–186.

Intraub, H. (1997). The representation of visual scenes. *Trends in Cognitive Sciences*, *1*(6), 217–222.

Intraub, H. (2002). Anticipatory spatial representation of natural scenes: Momentum without movement? *Visual Cognition*, *9*, 93–119.

Intraub, H. (2004). Anticipatory spatial representation of 3D regions explored by sighted observers and a deaf-and-blind observer. *Cognition*, *94*(1), 19–37.

Intraub, H. (2012). Rethinking visual scene perception. *Wiley Interdisciplinary Reviews: Cognitive Science*, *3*(1), 117–127.

Janzen, G., & Van Turennout, M. (2004). Selective neural representation of objects relevant for navigation. *Nature Neuroscience*, *7*(6), 673–677.

Johnson, M. R., Mitchell, K. J., Raye, C. L., D'Esposito, M., & Johnson, M. K. (2007). A brief thought can modulate activity in extrastriate visual areas: Top-down effects of refreshing just-seen visual stimuli. *NeuroImage*, *37*, 290–299.

Joubert, O. R., Rousselet, G., Fize, D., & Fabre-Thorpe, M. (2007). Processing scene context: Fast categorization and object interference. *Vision Research*, *47*, 3286–3297.

Kamitani, Y., & Tong, F. (2005). Decoding the visual and subjective contents of the human brain. *Nature Neuroscience*, *8*(5), 679–685.

Kay, K. N., Naselaris, T., Prenger, R. J., & Gallant, J. L. (2008). Identifying natural images from human brain activity. *Nature*, *452*(7185), 352–355.

Kim, J. G., & Biederman, I. (2011). Where do objects become scenes? *Cerebral Cortex*, *21*, 1738–1746.

Knierim, J. J., & Hamilton, D. A. (2010). Framing spatial cognition: Neural representations of proximal and distal frames of reference and their roles in navigation. *Physiological Reviews*, *91*, 1245–1279.

Konkle, T., & Oliva, A. (2012). A real-world size organization of object responses in occipito-temporal cortex. *Neuron*, *74*(6), 1114–1124.

Kourtzi, Z., & Kanwisher, N. (2000). Cortical regions involved in perceiving object shape. *Journal of Neuroscience*, *20*(9), 3310–3318.

Kravitz, D. J., Peng, C. S., & Baker, C. I. (2011). Real-world scene representations in high-level visual cortex: It's the spaces more than the places. *Journal of Neuroscience*, *31*(20), 7322–7333.

Kriegeskorte, N., Mur, M., Ruff, D., Kiani, R., Bodurka, J., Esteky, H., et al. (2008). Matching categorical object representations in inferior temporal cortex of man and monkey. *Neuron*, *60*(6), 1126–1141.

Kumaran, D., & Maguire, E. A. (2006). An unexpected sequence of events: Mismatch detection in the human hippocampus. *PLoS Biology*, *4*(12), e424.

Landis, T., Cummings, J. L., Benson, D. F., & Palmer, E. P. (1986). Loss of topographic familiarity. An environmental agnosia. *Archives of Neurology*, *43*, 132–136.

Lazebnik, S., Schmid, C., & Ponce, J. (2006). Beyond bags of features: Spatial pyramid matching for recognizing natural scene categories. *Computer Vision and Pattern Recognition*, *2*, 2169–2178.

Loftus, G. R., & Mackworth, N. H. (1978). Cognitive determinants of fixation location during picture viewing. *Journal of Experimental Psychology. Human Perception and Performance*, *4*(4), 565–572.

MacEvoy, S. P., & Epstein, R. A. (2007). Position selectivity in scene and object responsive occipitotemporal regions. *Journal of Neurophysiology*, *98*, 2089–2098.

MacEvoy, S. P., & Epstein, R. A. (2011). Constructing scenes from objects in human occipitotemporal cortex. *Nature Neuroscience*, *14*(10), 1323–1329.

Maguire, E. A. (2001). The retrosplenial contribution to human navigation: A review of lesion and neuroimaging findings. *Scandinavian Journal of Psychology*, *42*, 225–238.

Malach, R., Reppas, J. B., Benson, R. R., Kwong, K. K., Jiang, H., Kennedy, W. A., et al. (1995). Object-related activity revealed by functional magnetic resonance imaging in human occipital cortex. *Proceedings of the National Academy of Sciences of the United States of America*, *92*, 8135–8139.

Mendez, M. F., & Cherrier, M. M. (2003). Agnosia for scenes in topographagnosia. *Neuropsychologia*, *41*, 1387–1395.

Morgan, L. K., MacEvoy, S. P., Aguirre, G. K., & Epstein, R. A. (2011). Distances between real-world locations are represented in the human hippocampus. *Journal of Neuroscience*, *31*(4), 1238–1245.

Mullally, S. L., Intraub, H., & Maguire, E. A. (2012). Attenuated boundary extension produces a paradoxical memory advantage in amnesic patients. *Current Biology*, *22*(4), 261–268.

O'Keefe, J., & Burgess, N. (1996). Geometric determinants of the place fields of hippocampal neurons. *Nature*, *381*, 425–428.

O'Keefe, J., & Nadel, L. (1979). The hippocampus as a cognitive map. *Behavioral and Brain Sciences*, *2*, 487–494.

Oliva, A., Park, S., & Konkle, T. (2011). Representing, perceiving and remembering the shape of visual space. In L. R. Harris & M. Jenkin (Eds.), *Vision in 3D Environments* (pp. 107–134). Cambridge: Cambridge University Press.

Oliva, A., & Torralba, A. (2001). Modeling the shape of the scene: A holistic representation of the spatial envelope. *International Journal of Computer Vision*, *42*(3), 145–175.

Oliva, A., & Torralba, A. (2006). Building the gist of a scene: The role of global image features in recognition. *Progress in Brain Research: Visual Perception*, *155*, 23–36.

Palmer, S. E. (1975). The effects of contextual scenes on the identification of objects. *Memory & Cognition*, *3*, 519–526.

Park, S., Brady, T. F., Greene, M. R., & Oliva, A. (2011). Disentangling scene content from spatial boundary: Complementary roles for the PPA and LOC in representing real-world scenes. *Journal of Neuroscience*, *31*(4), 1333–1340.

Park, S., & Chun, M. M. (2009). Different roles of the parahippocampal place area (PPA) and retrosplenial cortex (RSC) in scene. *NeuroImage*, *47*(4), 1747–1756.

Park, S., Chun, M. M., & Johnson, M. K. (2010). Refreshing and integrating visual scenes in scene-selective cortex. *Journal of Cognitive Neuroscience*, *22*(12), 2813–2822.

Park, S., Intraub, H., Yi, D. J., Widders, D., & Chun, M. M. (2007). Beyond the edges of a view: Boundary extension in human scene-selective visual cortex. *Neuron*, *54*(2), 335–342.

Park, S., Konkle, T., & Oliva, A. (2014). Parametric coding of the size and clutter of natural scenes in the human brain. *Cerebral Cortex*, doi: 10.1093/cercor/bht418.

Potter, M. C. (1975). Meaning in visual scenes. *Science*, *187*, 965–966.

Potter, M. C. (1976). Short-term conceptual memory for pictures. *Journal of Experimental Psychology. Human Learning and Memory*, *2*(5), 509–522.

Potter, M. C., Staub, A., & O'Connor, D. H. (2004). Pictorial and conceptual representation of glimpsed pictures. *Journal of Experimental Psychology. Human Perception and Performance*, *30*, 478–489.

Quinn, P. C., & Intraub, H. (2007). Perceiving "outside the box" occurs early in development: Evidence for boundary extension in three- to seven-month-old infants. *Child Development*, *78*(1), 324–334.

Rauchs, G., Orban, P., Balteau, E., Schmidt, C., Degueldre, C., Luxen, A., et al. (2008). Partially segregated neural networks for spatial and contextual memory in virtual navigation. *Hippocampus*, *18*(5), 503–518.

Rosch, E. (1978). Principles of categorization. In E. Rosch & B. Lloyd (Eds.), *Cognition and categorization* (pp. 27–48). Hilldale, NJ: Lawrence Erlbaum Associates.

Schyns, P. G., & Oliva, A. (1994). From blobs to boundary edges: Evidence for time- and spatial-scale-dependent scene recognition. *Psychological Science*, *5*(4), 195–200.

Seamon, J. G., Schlegel, S. E., Hiester, P. M., Landau, S. M., & Blumenthal, B. F. (2002). Misremembering pictured objects: People of all ages demonstrate the boundary extension illusion. *American Journal of Psychology*, *115*(2), 151–167.

Summerfield, J. J., Hassabis, D., & Maguire, E. A. (2010). Differential engagement of brain regions within a "core" network during scene construction. *Neuropsychologia*, *48*(5), 1501–1509.

Takahashi, N., & Kawamura, M. (2002). Pure topographical disorientation—the anatomical basis of landmark agnosia. *Cortex*, *38*, 717–725.

Thorpe, S., Fize, D., & Marlot, C. (1996). Speed of processing in the human visual system. *Nature*, *381*(6582), 520–522.

Torralba, A., & Oliva, A. (2003). Statistics of natural image categories. *Network (Bristol, England)*, *14*(3), 391–412.

Torralba, A., Oliva, A., Castelhano, M., & Henderson, J. M. (2006). Contextual guidance of eye movements in real-world scenes: the role of global features on object search. *Psychological Review*, *113*(4), 766–786.

Turk-Browne, N. B., Scholl, B. J., Johnson, M. K., & Chun, M. M. (2010). Implicit perceptual anticipation triggered by statistical learning. *Journal of Neuroscience*, *30*, 11177–11187.

Turk-Browne, N. B., Simon, M. G., & Sederberg, P. B. (2012). Scene representations in parahippocampal cortex depend on temporal context. *Journal of Neuroscience*, *32*, 7202–7207.

Valenstein, E., Vowers, D., Verfaellie, M., Heilman, K. M., Day, A., & Watson, R. T. (1987). Retrosplenial amnesia. *Brain*, *110*, 1631–1646.

Vann, S. D., Aggleton, J. P., & Maguire, E. A. (2009). What does the retrosplenial cortex do? *Nature Reviews Neuroscience*, *10*(11), 792–802.

VanRullen, R., & Thorpe, S. J. (2001). The time course of visual processing: From early perception to decision making. *Journal of Cognitive Neuroscience*, *13*(4), 454–461.

Vinberg, J., & Grill-Spector, K. (2008). Representation of shapes, edges, and surfaces across multiple cues in the human visual cortex. *Journal of Neurophysiology*, *99*(3), 1380–1393.

Walther, D. B., Caddigan, E., Fei-Fei, L., & Beck, D. M. (2009). Natural scene categories revealed in distributed patterns of activity in the human brain. *Journal of Neuroscience*, *29*(34), 10573–10581.

Walther, D. B., Chai, B., Caddigan, E., Beck, D. M., & Fei-Fei, L. (2011). Simple line drawings suffice for functional MRI decoding of natural scene categories. *Proceedings of the National Academy of Sciences of the United States of America*, *108*(23), 9661–9666.

Ward, E. J., MacEvoy, S. P., & Epstein, R. A. (2010). Eye-centered encoding of visual space in scene-selective regions. *Journal of Vision, 10*(14), 1–12.

Yassa, M. A., & Stark, C. E. L. (2011). Pattern separation in the hippocampus. *Trends in Neurosciences*, *34*(10), 515–525.

4

Deconstructing Scene Selectivity in Visual Cortex

Reza Rajimehr, Shahin Nasr, and Roger Tootell

In high-order object-processing areas of the ventral visual pathway, discrete clusters of neurons ("modules") respond selectively to specific categories of complex images such as faces (Kanwisher, McDermott, & Chun, 1997; Tsao, Freiwald, Knutsen, Mandeville, & Tootell, 2003; Tsao, Moeller, & Freiwald, 2008), places/scenes (Aguirre, Zarahn, & D'Esposito, 1998; Epstein & Kanwisher, 1998), body parts (Downing, Jiang, Shuman, & Kanwisher, 2001; Grossman & Blake, 2002), and word forms (Cohen et al., 2000). On the other hand, stimuli of a common category often also share low-level visual cues, and correspondingly, it has been reported that many neurons in the inferior temporal (IT) cortex (which is the final stage of the ventral visual pathway) are selective for specific low-level properties, including surface curvature (Janssen, Vogels, Liu, & Orban, 2001; Kayaert, Biederman, & Vogels, 2005), Fourier descriptor shapes (Schwartz, Desimone, Albright, & Gross, 1983), simple geometry (Brincat & Connor, 2004; Kobatake & Tanaka, 1994), nonaccidental features (geons; Vogels, Biederman, Bar, & Lorincz, 2001), diagnostic features (Sigala & Logothetis, 2002), and color (Koida & Komatsu, 2007). Thus, any given category-selective response might be deconstructed into multiple low-level feature selectivities. In fact a recent theory suggests that overlapping continuous maps of simple features give rise to discrete modules that are selective for complex stimuli (Op de Beeck, Haushofer, & Kanwisher, 2008). Selectivity for low-level visual features may be particularly crucial in the processing of scene images. Scenes encompass a virtually infinite range of possible visual stimuli. Selectivity for such a wide range of stimuli may be constructed only by considering some low-level features that are common to images from the scene category. Here we review recent fMRI studies that have reported certain low-level preferences/biases in the scene-responsive areas of visual cortex.

Organization of Scene-Responsive Cortical Areas in Human and Nonhuman Primates

In humans, fMRI studies have described three visual cortical regions that are more active during the presentation of "places" (images of scenes or isolated houses) than during the presentation of other visual stimuli such as faces, objects, body parts, or scrambled scenes (Aguirre et al., 1998; Bar & Aminoff, 2003; Epstein & Kanwisher, 1998; Hasson, Harel, Levy, & Malach, 2003; Maguire, 2001) (figure 4.1, plate 7). These regions have been named for nearby anatomical landmarks as follows: (1) parahippocampal place area (PPA), (2) transverse occipital sulcus (TOS), and (3) retrosplenial cortex (RSC). A recent meta-analysis and comprehensive mapping of scene-related activations suggest that the three scene-responsive regions are actually centered near—but distinct from—the gyri/sulci for which they were originally named (Nasr et al., 2011).

The scene-responsive PPA is typically centered on the lips of the collateral sulcus and adjacent medial fusiform gyrus rather than on the parahippocampal gyrus per se. Although the size of the PPA varies when it is localized with different localizer stimuli, the peak activity is consistently located on the medial fusiform gyrus. More specifically, the fusiform gyrus is subdivided by a shallow sulcus (the middle fusiform sulcus) into a scene-responsive region on the medial fusiform gyrus (PPA) and a face-responsive region on the lateral fusiform gyrus (fusiform face area [FFA]) (Nasr et al., 2011).

The scene-responsive TOS (renamed the occipital place area by Dilks, Julian, Paunov, & Kanwisher, 2013) is typically centered on the nearby lateral occipital gyrus rather than within its namesake, the transverse occipital sulcus. This scene-responsive region lies immediately anterior and ventral to the retinotopically defined area V3A, in/near retinotopic areas V7, V3B, and/or LO-1 (Nasr et al., 2011).

The scene-responsive RSC is a discrete region consistently located in the fundus of the parieto-occipital sulcus, approximately 1 cm from the RSC as defined by Brodmann areas 26, 29, and 30. This scene-responsive region is located immediately adjacent to V1 in what would otherwise be the peripheral representation of dorsal V2 (Nasr et al., 2011).

The functional connectivity of these scene areas has been tested during the resting-state fMRI (Baldassano, Beck, & Fei-Fei, 2013). The RSC and TOS show differentiable functional connections with the anterior-medial and posterior-lateral parts of the PPA, respectively. Each of these areas is also functionally connected with specific parts of the cortex. The RSC shows connections with the superior frontal sulcus (Brodmann areas 8/9) and the peripheral representation of early visual areas V1 and V2. The TOS shows connections with the intraparietal sulcus, the lateral occipital complex, and retinotopic early visual areas. The PPA shows connections with the lateral occipital complex and the peripheral representation of early visual areas V1 and V2.

Corresponding (presumptively homologous) scene-responsive regions have been identified by use of fMRI in awake macaque monkeys (Nasr et al., 2011; Rajimehr, Devaney, Bilenko, Young, & Tootell, 2011) (figure 4.1, plate 7). These studies used identical stimuli and largely overlapping fMRI procedures in human and monkey scans so that a relatively direct comparison between human and monkey maps was possible. Mirroring the arrangement of the human FFA and PPA (which are adjacent to each other in cortex), the presumptive monkey homologue of the human PPA (mPPA) is located adjacent to the most prominent face patch in the IT cortex. This location is immediately anterior to area TEO. The monkey TOS (mTOS) includes the region predicted from the human maps (macaque V4d), extending posteriorly into V3A. A possible monkey homologue of the human RSC lies in the medial bank, near the peripheral V1.

In addition to mPPA, a recent study has reported two other "place patches" in macaque ventral temporal cortex (Kornblith, Cheng, Ohayon, & Tsao, 2013). These patches, the lateral place patch (LPP) and the medial place patch (MPP), are located in the occipitotemporal sulcus and the parahippocampal gyrus, respectively. The LPP contains a large concentration of scene-selective single units, with individual units coding specific scene parts. Based on microstimulation, the LPP is connected with the MPP and with extrastriate visual areas V4v and DP.

Retinotopic Selectivity in Scene-Responsive Areas

Early electrophysiological studies suggested that neurons in the IT cortex have large receptive fields (>20°) (Desimone & Gross, 1979; Richmond, Wurtz, & Sato, 1983). Those studies emphasized that the positional information is lost at progressively higher stages of the ventral visual pathway, and neurons become selective for visual features and objects independent of their locations in the visual field (Ito, Tamura, Fujita, & Tanaka, 1995; Lueschow, Miller, & Desimone, 1994). However, more recent studies of the IT cortex have reported the presence of small receptive fields (<5°) even in the anterior IT cortex (DiCarlo & Maunsell, 2003; Op De Beeck & Vogels, 2000). In fact there is a wide distribution of receptive field sizes in the IT cortex, ranging from 3° to 25° with a mean size of 10° (Op De Beeck & Vogels, 2000). These data are consistent with the idea that representations in the IT cortex are position dependent. This position sensitivity could be considered a low-level selectivity for object-selective IT neurons.

Analogously, early human fMRI studies distinguished between retinotopic and nonretinotopic cortex (e.g., Grill-Spector et al., 1998; Halgren et al., 1999). Those studies described retinotopic maps in occipital visual areas such as V1, V2, V3, V3A/B, hV4, and V5/hMT+ (e.g., Brewer, Liu, Wade, & Wandell, 2005; DeYoe et al., 1996; Engel, Rumelhart, Wandell, & Lee, 1994; Huk, Dougherty, & Heeger, 2002; Sereno et al., 1995; Tootell et al., 1997) but failed to find consistent retinotopy in higher-level areas of the ventral visual pathway—perhaps due to technical limitations. With technical advancements in neuroimaging and better stimulus designs, recent fMRI studies have reported retinotopic maps located beyond (anterior to) V4. Such maps have been identified in object-selective lateral occipital cortex (Larsson & Heeger, 2006; Sayres & Grill-Spector, 2008) and within regions selective for object categories such as body parts (Weiner & Grill-Spector, 2011). Distributed positional information from multivoxel pattern analysis has also been reported in almost all category-selective

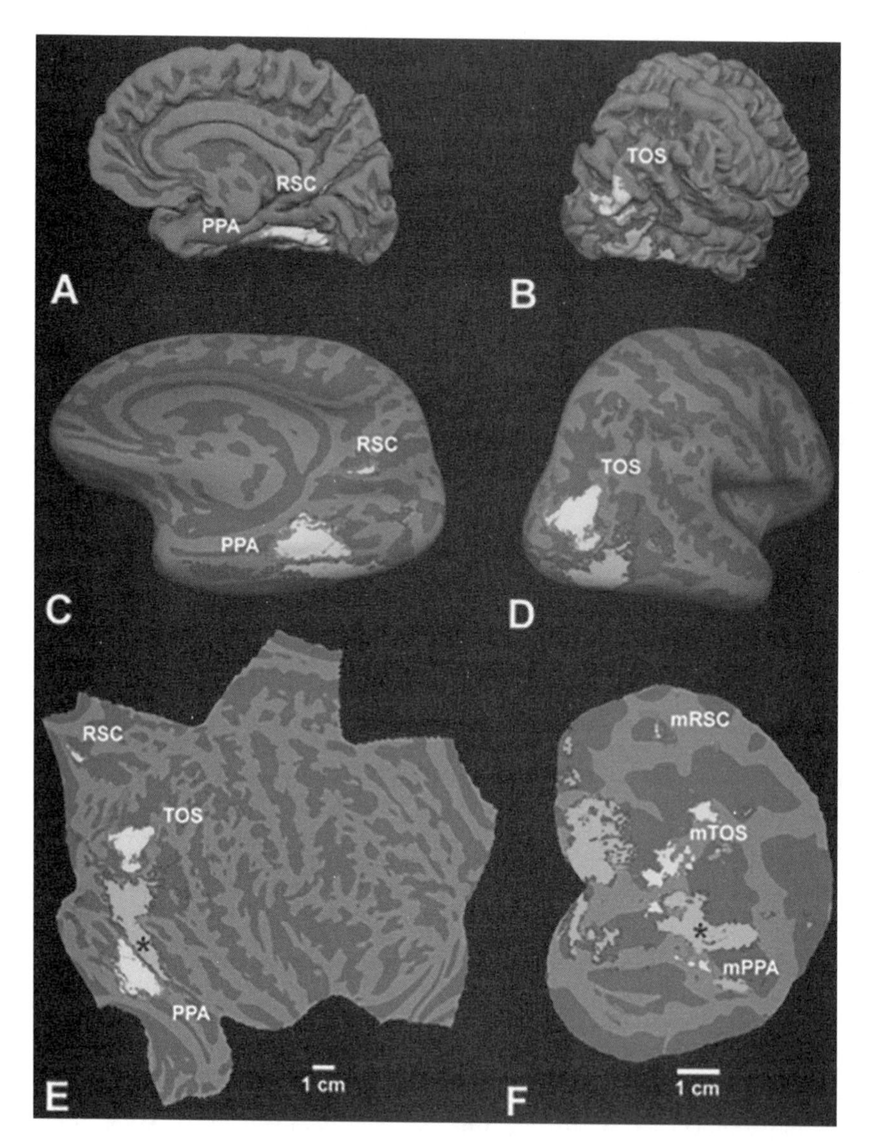

Figure 4.1 (plate 7)
Overall view of scene-responsive areas in human and monkey visual cortex. Both species fixated the center of a screen during block-designed presentation of identical scene versus face-localizing stimuli. In the human data (panels A–E), relatively higher activity to scenes versus faces is shown in red/yellow versus blue/cyan, respectively (minimum = $p < 10^{-10}$; maximum = $p < 10^{-30}$). The human map is a group average of both the functional and the anatomical data (n = 10) in cortical surface format. The right hemisphere is illustrated. Panels A and B show the medial and lateral-posterior views of folded cortex, respectively. Panels C and D show corresponding views of inflated cortex, and panel E shows the flattened view. For comparison, panel F shows the flattened activity map from a macaque monkey viewing the same stimuli (minimum = $p < 10^{-5}$; maximum = $p < 10^{-10}$). In both species presumptive corresponding scene-responsive areas are named in white (preceded by "m" in the macaque map). Adapted from Nasr et al. (2011).

areas (Carlson, Hogendoorn, Fonteijn, & Verstraten, 2011; Cichy, Chen, & Haynes, 2011; Kravitz, Kriegeskorte, & Baker, 2010; Schwarzlose, Swisher, Dang, & Kanwisher, 2008). Thus, low-level selectivity for the retinotopic location of visual stimuli is preserved in higher-level areas in human IT cortex.

Using retinotopic mapping combined with an attentional tracking paradigm, Arcaro and colleagues reported two new retinotopic maps anterior to the VO cluster within the posterior parahippocampal cortex (PHC), referred to as PHC-1 and PHC-2 (Arcaro, McMains, Singer, & Kastner, 2009). Each PHC area contains a complete representation of the contralateral visual field with a bias for stimuli in the upper visual field (see also Schwarzlose et al., 2008). Both areas are heavily overlapped with the functionally defined area PPA (Arcaro et al., 2009); this suggests a position-dependent coding of scenes in the PPA. The scene-responsive TOS also has retinotopic selectivity because it is located within the retinotopic extrastriate cortex in both humans and macaques (Nasr et al., 2011).

Scene-related areas, including the PPA and TOS, also show a strong preference for stimuli presented in the peripheral visual field. In a series of experiments Malach and colleagues reported an association between category selectivity and eccentricity bias in high-order object areas; face areas were associated with central visual-field bias, whereas scene areas were associated with peripheral visual-field bias (Hasson et al., 2003; Levy, Hasson, Avidan, Hendler, & Malach, 2001; Malach, Levy, & Hasson, 2002). Retinotopic eccentricity might be an organizing principle of object representations in these areas, and it can be tightly linked to acuity demands (Hasson, Levy, Behrmann, Hendler, & Malach, 2002). It is conceivable that scene-related processes such as spatial navigation and texture segregation depend crucially on large-scale integration, and thus, these functions might be better served by a strong association with peripheral, low-magnification representations.

A Preference for High Spatial Frequencies in Scene-Responsive Areas

In the lower-tier (occipital) visual cortex (e.g., in V1, V2, and V3), the sensitivity to spatial frequency covaries systematically with the retinotopic representation of visual field eccentricity. That is, the foveal/parafoveal cortex in these areas shows a preference for higher spatial frequencies (Henriksson, Nurminen, Hyvärinen, & Vanni, 2008; Sasaki et al., 2001). Recently, the sensitivity to spatial frequency has been tested in the higher-tier areas of the ventral visual pathway (Awasthi, Sowman, Friedman, & Williams, 2013; Rajimehr et al., 2011; Zeidman, Mullally, Schwarzkopf, & Maguire, 2012). One study (Rajimehr et al., 2011) manipulated the spatial frequency in a variety of stimuli and found a significant preference for high spatial frequencies in the scene-responsive area PPA. The high-spatial-frequency bias in the PPA was demonstrated using high-pass-filtered scene, face, and even checkerboard stimuli (figure 4.2, plate

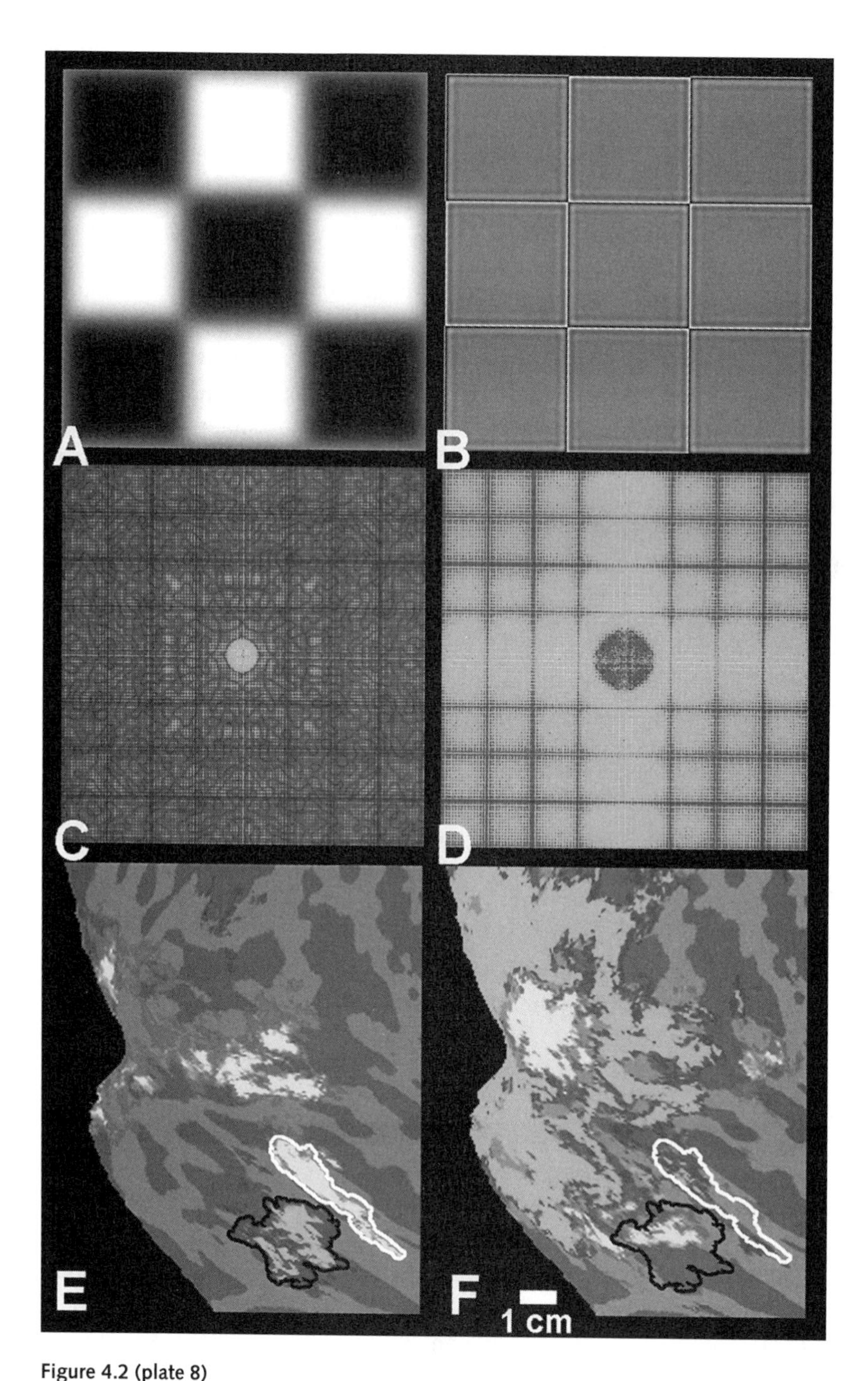

Figure 4.2 (plate 8)

High-pass-filtered checkerboard images selectively activate the PPA. Panels A and B show examples of low-spatial-frequency (low-SF) and high-spatial-frequency (high-SF) checkerboards. Panels C and D show the FFT amplitude spectra of low-SF and high-SF checkerboards. Panel E shows the FFA (indicated by white border) and PPA (indicated by black border), localized based on a comparison between faces and places, in the averaged map of four human subjects. The group-averaged activity map is displayed on a flattened view of the right occipitotemporal cortex. Panel F shows the comparison of activity between high-SF (yellow/red) and low-SF (cyan/blue) checkerboards. This comparison revealed a high-SF bias within the PPA. If anything, the opposite bias was found in parts of the FFA. The maps are significant at the threshold of $p < 10^{-2}$. Adapted from Rajimehr et al. (2011).

8). This bias was more prominent in the posterior-lateral part of the PPA. The PPA also showed a higher response to unfiltered natural scenes that were dominated by high spatial frequencies. This study (Rajimehr et al., 2011) used identical stimuli in monkeys and found that the mPPA (the apparent homologue of PPA in monkeys) also has a preference for high spatial frequencies.

An image analysis suggests that scenes have more spatial discontinuities (in the form of high-spatial-frequency components) compared to other object categories, such as faces (Rajimehr et al., 2011). Thus, a low-level sensitivity to high spatial frequencies in the PPA can be particularly useful for detecting edges, object borders, and scene details during spatial perception and navigation. Furthermore, there may be an evolutionary advantage for the PPA to be preferentially tuned for high spatial frequencies (e.g., in facilitating the detection of food/predators in visually complex environments).

A Cardinal Orientation Bias in Scene-Responsive Areas

Human vision is more sensitive to contours at cardinal (horizontal and vertical) orientations compared to oblique orientations, a phenomenon called the "oblique effect" (Appelle, 1972; Mach, 1861). Because the oblique effect is linked to stimulus orientation, and orientation-selective cells are common in V1, prior psychophysical and physiological experiments have often hypothesized a neural correlate of the perceptual oblique effect in lower-level retinotopic visual cortex (Vogels & Orban, 1985). However, fMRI studies in humans have not reported a consistent activity bias for cardinal orientations in V1 (Freeman, Brouwer, Heeger, & Merriam, 2011; but see Furmanski & Engel, 2000; Swisher et al., 2010).

A recent study suggests that the oblique effect may be related to scene processing (Nasr & Tootell, 2012). The link between the oblique effect and scene processing is supported by ecological evidence. Image statistics confirm that many scenes are dominated by cardinal orientations (Torralba & Oliva, 2003). Such a statistical bias is present not only in carpentered environments (such as cityscapes, buildings, and indoor scenes) but also in some natural scenes, often due to the orthogonal influences of gravity and/or phototropism. Consistent with this idea, the scene-responsive area PPA shows a stronger fMRI response to cardinal (compared to oblique) orientations (Nasr & Tootell, 2012) (figure 4.3, plate 9). This low-level orientation bias in the PPA can be observed even for simple geometrical stimuli such as arrays of overlapping squares or arrays of line segments (Nasr & Tootell, 2012).

Conclusion

Here we reviewed evidence for selectivity to low-level visual features in the scene-responsive areas, particularly in the PPA. It is important to define such features for

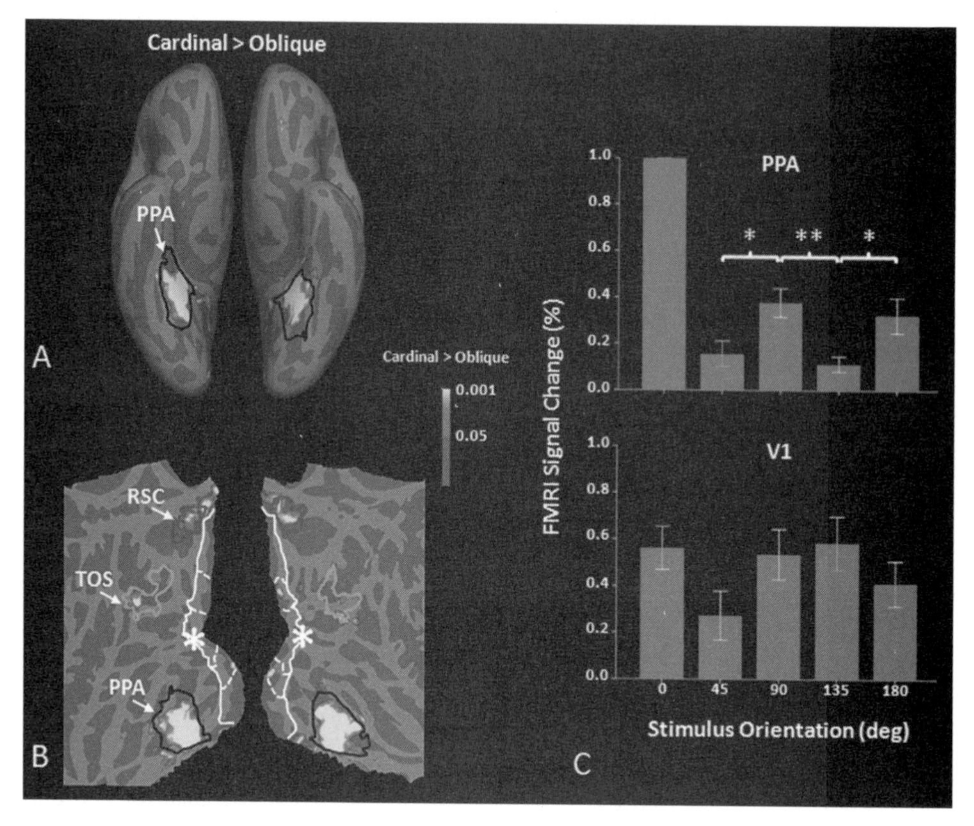

Figure 4.3 (plate 9)
Scenes with dominant power at cardinal orientations selectively activate the PPA. Panel A shows group-averaged fMRI activity ($n = 13$; random-effects analysis) for spatially filtered scenes with dominant power at cardinal versus oblique orientations. The map shows a ventral view of the inflated (panel A) and flattened (panel B) cortical surfaces. The borders of the PPA, TOS, RSC, and V1 are shown using black, green, blue, and white lines, respectively. White dashed lines indicate the peripheral visual field representation, and white asterisks indicate the foveal representation. Panel C shows the region-of-interest analysis in the PPA and V1. Orientation differences were significant only in the PPA ($*p < 0.05$; $**p < 0.01$). Adapted from Nasr et al. (2012).

both practical and conceptual reasons. At a practical level it is important to define such variables to avoid uncontrolled stimulus influences on otherwise carefully controlled tests of higher-order variables. More conceptually, such lower-level variables may well serve as intermediate "building blocks" for higher-order selectivity and thus clarify the nature of further higher-order variables. The latter is particularly important because after its initial characterization as a scene-selective area, the PPA has been reinterpreted as selective for a wide range of other higher-order categories, including tools (Chao, Haxby, & Martin, 1999), single houses (Tootell et al., 2008), inanimate objects (Kriegeskorte et al., 2008), big objects (Konkle & Oliva, 2012; Troiani, Stigliani, Smith, & Epstein, 2014), object ensembles, and surface textures (Cant & Xu, 2012). Due to this wide range of stimulus selectivities in the PPA, we have used the term "scene-responsive" (rather than "scene-selective") when referring to the PPA and other scene areas.

Evidence to date demonstrates that multiple lower-level variables influence the fMRI response properties in the PPA. Retinotopically, studies have reported that this area shows a bias for stimuli in the peripheral (as opposed to foveal) visual field (e.g., Levy et al., 2001) and a bias for upper visual field (Arcaro et al., 2009). It can be argued that to the extent that a response is specific to portions of the visual field, it cannot be strictly selective for a given visual category per se.

More globally, it has been shown that the PPA is also selective for high spatial frequencies and cardinal orientations. It is possible that PPA and other scene-responsive areas also show selectivity to other low-level features such as specific line junctions or specific shapes. For instance, the fact that the PPA responds more strongly to cubes than to spheres (Rajimehr et al., 2011) may be partly due to the presence of right angles in the "cube" stimuli. Future studies would be needed to test this and other low-level feature selectivities in these areas. Demonstration of low-level selectivity in the scene areas makes the single-cell studies of scene processing more tractable, as these low-level features can be parametrically manipulated in well-defined stimulus spaces.

Acknowledgments

This research was supported by National Institutes of Health (NIH) Grants R01 MH67529 and R01 EY017081 to R.B.H.T., the Martinos Center for Biomedical Imaging, the NCRR, and the MIND Institute.

References

Aguirre, G. K., Zarahn, E., & D'Esposito, M. (1998). An area within human ventral cortex sensitive to "building" stimuli: Evidence and implications. *Neuron*, *21*, 373–383.

Appelle, S. (1972). Perception and discrimination as a function of stimulus orientation: The oblique effect in man and animals. *Psychological Bulletin*, *78*, 266–278.

Arcaro, M. J., McMains, S. A., Singer, B. D., & Kastner, S. (2009). Retinotopic organization of human ventral visual cortex. *Journal of Neuroscience*, *29*(34), 10638–10652.

Awasthi, B., Sowman, P. F., Friedman, J., & Williams, M. A. (2013). Distinct spatial scale sensitivities for early categorization of faces and places: Neuromagnetic and behavioral findings. *Frontiers in Human Neuroscience*, *7*(91).

Baldassano, C., Beck, D. M., & Fei-Fei, L. (2013). Differential connectivity within the parahippocampal place area. *NeuroImage*, *75*, 228–237.

Bar, M., & Aminoff, E. M. (2003). Cortical analysis of visual context. *Neuron*, *38*, 347–358.

Brewer, A. A., Liu, J., Wade, A. R., & Wandell, B. A. (2005). Visual field maps and stimulus selectivity in human ventral occipital cortex. *Nature Neuroscience*, *8*(8), 1102–1109.

Brincat, S. L., & Connor, C. E. (2004). Underlying principles of visual shape selectivity in posterior inferotemporal cortex. *Nature Neuroscience*, *7*(8), 880–886.

Cant, J. S., & Xu, Y. (2012). Object ensemble processing in human anterior-medial ventral visual cortex. *Journal of Neuroscience*, *32*(22), 7685–7700.

Carlson, T., Hogendoorn, H., Fonteijn, H., & Verstraten, F. A. (2011). Spatial coding and invariance in object-selective cortex. *Cortex*, *47*(1), 14–22.

Chao, L. L., Haxby, J. V., & Martin, A. (1999). Attribute-based neural substrates in temporal cortex for perceiving and knowing about objects. *Nature Neuroscience*, *2*(10), 913–919.

Cichy, R. M., Chen, Y., & Haynes, J. D. (2011). Encoding the identity and location of objects in human LOC. *NeuroImage*, *54*(3), 2297–2307.

Cohen, L., Dehaene, S., Naccache, L., Lehéricy, S., Dehaene-Lambertz, G., Hénaff, M. A., et al. (2000). The visual word form area: Spatial and temporal characterization of an initial stage of reading in normal subjects and posterior split-brain patients. *Brain*, *123*(2), 291–307.

Desimone, R., & Gross, C. G. (1979). Visual areas in the temporal cortex of the macaque. *Brain Research*, *178*(2), 363–380.

DeYoe, E. A., Carman, G. J., Bandettini, P., Glickman, S., Wieser, J., Cox, R., et al. (1996). Mapping striate and extrastriate visual areas in human cerebral cortex. *Proceedings of the National Academy of Sciences of the United States of America*, *93*(2), 2382–2386.

DiCarlo, J. J., & Maunsell, J. H. (2003). Anterior inferotemporal neurons of monkeys engaged in object recognition can be highly sensitive to object retinal position. *Journal of Neurophysiology*, *89*(6), 3264–3278.

Dilks, D. D., Julian, J. B., Paunov, A. M., & Kanwisher, N. (2013). The occipital place area (OPA) is causally and selectively involved in scene perception. *Journal of Neuroscience*, *33*(4), 1331–1336.

Downing, P. E., Jiang, Y., Shuman, M., & Kanwisher, N. (2001). A cortical area selective for visual processing of the human body. *Science*, *293*(5539), 2470–2473.

Engel, S. A., Rumelhart, D. E., Wandell, B. A., & Lee, A. T. (1994). fMRI of human visual cortex. *Nature*, *369*, 525.

Epstein, R. A., & Kanwisher, N. (1998). A cortical representation of the local visual environment. *Nature*, *392*(6676), 598–601.

Freeman, J., Brouwer, G. J., Heeger, D. J., & Merriam, E. P. (2011). Orientation decoding depends on maps, not columns. *Journal of Neuroscience*, *31*(13), 4792–4804.

Furmanski, C. S., & Engel, S. A. (2000). An oblique effect in human primary visual cortex. *Nature Neuroscience*, *3*(6), 535–536.

Grill-Spector, K., Kushnir, T., Hendler, T., Edelman, S., Itzchak, Y., & Malach, R. (1998). A sequence of object-processing stages revealed by fMRI in the human occipital lobe. *Human Brain Mapping*, *6*(4), 316–328.

Grossman, E. D., & Blake, R. (2002). Brain areas active during visual perception of biological motion. *Neuron*, *35*(6), 1167–1175.

Halgren, E., Dale, A. M., Sereno, M. I., Tootell, R. B. H., Marinkovic, K., & Rosen, B. R. (1999). Location of human face-selective cortex with respect to retinotopic areas. *Human Brain Mapping*, *7*(1), 29–37.

Hasson, U., Harel, M., Levy, I., & Malach, R. (2003). Large-scale mirror-symmetry organization of human occipito-temporal object areas. *Neuron*, *37*, 1027–1041.

Hasson, U., Levy, I., Behrmann, M., Hendler, T., & Malach, R. (2002). Eccentricity bias as an organizing principle for human high-order object areas. *Neuron*, *34*(3), 479–490.

Henriksson, L., Nurminen, L., Hyvärinen, A., & Vanni, S. (2008). Spatial frequency tuning in human retinotopic visual areas. *Journal of Vision, 8*(10), 1–13.

Huk, A. C., Dougherty, R. F., & Heeger, D. J. (2002). Retinotopy and functional subdivision of human areas MT and MST. *Journal of Neuroscience*, *22*(16), 7195–7205.

Ito, M., Tamura, H., Fujita, I., & Tanaka, K. (1995). Size and position invariance of neuronal responses in monkey inferotemporal cortex. *Journal of Neurophysiology*, *73*(1), 218–226.

Janssen, P., Vogels, R., Liu, Y., & Orban, G. A. (2001). Macaque inferior temporal neurons are selective for three-dimensional boundaries and surfaces. *Journal of Neuroscience*, *21*(23), 9419–9429.

Kanwisher, N., McDermott, J., & Chun, M. M. (1997). The fusiform face area: A module in human extrastriate cortex specialized for face perception. *Journal of Neuroscience*, *17*(11), 4302–4311.

Kayaert, G., Biederman, I., & Vogels, R. (2005). Representation of regular and irregular shapes in macaque inferotemporal cortex. *Cerebral Cortex*, *15*(9), 1308–1321.

Kobatake, E., & Tanaka, K. (1994). Neuronal selectivities to complex object features in the ventral visual pathway of the macaque cerebral cortex. *Journal of Neurophysiology*, *71*(3), 856–867.

Koida, K., & Komatsu, H. (2007). Effects of task demands on the responses of color-selective neurons in the inferior temporal cortex. *Nature Neuroscience*, *10*, 108–116.

Konkle, T., & Oliva, A. (2012). A real-world size organization of object responses in occipito-temporal cortex. *Neuron*, *74*(6), 1114–1124.

Kornblith, S., Cheng, X., Ohayon, S., & Tsao, D. Y. (2013). A network for scene processing in the macaque temporal lobe. *Neuron*, *79*(4), 766–781.

Kravitz, D. J., Kriegeskorte, N., & Baker, C. I. (2010). High-level visual object representations are constrained by position. *Cerebral Cortex*, *20*(12), 2916–2925.

Kriegeskorte, N., Mur, M., Ruff, D., Kiani, R., Bodurka, J., Esteky, H., et al. (2008). Matching categorical object representations in inferior temporal cortex of man and monkey. *Neuron*, *60*(6), 1126–1141.

Larsson, J., & Heeger, D. J. (2006). Two retinotopic visual areas in human lateral occipital cortex. *Journal of Neuroscience*, *26*(51), 13128–13142.

Levy, I., Hasson, U., Avidan, G., Hendler, T., & Malach, R. (2001). Center-periphery organization of human object areas. *Nature Neuroscience*, *4*(5), 533–539.

Lueschow, A., Miller, E. K., & Desimone, R. (1994). Inferior temporal mechanisms for invariant object recognition. *Cerebral Cortex*, *4*(5), 523–531.

Mach, E. (1861). Über das Sehen von Lagen und Winkeln durch die Bewegung des Auges. *Sitzungsberichte der Kaiserlichen Akademie der Wissenschaften*, *43*(2), 215–224.

Maguire, E. A. (2001). The retrosplenial contribution to human navigation: A review of lesion and neuroimaging findings. *Scandinavian Journal of Psychology*, *42*, 225–238.

Malach, R., Levy, I., & Hasson, U. (2002). The topography of high-order human object areas. *Trends in Cognitive Sciences*, *6*(4), 176–184.

Nasr, S., Liu, N., Devaney, K. J., Yue, X., Rajimehr, R., Ungerleider, L. G., et al. (2011). Scene-selective cortical regions in human and non-human primates. *Journal of Neuroscience*, *31*(39), 13771–13785.

Nasr, S., & Tootell, R. B. H. (2012). A cardinal orientation bias in scene-selective visual cortex. *Journal of Neuroscience*, *32*(43), 14921–14926.

Op de Beeck, H., Haushofer, J., & Kanwisher, N. G. (2008). Interpreting fMRI data: Maps, modules and dimensions. *Nature Reviews Neuroscience*, *9*(2), 123–135.

Op De Beeck, H., & Vogels, R. (2000). Spatial sensitivity of macaque inferior temporal neurons. *Journal of Comparative Neurology*, *426*(4), 505–518.

Rajimehr, R., Devaney, K. J., Bilenko, N. Y., Young, J. C., & Tootell, R. B. H. (2011). The parahippocampal place area responds preferentially to high spatial frequencies in humans and monkeys. *PLoS Biology*, *9*(4), e1000608.

Richmond, B. J., Wurtz, R. H., & Sato, T. (1983). Visual responses of inferior temporal neurons in awake rhesus monkey. *Journal of Neurophysiology*, *50*(6), 1415–1432.

Sasaki, Y., Hadjikhani, N., Fischl, B., Liu, A. K., Marret, S., Dale, A. M., et al. (2001). Local and global attention are mapped retinotopically in human occipital cortex. *Proceedings of the National Academy of Sciences of the United States of America*, *98*(4), 2077–2082.

Sayres, R., & Grill-Spector, K. (2008). Relating retinotopic and object-selective responses in human lateral occipital cortex. *Journal of Neurophysiology*, *100*(1), 249–267.

Schwartz, E. L., Desimone, R., Albright, T. D., & Gross, C. G. (1983). Shape recognition and inferior temporal neurons. *Proceedings of the National Academy of Sciences of the United States of America*, *80*(18), 5776–5778.

Schwarzlose, R. F., Swisher, J. D., Dang, S., & Kanwisher, N. (2008). The distribution of category and location information across object-selective regions of visual cortex. *Proceedings of the National Academy of Sciences of the United States of America*, *105*(11), 4447–4452.

Sereno, M. I., Dale, A. M., Reppas, J. B., Kwong, K. K., Belliveau, J. W., Brady, T. J., et al. (1995). Borders of multiple visual areas in human revealed by functional magnetic resonance imaging. *Science*, *268*, 889–893.

Sigala, N., & Logothetis, N. K. (2002). Visual categorization shapes feature selectivity in the primate temporal cortex. *Nature*, *415*(6869), 318–320.

Swisher, J. D., Gatenby, J. C., Gore, J. C., Wolfe, B. A., Moon, C. H., Kim, S. G., et al. (2010). Multiscale pattern analysis of orientation-selective activity in the primary visual cortex. *Journal of Neuroscience*, *30*(1), 325–330.

Tootell, R. B. H., Devaney, K. J., Young, J. C., Postelnicu, G., Rajimehr, R., & Ungerleider, L. G. (2008). fMRI mapping of a morphed continuum of 3D shapes within inferior temporal cortex. *Proceedings of the National Academy of Sciences of the United States of America*, *105*(9), 3605–3609.

Tootell, R. B. H., Mendola, J. D., Hadjikhani, N., Ledden, P. J., Liu, A. K., Reppas, J. B., et al. (1997). Functional analysis of V3A and related areas in human visual cortex. *Journal of Neuroscience*, *17*(18), 7060–7078.

Torralba, A., & Oliva, A. (2003). Statistics of natural image categories. *Network (Bristol, England)*, *14*(3), 391–412.

Troiani, V., Stigliani, A., Smith, M. E., & Epstein, R. A. (2014). Multiple object properties drive scene-selective regions. *Cerebral Cortex*, *24*(4), 883–897.

Tsao, D. Y., Freiwald, W. A., Knutsen, T. A., Mandeville, J. B., & Tootell, R. B. H. (2003). The representation of faces and objects in macaque cerebral cortex. *Nature Neuroscience*, *6*, 989–995.

Tsao, D. Y., Moeller, S., & Freiwald, W. A. (2008). Comparing face patch systems in macaques and humans. *Proceedings of the National Academy of Sciences of the United States of America*, *105*(49), 19514–19519.

Vogels, R., Biederman, I., Bar, M., & Lorincz, A. (2001). Inferior temporal neurons show greater sensitivity to nonaccidental than to metric shape differences. *Journal of Cognitive Neuroscience*, *13*(4), 444–453.

Vogels, R., & Orban, G. A. (1985). The effect of practice on the oblique effect in line orientation judgments. *Vision Research*, *25*(11), 1679–1687.

Weiner, K. S., & Grill-Spector, K. (2011). Not one extrastriate body area: Using anatomical landmarks, hMT+, and visual field maps to parcellate limb-selective activations in human lateral occipitotemporal cortex. *NeuroImage*, *56*(4), 2183–2199.

Zeidman, P., Mullally, S. L., Schwarzkopf, D. S., & Maguire, E. A. (2012). Exploring the parahippocampal cortex response to high and low spatial frequency spaces. *Neuroreport*, *23*(8), 503–507.

5

The Neurophysiology of Attention and Object Recognition in Visual Scenes

Daniel I. Brooks,* Heida Maria Sigurdardottir,* and David L. Sheinberg

If you are like most academics and scholars, you start your day with a cup of coffee. This task (Land & Hayhoe, 2001), among other things, requires you to locate the correct cupboard in the kitchen, open it, search for the can of coffee grounds among the surrounding clutter, and operate the coffee brewer by finding and pushing the correct buttons on the coffee maker. Depending on your view and current goal, the kitchen with its furniture and appliances, the cupboard full of miscellaneous objects, and the coffee brewer with its all too many buttons can be thought of as scenes in which you are searching for and recognizing a particular target object among other irrelevant and distracting ones.

Everyone knows what a scene is, to twist a famous quote by William James, but the concept is nonetheless surprisingly hard to define. When talking about scenes, it is often implied that they depict real-world environments consisting of a background and several objects in a particular spatial arrangement (Henderson & Hollingworth, 1999). Whereas there exists a long list of reports on human scene perception in the aforementioned sense, visual stimuli traditionally used in neurophysiology are rather sparse and artificial in comparison. When reviewing some of the literature on the neurophysiology of searching through and recognizing objects in visual scenes, we have therefore stretched the scene concept to include not only naturalistic environments but also simplified arrays of two or more visual elements.

The evidence presented here will be drawn mostly, although not exclusively, from the visual system of the macaque, one of the animal models most anatomically and functionally comparable with the human visual system (for a detailed overview of the macaque visual system, see, e.g., Van Essen, Anderson, & Felleman, 1992). Monkeys rarely search for coffee, but they nonetheless are often given the task of searching for particular things in visual scenes in order to gain other liquid rewards, such as drops of juice. Even though this can be considered rather unusual behavior, macaques in the wild take on many analogous tasks in their everyday environments, such as foraging for food and finding shelter, that require finding objects of interest in a complex scene.

Here, we review some of the neural processing that a scene image likely undergoes as it passes through the visual system, leading to the identification of an object in the scene. Our attempt, no doubt, is oversimplified; we focus on only particular processing steps and brain regions, but in reality the steps taken are likely to be both stimulus and task dependent and might involve several iterations where information is sent back and forth between low- and high-level visual regions (see, e.g., Hochstein & Ahissar, 2002; Peters & Payne, 1993; Tsotsos, 2008). Much is yet to be learned about the electrophysiology of searching for and recognizing objects in scenes.

Researchers often go to great lengths to eliminate any temporal and spatial correlations in the stimuli that they use in experiments because they do not want participants to be able to guess what they are going to be shown based either on what they just saw or the surrounding visual elements. The natural viewing of scenes, on the other hand, imposes all sorts of contingencies in both space and time (Attneave, 1954; Field, 1987; Kersten, 1987). If you are at the beach now, you are going to be at the beach one second later as well (probably even seeing almost the same view as before), and one grain of sand is most likely going to be next to another grain of sand. Change in the real world is slow. Similarly, the spectrum of naturalistic visual input is not white—that is, not all frequencies are equally presented in natural scenes; low temporal and spatial frequencies are more abundant than high frequencies (Dong & Atick, 1995; Field, 1987).

What this means is that natural scenes are full of redundancies. It might not be a particularly good strategy to use a large part of your computational resources to essentially code for the same information over and over again (Attneave, 1954; Simoncelli & Olshausen, 2001). Computer scientists know this and have developed methods that make use of the correlational structure of images to compress them into a more easily transferable format. The primate visual system, apparently, also makes use of such contingencies to more efficiently code visual information (but see, e.g., Barlow, 2001, for a critical look at the idea that redundancy in natural viewing is used mainly for compressive coding in perceptual systems).

This apparently happens before light even enters the eye. Primates have eyes that can move and thus possess the ability to influence their own visual input. The eyes are never completely still, even when an attempt is made to keep them fixated at a single spot (Ratliff & Riggs, 1950). Computational modeling work indicates that such fixational instability changes the statistics of neuronal firing in the retina, where the responses of retinal ganglion cells with nonoverlapping receptive fields become uncorrelated when images with spectral densities like those of natural images are fixated (Rucci, 2008). Most retinal ganglion cells project to the lateral geniculate nucleus (LGN), a subcortical structure in the thalamus, which then projects mainly to the primary visual cortex (V1) (see figure 5.1, plate 10). Electrophysiological recordings have shown that the linear filtering properties of neurons in the LGN also lead to a

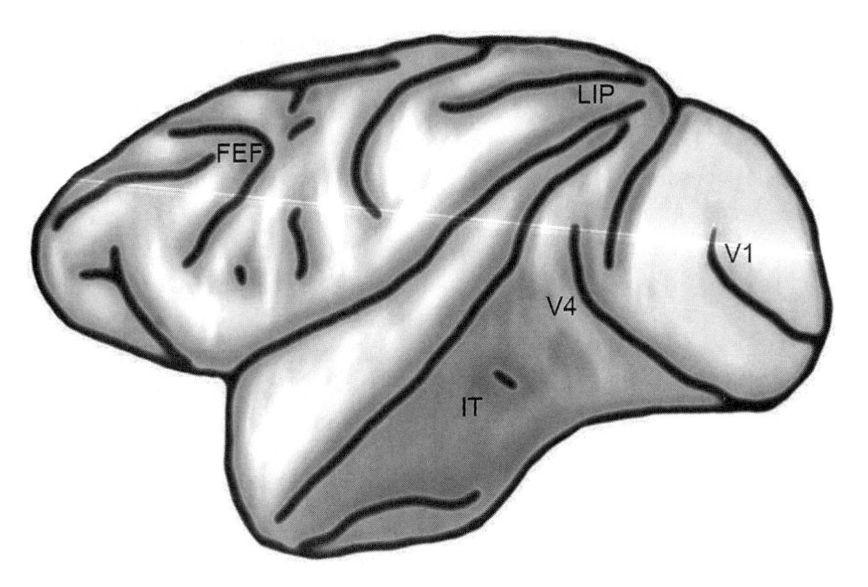

Figure 5.1 (plate 10)
The macaque brain. Neurons in the primary visual cortex (V1) were mainly thought to detect simple features, such as lines of particular orientations, within a very small part of visual space but are now known to integrate visual information from a much larger area (Allman et al., 1985; Hubel & Wiesel, 1959, 1968). Visual information originating from V1 reaches both a frontoparietal network, including the lateral intraparietal area (LIP) and the frontal eye fields (FEF), and the ventral visual stream, including the fourth visual area (V4) and the inferior temporal cortex (IT). Cells in LIP can respond to visual information, have memory-related responses, respond before, during, or after a saccadic eye movement, or have complex combinations of these types of responses (Barash, Bracewell, Fogassi, Gnadt, & Andersen, 1991). These neurons have a spatial receptive or response field, so in order to evoke a neural response, the thing that is visually presented, attended, looked at, or memorized, needs to have been within a particular spatial location (Barash et al., 1991; Blatt et al., 1990). Similar to LIP, the FEF also contain neurons with visual responses, oculomotor responses, and a combination of the two (Bruce & Goldberg, 1985). V4 and especially IT likely represent the final output stages of object processing in the primate brain. Unlike the frontoparietal system, IT neurons often have very large receptive fields (Gross, Bender, & Rocha-Miranda, 1969; but see Rolls et al., 2003). These cells might therefore not carry as much information on the location of an object, but instead they selectively respond to the complex features of objects of interest (Logothetis & Sheinberg, 1996).

similar temporal decorrelation or whitening of visual information in natural scenes (Dan, Atick, & Reid, 1996).

The whitening of the signal reduces redundancies found in real-world scenes, but it might also have an even more important effect. Because low temporal and spatial frequencies are abundant in natural scenes, the whitening of this input effectively emphasizes the throughput of high-frequency information—fast changes—while deemphasizing low-frequency information—slow changes. A quick change over time can often signify the appearance of a new object in the scene, and a quick change over space is quite likely to happen at object boundaries. Visual information from a scene that reaches the primary visual cortex is therefore already preprocessed in a manner that accentuates features that are important for finding and segmenting objects in scenes.

It is good to keep in mind that just because a brain region, such as the LGN, appears to behave differently under natural conditions than when artificial reduced visual stimuli are used, that does not necessarily indicate that it "knows" anything about the naturalness of its visual inputs. Even in the primary visual cortex (V1), local field potentials and multiunit firing rates are not detectably different for natural movies and noise stimuli with a frequency distribution similar to that of natural scenes (Kayser, Salazar, & Konig, 2003). Neural activity in V1 for simple stimuli such as bars and gratings, on the other hand, significantly differs from several more complex or naturalistic inputs such as natural movies (Kayser et al., 2003), pink noise (Kayser et al., 2003), random textures (Lehky, Sejnowski, & Desimone, 1992), and three-dimensional surfaces (Lehky et al., 1992). The V1 might therefore not be adapted to processing natural scenes as such but to visual input with temporal and spatial statistics or complexities similar to those of real-world environments.

One aspect of real-world environments to which V1 might be adapted is that natural input is not confined to a small spot, as it often is in visual experiments, but encompasses the entire visual field. Natural scenes are therefore bound to stimulate both the classical receptive field of a V1 neuron, a small region within the visual field where stimuli can directly drive the neuron, and the nonclassical receptive field, a region that does not evoke a neural response when stimulated alone but can selectively modulate the neuron's responses to other stimuli within the classical receptive field (Allman, Miezin, & McGuinness, 1985).

Under such more naturalistic viewing conditions the nonlinear interactions between the classical and nonclassical receptive fields of V1 neurons not only produce a sparse code that is energy efficient and minimizes redundancy (Vinje & Gallant, 2000) but can also be sensitive to scene structure such as whether the classical and nonclassical receptive fields are part of a continuous contour (Guo, Robertson, Mahmoodi, & Young, 2005). Such interactions are thought to facilitate contour integration (Guo et al., 2005), which again is important for segmenting objects from the background and from each other. Contours also serve as cues for visual object recognition, as attested by the fact that people can recognize line drawings of objects with relative ease.

In general, neurons in V1 might respond to the greatest extent when the visual features within the classical and nonclassical receptive fields are dissimilar, such as when there are orthogonal orientations inside and outside the classical receptive field (Guo et al., 2005; Knierim & van Essen, 1992; Sillito, Grieve, Jones, Cudeiro, & Davis, 1995). Again, this would emphasize the throughput of fast changes in visual scenes more than slow changes. Such contrast between visual features, like a yellow beach ball in a sea of blue, is salient to a human observer and an important cue for finding objects in scenes. We turn next to higher-level brain regions that might use information from V1 about feature contrast as well as other relevant data such as reward history or the goal of the current task to prioritize spatial locations in a visual scene. This task is thought to be accomplished mainly by regions within the so-called dorsal visual stream and its interconnectivity with areas of frontal cortex involved in the allocation of attention.

The classic characterization of the primate visual system includes a broad segregation into two main processing streams, both of which get input from the V1. (For a detailed review on parallel visual processing, see Goodale & Milner, 1992; Nassi & Callaway, 2009; Ungerleider & Haxby, 1994; Ungerleider & Mishkin, 1982.) The ventral ("what") stream is theorized to accomplish the recognition of complex forms and, ultimately, objects, whereas the dorsal ("where" or "how") stream has been implicated in the control of visual attention and spatial guidance of appropriate actions. These streams are at least partially anatomically and functionally separable, but they can be viewed as forming a widely interconnected system that guides the eyes and attention to visually salient or otherwise potentially important features/objects that then can be further scrutinized, recognized, and categorized.

As the dorsal stream projecting from V1 enters the parietal lobe, information is integrated into a frontoparietal network including the lateral intraparietal area (LIP) in the intraparietal sulcus and the frontal eye fields (FEF) in the prearcuate sulcus of the prefrontal cortex (figure 5.1, plate 10). These two regions, as well as a subcortical midbrain structure called the superior colliculus (SC), have been the focus of most research on the neuroanatomical substrates of visual search because each of them is thought to participate in guiding the eyes and attention to important objects or locations within visual scenes (Baluch & Itti, 2011; Bisley & Goldberg, 2010; Fecteau & Munoz, 2006).

One conceptual framework for this network of structures is that together they form a priority map of visual space in which each location in a scene is weighed or prioritized according to both its visual salience and probable task relevance (Baluch & Itti, 2011; Bisley & Goldberg, 2010; Fecteau & Munoz, 2006; Goldberg, Bisley, Powell, & Gottlieb, 2006; Itti & Koch, 2001; Thompson & Bichot, 2005). A saliency map, inspired by the psychological work of Treisman and Gelade (1980), was originally proposed as a computational model that uses the distribution of features such as color, brightness, and orientation to make a single representation of visual conspicuity (Itti & Koch, 2000; Itti, Koch, & Niebur, 1998; Wolfe, 1994).

As an example of this phenomenon, consider the results of Thomas and Paré (2007). In that study monkeys were required to make an eye movement to an oddball stimulus (a red target in an array of green distracters or a green target among red distracters). Cells in the lateral intraparietal area initially responded indiscriminately to a target and a distracter (i.e., whether the object was green or red, or whether it was task relevant or irrelevant), but the response of cells shifted to signal whether or not the oddball target was in or out of their receptive fields. Thus, neural activity in the lateral intraparietal area might gradually develop to represent the visual priority of the location occupied by an object in a scene. Similar effects can be observed in the frontal eye fields and the superior colliculus (for reviews on these regions acting as priority maps, see Baluch & Itti, 2011; Bisley & Goldberg, 2010; Fecteau & Munoz, 2006).

It is a good guess that something that stands out from the background—because it is brighter, darker, of a different color, or differently oriented than the surrounding elements—is a candidate object that might need to be recognized and is thus worth scrutinizing further. However, it is also likely that in a complex environment, dozens of relatively unimportant objects will stand out from the background while a single object (with certain features) is the only behaviorally relevant object. If you are looking for something in particular, such as a red parasol in our hypothetical beach scene, your visual system can prioritize this type of object over objects with different features (green things, square things, etc.) because they are not immediately relevant (see, e.g., Wolfe, 2007; Wolfe & Horowitz, 2004). This integration of top-down information with bottom-up saliency likely also happens in the aforementioned brain regions. Because a priority map weights locations based on such information, it can, at least theoretically, be used to select potential objects to which attention and eye gaze should be directed.

For an example of visual selection after attentional prioritization, let us consider the frontal eye fields. The frontal eye fields bear this name because electrical microstimulation within the region readily evokes saccadic eye movements, the endpoints of which depend on which subregion within the frontal eye fields is stimulated (Bruce, Goldberg, Bushnell, & Stanton, 1985). When the frontal eye fields are stimulated at a subthreshold current level—too low to cause eye movements—this modulated activity can nonetheless have a measurable effect on behavior. Moore and Fallah (2004) trained monkeys to detect a change in a small peripheral target in the presence of a continuously flashing distracting stimulus. The researchers then stimulated a subregion of the monkeys' frontal eye fields that at higher current levels would have evoked saccades to the target location. When this was done at the time of the target change, it improved the monkeys' target detection to a degree comparable with removing the distracter altogether. The frontal eye fields therefore causally contribute to the distribution of visuospatial attention.

A single target and a single distracter make a poor scene, if such a display can be considered a scene at all. It is nonetheless likely that activity in the frontal eye fields and the lateral intraparietal area aids target detection in complex and cluttered scenes by giving priority and guiding attention to the location of a potentially important object. Attending to a location means, among other things, that finer visual details can be sampled from an object that occupies that location (Montagna, Pestilli, &

Carrasco, 2009), details that might be relevant for their successful recognition. Further, it is likely that prioritizing a location directly affects the selection of visual features or candidate objects in the ventral visual pathway. In cluttered scenes multiple visual objects might compete for detailed representation in the ventral visual pathway. Signals from the frontal eye fields and the lateral intraparietal area could bias this competition (Desimone & Duncan, 1995) in favor of an object that occupies a high-priority location.

In order for this to be plausible, brain areas that implement a priority map must somehow communicate with ventral regions that are known to be important for object recognition. In other words, attentional selection should affect neural responses to objects in the ventral stream, and the prioritizing of the locations of candidate objects should precede object recognition.

There are now quite a lot of data that confirm that regions such as the lateral intraparietal area and the frontal eye field have reciprocal structural connections to some regions in the ventral visual stream, such as V4 and parts of the inferior temporal cortex (see figure 5.1, plate 10; Blatt, Andersen, & Stoner, 1990; Distler, Boussaoud, Desimone, & Ungerleider, 1993; Lewis & Van Essen, 2000; Schall, Morel, King, & Bullier, 1995; Stanton, Bruce, & Goldberg, 1995; Webster, Bachevalier, & Ungerleider, 1994) that respond selectively to moderately or highly complex visual features or even whole objects and are thought important for visual object recognition (Logothetis & Sheinberg, 1996). Because these connections exist, the lateral intraparietal area and the frontal eye fields must, at least under some circumstances, exchange information with regions in the ventral visual stream. It is plausible that this happens when the situation requires attentional selection.

Attentional selection strongly modulates object responses in the ventral stream. This is demonstrated by the work of Chelazzi and colleagues (1993; 1998; 2001), who presented monkeys with visual search displays containing two or more objects while they recorded neural activity in lower-level (V4) and higher-level (inferior temporal cortex) regions of the ventral visual stream (see also, e.g., Moran & Desimone, 1985; Reynolds, Chelazzi, & Desimone, 1999). One of the objects was known to elicit a strong response from the neuron being recorded from (an effective stimulus), whereas another one of the objects tended to elicit a poor or even no response from the same neuron (an ineffective stimulus); note that each neuron can have a unique stimulus preference, so an object that effectively drives one neuron is not necessarily an effective stimulus for another neuron. The monkey was shown one of the two objects, had to keep it in memory, and was then required to find it again in a following stimulus array or scene.

When both objects were present in the scene, the activity of the neuron nonetheless changed to reflect which object the monkey had to attend to, so its activity tended to be greater if the searched-for target was the effective stimulus, and neural activity was lower if the target was the ineffective stimulus. This observation supports the idea that visual attention biases the responses of these object-selective neurons to process mainly one object even when many other objects are present in the neurons' receptive fields (Desimone & Duncan, 1995; Reynolds et al., 1999), a situation that often arises in real-world visual scenes.

The supposition is that biasing signals from spatial priority maps (in the lateral intraparietal area and the frontal eye fields) guide the initial selection of the object that then gets primary representation in brain regions that partake in object recognition (the ventral visual stream). In order for this to be feasible it is quite important that these priority maps can be computed quickly so that object processing and recognition will be little delayed, even in cluttered scenes with many objects. Electrophysiological work has indeed shown that visual response latency of neurons in the lateral intraparietal area (Tanaka, Nishida, Aso, & Ogawa, 2013) and the frontal eye fields (Schmolesky et al., 1998) can be very short (e.g., frontal eye fields ~50 ms to ~100 ms), whereas latencies in several areas within the ventral stream tend to be longer (e.g., V4 ~70 ms to ~160 ms, TE1 80 ms to 200 ms) (Baylis, Rolls, & Leonard, 1987; Kiani, Esteky, & Tanaka, 2005; Schmolesky et al., 1998; Tamura & Tanaka, 2001).

The response latency of neurons associated with object recognition or spatial attention is nonetheless somewhat variable and might depend on the task, the stimuli being shown (see, e.g., Tanaka et al., 2013), and can even vary from trial to trial. In addition, the fact that one event precedes another does not imply that the former has a causal effect on the latter. Dual-area studies, in which neural activity is either simultaneously recorded from two brain regions or manipulated in one region but recorded in another, have the potential of eliminating these concerns. These challenging experiments can answer critical questions related to the timing of visual selection processes and more closely map the interaction between different processing systems.

Armstrong, Fitzgerald, and Moore (2006) electrically stimulated neurons in the frontal eye fields at current levels not strong enough to evoke a saccade but nonetheless known to affect the distribution of covert attention (Moore & Fallah, 2004). At the same time, they recorded neural activity in the ventral stream region V4. When they stimulated in this way, Armstrong and colleagues were able to enhance the V4 representation of an object if it was located at the endpoint of the saccade that the stimulation would have evoked at higher current levels. This was true even when another object was present in the scene and also occupied a nearby location within a V4 neuron's receptive field, as had previously been demonstrated by manipulating the distribution of attention (Chelazzi et al., 1993, 1998, 2001; Moran & Desimone, 1985; Reynolds et al., 1999). This indicates that neural activity in the frontal eye fields can have a causal effect on visual representations in the ventral stream and might work by prioritizing object processing in a particular location in the visual field.

Despite the apparent causality between the stimulation of the frontal eye fields and ventral stream processing, it was still unclear if activity in the frontal eye fields necessarily preceded that of the ventral stream during active visual search through more complex multiobject scenes. By simultaneously recording from the frontal eye fields and the inferior temporal cortex, Monosov, Sheinberg, and Thompson (2010) attempted to distinguish between an early selection process, in which spatial selection precedes object identification, and a late selection process, in which object identification precedes spatial selection.

In their experiment, monkeys were rewarded for identifying a target object presented with an array of several distracter objects while the monkey maintained central fixation. The target on each trial was known to be either effective or ineffective at driving the responses of an inferior temporal neuron being recorded from. The monkeys were not allowed to look directly to any of the objects, but the target location was presumably assigned a high priority, leading to a covert attentional shift. In this task information on visuospatial selection related to target location could be read out from neurons in the frontal eye fields before information about object identity useful for recognition was available in inferior temporal neurons.

These results are in agreement with those of Zhou and Desimone (2011), who simultaneously recorded from both the frontal eye fields and V4. In their task, monkeys were presented with a scene that contained a single searched-for target object and several distracter objects, some of which matched the target in color or shape. Neurons in both the frontal eye fields and V4 tended to respond more to objects that shared features with the target object (and were thus potentially task-relevant), but this feature-based attentional enhancement developed faster in the frontal eye fields than in V4. Feature-based attentional effects in V4 could therefore be caused by top-down selection from the frontal eye fields.

What about the lateral intraparietal area? Buschman and Miller (2007) recorded the activity of neurons in the lateral intraparietal area and the frontal eye fields while monkeys searched for a target within a visual scene. In some cases the target was not particularly conspicuous because it shared some features with the distracters (similar to the experiment by Zhou & Desimone, 2011); the monkeys therefore needed to find and identify the target based on its remembered appearance. In other cases the target was of a color different from that of the distracters and was therefore almost immediately noticeable based on bottom-up feature contrast. Buschman and Miller (2007) found that, in the former case, neurons of the frontal eye fields signaled the location of the target at an earlier time point than neurons in the lateral intraparietal area. In the latter case, when the target was visually salient, neurons in the lateral intraparietal area "found" the target faster than neurons in the frontal eye fields.

The results of the experiments described above are consistent with the idea that visual recognition mechanisms in the ventral stream are biased toward certain objects in visual scenes because they occupy a location that has already been assigned a high priority by regions such as the lateral intraparietal area and the frontal eye fields. The two regions might both implement a priority map, and these priority maps might very well update each other through the strong connections that exist between these cortical areas (Ferraina, Pare, & Wurtz, 2002; Lewis & Van Essen, 2000; Stanton et al., 1995). They might nonetheless also divide some of their responsibilities, so that priority signals based on bottom-up saliency could possibly arrive first to the ventral stream from the lateral intraparietal area, whereas priority based on task-relevant yet inconspicuous features might initially be sent from the frontal eye fields. If objects compete for representation in the ventral visual stream when complex multiobject scenes are viewed, then weighing them by their assigned priority can bias this competition in favor of those likely to be behaviorally important.

So far we have talked about how locations of objects might be prioritized so that neural machinery in the ventral stream is mostly used to process objects that are deemed most important at any given time. But how does the ventral stream actually behave when objects need to be found and recognized in realistic scenes?

When objects are presented alone, selectivity of inferior temporal neurons is remarkably insensitive to retinal location (Rolls, Aggelopoulos, & Zheng, 2003). Individual neurons, however, do tend to respond to several different objects, so if these objects were all presented to such a neuron at once, the resulting neural activity might be expected to end in a total cacophony.

This, however, does not actually seem to occur. Rolls and colleagues (2003) visually presented an effective object (known to drive the neuron) at various eccentricities away from the center of the visual field. Inferior temporal neurons tended to respond robustly to the object when it was shown at the center of gaze, regardless of whether it was presented alone on a blank background or as part of a scene. The neural firing rate, however, sharply fell off with increased object eccentricity in the complex scene but stayed relatively high even when the effective object was shown in the periphery of the visual field on a blank background.

The interpretation of Rolls and colleagues (2003) is that receptive fields of inferior temporal neurons shrink in complex natural scenes and that neural output is weighted toward objects closest to the fovea. However, the center of gaze often, but not always, coincides with the center of visuospatial attention. Although a central location might always be given a relatively high attentional priority, the priority given to a single object on a blank background, even a peripheral one, is also likely to be high because there are no other potential objects that compete with it. It is quite likely that a single object on a blank background will immediately attract the monkey's attention, and under most normal circumstances, the object would also become the target of the monkey's next saccade. In a complex scene the priority given to the object in question is likely to be comparatively lower because multiple other stimuli are present and are also given some weight.

Another possible way of thinking about this phenomenon, therefore, is that inferior temporal neurons process mainly the object that is at the center of attention. This is

A

B
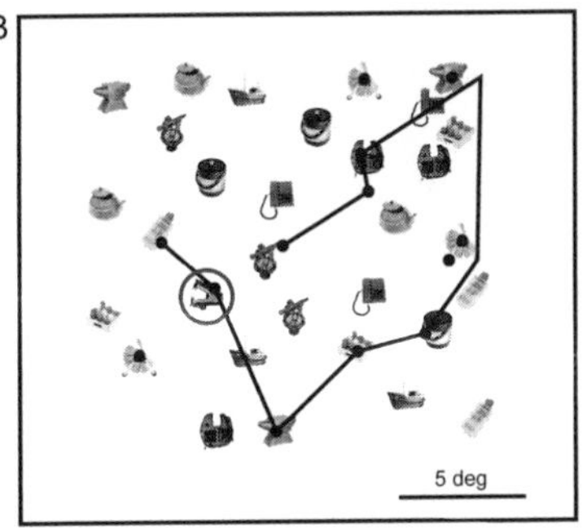

Figure 5.2 (plate 11)
Paradigms for studying visual search in neurophysiological studies. The range of tasks used to understand the physiological contributions of particular neural areas involved in visual search constitutes a large class of related, but distinct, experimental paradigms. Many of these focus on the ability to select a region of space in the context of a simple environment: the subject is presented with arrays of simple stimuli, such as colored patches or lines in various orientations, that form the basis for either pop-out or conjunctive search displays (Treisman & Gelade, 1980). Here, the recognition of an object is not a critical component of the task. This broad class of experimental studies has primarily investigated the contributions of the frontoparietal network with the objective of understanding the physiological process of attentional selection. Some other tasks are designed to investigate object recognition in a scene with other distracting objects or complex backgrounds. These tasks can be used to investigate either the dorsal or ventral visual streams and, sometimes, directly look at the interaction of these two pathways. The visual search for a realistic object embedded within a natural scene is very little used because of the complexities involved in teaching such a task to nonhuman primates and the difficulty of analyzing neural responses from an animal engaged in an active search through a complex scene. Two examples of this kind are depicted. (A) Sheinberg and Logothetis (2001) trained monkeys to search for and then categorize objects embedded in real-world scenes. Each category of possible objects was arbitrarily assigned to either a left or a right lever press (e.g., pull the left lever whenever you find a monkey in the scene). A saccadic scan path from a single trial is shown; the search target in this case was an image of a monkey. (B) In the experiment by Mruczek and Sheinberg (2007) monkeys had to search for any one of a number of possible target objects among either familiar or unfamiliar task-irrelevant objects. Once the target object was found, the monkey pushed one of two buttons, only one of which had been arbitrarily deemed to be the correct choice for that particular object.

in alignment with the results of Sheinberg and Logothetis (2001), who found that inferior temporal neurons represent an object in a complex scene before it is actually fixated, as long as it is the target of the next fixation and has thus presumably become the center of attention.

Sheinberg and Logothetis (2001) recorded from inferior temporal neurons while monkeys searched for and identified objects in a complex visual scene (figure 5.2A, plate 11). One of dozens of possible target objects was either presented alone on a blank background or blended into a complex realistic scene. In the latter case the monkey had to make several saccades within the scene before he would find and recognize the target, just as one would have to do when searching for an object in a cluttered real-world environment.

Under these conditions, Sheinberg and Logothetis (2001) found that responses of inferior temporal neurons appeared to be linked to the monkeys' recognition performance, even in the complex scenes. They also saw that the object selectivity of inferior temporal neurons was similar, regardless of whether a target object was presented alone or in a complex scene. In both cases the neurons often responded vigorously to a few objects and almost not at all to many others, and the objects that effectively drove the neurons when presented on the blank background also tended to be the preferred objects in the complex scene. This object selectivity did not unfold until just before the monkey made a saccadic eye movement to the target object in a scene, presumably because this was the time when the monkey noticed that the object was there and could recognize it. Once the monkeys seemingly attended to the target's location, these neurons behaved almost as if the rest of the complex environment had been filtered out. In the complex scene condition it was not unusual for the eyes to be within a few degrees of an effective target object without eliciting a response from a neuron if the monkey's search behavior indicated that it was not actually noticed.

Effectively filtering out several salient objects that nonetheless happen to be irrelevant for one's current goal might require some learning. Visual object representations in the inferior temporal cortex do appear to be plastic and develop as a function of experience with particular objects. In this region neural representations for familiar objects appear to be more selective than neural representations for unfamiliar objects (Freedman, Riesenhuber, Poggio, & Miller, 2006; Kobatake, Wang, & Tanaka, 1998; Woloszyn & Sheinberg, 2012), which is likely related to the increased ability to recognize objects that one has encountered many times before.

However, these studies were conducted by presenting isolated objects at fixation. In real-world scenes a task-relevant object often needs to be found and recognized even in the presence of several unimportant objects that are best left ignored, such as when you need to find the key to your house among your car key, your office key, and the key to your garage. Familiarity with such irrelevant and distracting objects appears to promote more efficient visual search through an environment. For example, Mruczek and Sheinberg (2005) found that people search more effectively through scenes with familiar distracting objects than through scenes with distracting objects that are unfamiliar to them.

Mruczek and Sheinberg (2007) then tested whether such familiarity with the contents of a scene affected object processing in the inferior temporal cortex. They trained monkeys to search for a particular object, effective at driving neural responses, hidden among either familiar objects (which had been seen hundreds of times) or novel objects (which were new for each particular testing session; see figure 5.2B, plate 11). Mruczek and Sheinberg (2007) found that effective target objects could be detected by individual inferior temporal cortical neurons at greater eccentricities when the distracting objects were familiar than when they were unfamiliar. Furthermore, this enlarged functional visual receptive field was predictive of the monkeys' performance, which showed greater efficiency at finding objects among familiar distracters compared to novel ones, just as had been shown for human observers (Mruczek & Sheinberg, 2005).

Interestingly, the largest differences in firing rate for the target object in the presence of novel and familiar distracting objects were found when the target was shown in the periphery, not in the center of the visual field. This is reminiscent of the results by Rolls et al. (2003), who found that objects in the visual periphery were less effective for driving the responses of inferior temporal neurons when presented in a scene than when presented alone, whereas the difference between object responses with and without a scenic background was not as apparent when the object was at the center of gaze. It is possible that top-down priority signals are used to down-weight familiar distracters so that a peripheral target object becomes more noticeable, attracts more attention, and is thus more likely to be processed by inferior temporal neurons. A relevant object in a familiar scene might be treated by inferior temporal neurons almost as if the object were the only thing present.

The inferior temporal cortex likely plays a role in object recognition in visual scenes, but it might not always be crucial for this task. Meyers et al. (2010) recorded neural activity in the anterior part of the inferior temporal cortex while monkeys viewed natural scenes, some of which contained animals and some that did not. Meyers and colleagues (2010) then looked to see if the neural data contained enough information to judge whether the monkey had been looking at a scene with or without an animal. Although decoding accuracy was above what would have been expected by mere chance, the information useful for this classification appeared quite late in the responses of inferior temporal neurons (around 125 ms after the visual onset of a scene). When given the choice of two scenes only one of which contains an animal, both people (Kirchner & Thorpe, 2006) and monkeys (Girard, Jouffrais, & Kirchner, 2008) can reliably make a saccade to the animal scene before any information about object category is present in inferior temporal neural responses to similar scenes. Such rapid object recognition in natural scenes seems to rely on regions other than the inferior temporal cortex.

This kind of object recognition in visual scenes might not even depend on selective attention. Current theories of visual search (Wolfe, Vo, Evans, & Greene, 2011) have now begun to acknowledge that searching through scenes might involve both a selective and a nonselective path. Whereas the former requires the selection of individual objects, the latter makes use of summary statistics from the whole scene. This global information might in some cases suffice for simple categorization, such as distinguishing between an animal and a nonanimal and could provide the basis for a scene-based guidance of visual search.

Last, we must consider the identification and processing of spatial contexts themselves. Kornblith and colleagues (2013) recently reported electrophysiological data from two newly identified ventral stream areas termed the lateral place patch (LPP) and the medial place patch (MPP), a possible macaque homologue to the human parahippocampal place area (PPA). These scene-processing regions contain cells that are strongly responsive not just to individual objects (as are neurons in the inferior temporal cortex) or the spatial distribution of object salience or priority (as are neurons in the frontoparietal attention systems) but instead respond to the environmental structure and texture of both familiar and unfamiliar scenes. Stimulation studies done in concert with the neural recordings suggest that these regions lie within a neural network that feeds back to other regions that process visual information, such as V4. Interestingly, these scene-based responses were observed even while the monkey engaged in a passive fixation task, suggesting that active search or exploration through the environment was not necessary to activate brain regions that process scene stimuli. Performing recordings in these scene-selective areas while animals engage in active behavioral tasks may further elucidate the contribution of this processing to search and recognition.

This chapter has attempted to follow a scene as it passes through the visual hierarchy until an object in the scene is found and successfully recognized. We have focused on two divergent yet highly interactive neural systems, a frontoparietal network and the ventral visual stream, one of which allocates and directs visual attention to important features of the environment while the other processes and identifies the objects in that environment. This division of labor by the two systems is supported by the work reviewed here as well as by behavioral work that indicates that spatial selection and target identification are separable parts of finding objects in visual scenes (Ghorashi, Enns, Klein, & Di Lollo, 2010).

Although some progress has recently been made in understanding how the visual system finds and identifies objects in complex environments, several aspects of this work make it difficult to generalize from these experimental results to the more human undertaking of "scene processing."

First, because of the difficulty of analyzing data from experiments with sufficiently complex stimuli, a large majority of electrophysiological work makes use of simplified stimuli that are easier to control but are quite impoverished in comparison to the richness of visual input we encounter in our everyday lives. Because neural responses can be highly nonlinear, it is almost impossible to know how well experimental results derived from highly artificial conditions translate to the types of environments that the visual system actually evolved, developed, and learned to process.

Second, although most human observers no doubt recognize a high-resolution photograph of a "beach scene" as a real-world place with three-dimensional structure and individuated objects, and as a setting for particular memories consistent with individual experiences, the animals studied in the research lab arguably lack the former of these abilities and definitely lack the latter. How this translation of a two-dimensional photographic representation into a three-dimensional environment with shared experiences changes the outcome or process of neural activity is difficult to say.

Last, even pictures of real-world scenes are in many cases a poor approximation of the genuine thing because the real world is ever changing. When searching for a friendly face while walking through a crowd of people, we are not only moving within the environment but the environment is also moving, in chaotic ways, around us. How the visual system keeps track of contexts with many moving objects and still manages to search in an efficient and effective manner—and how what has been discovered about searching for and recognizing objects in static scenes translates into dynamic scenes—is for the most part still an open question. It is also the case that we often search in service of some larger behavioral goal (e.g., "make a cup of coffee"), but it is rarer to search per se with no larger goal in mind (e.g., "locate the sugar on the shelf"). How these priority maps dynamically change and update to reflect the nature of our shifting behavioral goals is still challenging to explain.

These open issues deal mainly with the interpretative difficulty in studying the very human experience of "scene perception" in nonhuman animals. In some ways a comparative psychology and neurobiology of scene perception will always suffer from the difficulty of translating naturalistic vision into the laboratory. But despite this issue, the last decade has seen enormous progress in understanding how the brain makes sense of complex environments and the objects within them and has seen the discovery of multiple homologous human and animal pathways.

Thus, by designing experiments that require animals to search through complex environments, we can better understand how visual processing occurs through the dynamic interaction between multiple brain regions within the more naturalistic context of a visual scene.

Acknowledgments

This work was supported by the National Science Foundation (NSF 0827427 to DLS) and the National Institutes of Health (5F32EY021692 to DIB and R01 EY014681 to DLS).

Note

*These authors contributed equally to this work and are listed alphabetically.

References

Allman, J., Miezin, F., & McGuinness, E. (1985). Stimulus specific responses from beyond the classical receptive field: Neurophysiological mechanisms for local-global comparisons in visual neurons. *Annual Review of Neuroscience*, *8*, 407–430.

Armstrong, K. M., Fitzgerald, J. K., & Moore, T. (2006). Changes in visual receptive fields with microstimulation of frontal cortex. *Neuron*, *50*(5), 791–798.

Attneave, F. (1954). Some informational aspects of visual perception. *Psychological Review*, *61*(3), 183–193.

Baluch, F., & Itti, L. (2011). Mechanisms of top-down attention. *Trends in Neurosciences*, *34*(4), 210–224.

Barash, S., Bracewell, R. M., Fogassi, L., Gnadt, J. W., & Andersen, R. A. (1991). Saccade-related activity in the lateral intraparietal area. I. Temporal properties; comparison with area 7a. *Journal of Neurophysiology*, *66*(3), 1095–1108.

Barlow, H. (2001). Redundancy reduction revisited. *Network*, *12*(3), 241–253.

Baylis, G. C., Rolls, E. T., & Leonard, C. M. (1987). Functional subdivisions of the temporal lobe neocortex. *Journal of Neuroscience*, *7*(2), 330–342.

Bisley, J. W., & Goldberg, M. E. (2010). Attention, intention, and priority in the parietal lobe. *Annual Review of Neuroscience*, *33*, 1–21.

Blatt, G. J., Andersen, R. A., & Stoner, G. R. (1990). Visual receptive field organization and cortico-cortical connections of the lateral intraparietal area (area LIP) in the macaque. *Journal of Comparative Neurology*, *299*(4), 421–445.

Bruce, C. J., & Goldberg, M. E. (1985). Primate frontal eye fields. I. Single neurons discharging before saccades. *Journal of Neurophysiology*, *53*(3), 603–635.

Bruce, C. J., Goldberg, M. E., Bushnell, M. C., & Stanton, G. B. (1985). Primate frontal eye fields. II. Physiological and anatomical correlates of electrically evoked eye movements. *Journal of Neurophysiology*, *54*(3), 714–734.

Buschman, T. J., & Miller, E. K. (2007). Top-down versus bottom-up control of attention in the prefrontal and posterior parietal cortices. *Science*, *315*(5820), 1860–1862.

Chelazzi, L., Duncan, J., Miller, E. K., & Desimone, R. (1998). Responses of neurons in inferior temporal cortex during memory-guided visual search. *Journal of Neurophysiology*, *80*(6), 2918–2940.

Chelazzi, L., Miller, E. K., Duncan, J., & Desimone, R. (1993). A neural basis for visual search in inferior temporal cortex. *Nature*, *363*(6427), 345–347.

Chelazzi, L., Miller, E. K., Duncan, J., & Desimone, R. (2001). Responses of neurons in macaque area V4 during memory-guided visual search. *Cerebral Cortex*, *11*(8), 761–772.

Dan, Y., Atick, J. J., & Reid, R. C. (1996). Efficient coding of natural scenes in the lateral geniculate nucleus: Experimental test of a computational theory. *Journal of Neuroscience*, *16*(10), 3351–3362.

Desimone, R., & Duncan, J. (1995). Neural mechanisms of selective visual attention. *Annual Review of Neuroscience*, *18*, 193–222.

Distler, C., Boussaoud, D., Desimone, R., & Ungerleider, L. G. (1993). Cortical connections of inferior temporal area TEO in macaque monkeys. *Journal of Comparative Neurology*, *334*(1), 125–150.

Dong, D. W., & Atick, J. J. (1995). Statistics of natural time-varying images. *Network (Bristol, England)*, *6*(3), 345–358.

Fecteau, J. H., & Munoz, D. P. (2006). Salience, relevance, and firing: A priority map for target selection. *Trends in Cognitive Sciences*, *10*(8), 382–390.

Ferraina, S., Pare, M., & Wurtz, R. H. (2002). Comparison of cortico-cortical and cortico-collicular signals for the generation of saccadic eye movements. *Journal of Neurophysiology*, *87*(2), 845–858.

Field, D. J. (1987). Relations between the statistics of natural images and the response properties of cortical cells. *Journal of the Optical Society of America. A, Optics, Image Science, and Vision*, *4*(12), 2379–2394.

Freedman, D. J., Riesenhuber, M., Poggio, T., & Miller, E. K. (2006). Experience-dependent sharpening of visual shape selectivity in inferior temporal cortex. *Cerebral Cortex*, *16*(11), 1631–1644.

Ghorashi, S., Enns, J. T., Klein, R. M., & Di Lollo, V. (2010). Spatial selection and target identification are separable processes in visual search. *Journal of Vision (Charlottesville, Va.)*, *7*, 1–12.

Girard, P., Jouffrais, C., & Kirchner, C. (2008). Ultra-rapid categorisation in non-human primates. *Animal Cognition*, *11*(3), 485–493.

Goldberg, M. E., Bisley, J. W., Powell, K. D., & Gottlieb, J. (2006). Saccades, salience and attention: The role of the lateral intraparietal area in visual behavior. *Progress in Brain Research*, *155*, 157–175.

Goodale, M. A., & Milner, A. D. (1992). Separate visual pathways for perception and action. *Trends in Neurosciences*, *15*(1), 20–25.

Gross, C. G., Bender, D. B., & Rocha-Miranda, C. E. (1969). Visual receptive fields of neurons in inferotemporal cortex of the monkey. *Science*, *166*(3910), 1303–1306.

Guo, K., Robertson, R. G., Mahmoodi, S., & Young, M. P. (2005). Centre-surround interactions in response to natural scene stimulation in the primary visual cortex. *European Journal of Neuroscience*, *21*(2), 536–548.

Henderson, J. M., & Hollingworth, A. (1999). High-level scene perception. *Annual Review of Psychology*, *50*, 243–271.

Hochstein, S., & Ahissar, M. (2002). View from the top: Hierarchies and reverse hierarchies in the visual system. *Neuron*, *36*(5), 791–804.

Hubel, D. H., & Wiesel, T. N. (1959). Receptive fields of single neurones in the cat's striate cortex. *Journal of Physiology*, *148*(3), 574–591.

Hubel, D. H., & Wiesel, T. N. (1968). Receptive fields and functional architecture of monkey striate cortex. *Journal of Physiology*, *195*(1), 215–243.

Itti, L., & Koch, C. (2000). A saliency-based search mechanism for overt and covert shifts of visual attention. *Vision Research*, *40*(10–12), 1489–1506.

Itti, L., & Koch, C. (2001). Computational modelling of visual attention. *Nature Reviews. Neuroscience*, *2*(3), 194–203.

Itti, L., Koch, C., & Niebur, E. (1998). A model of saliency-based visual attention for rapid scene analysis. *IEEE Transactions on Pattern Analysis and Machine Intelligence*, *20*, 1254–1259.

Kayser, C., Salazar, R. F., & Konig, P. (2003). Responses to natural scenes in cat V1. *Journal of Neurophysiology*, *90*(3), 1910–1920.

Kersten, D. (1987). Predictability and redundancy of natural images. *Journal of the Optical Society of America. A, Optics, Image Science, and Vision*, *4*(12), 2395–2400.

Kiani, R., Esteky, H., & Tanaka, K. (2005). Differences in onset latency of macaque inferotemporal neural responses to primate and non-primate faces. *Journal of Neurophysiology*, *94*(2), 1587–1596.

Kirchner, H., & Thorpe, S. J. (2006). Ultra-rapid object detection with saccadic eye movements: Visual processing speed revisited. *Vision Research*, *46*(11), 1762–1776.

Knierim, J. J., & van Essen, D. C. (1992). Neuronal responses to static texture patterns in area V1 of the alert macaque monkey. *Journal of Neurophysiology*, *67*(4), 961–980.

Kobatake, E., Wang, G., & Tanaka, K. (1998). Effects of shape-discrimination training on the selectivity of inferotemporal cells in adult monkeys. *Journal of Neurophysiology*, *80*(1), 324–330.

Kornblith, S., Cheng, X., Ohayon, S., & Tsao, D. Y. (2013). A network for scene processing in the macaque temporal lobe. *Neuron*, *79*(4), 766–781.

Land, M. F., & Hayhoe, M. (2001). In what ways do eye movements contribute to everyday activities? *Vision Research*, *41*(25–26), 3559–3565.

Lehky, S. R., Sejnowski, T. J., & Desimone, R. (1992). Predicting responses of nonlinear neurons in monkey striate cortex to complex patterns. *Journal of Neuroscience*, *12*(9), 3568–3581.

Lewis, J. W., & Van Essen, D. C. (2000). Corticocortical connections of visual, sensorimotor, and multimodal processing areas in the parietal lobe of the macaque monkey. *Journal of Comparative Neurology*, *428*(1), 112–137.

Logothetis, N. K., & Sheinberg, D. L. (1996). Visual object recognition. *Annual Review of Neuroscience*, *19*, 577–621.

Meyers, E., Embark, H., Freiwald, W., Serre, T., Kreiman, G., & Poggio, T. (2010). *Examining high level neural representations of cluttered scenes*. MIT-CSAIL-TR-2010-034 / CBCL-289, Massachusetts

Institute of Technology, Cambridge, MA, July 29, 2010. Available at http://cbcl.mit.edu/publications/ps/Meyers_MIT-CSAIL-TR-2010-034-CBCL-289.pdf.

Monosov, I. E., Sheinberg, D. L., & Thompson, K. G. (2010). Paired neuron recordings in the prefrontal and inferotemporal cortices reveal that spatial selection precedes object identification during visual search. *Proceedings of the National Academy of Sciences of the United States of America*, *107*(29), 13105–13110.

Montagna, B., Pestilli, F., & Carrasco, M. (2009). Attention trades off spatial acuity. *Vision Research*, *49*(7), 735–745.

Moore, T., & Fallah, M. (2004). Microstimulation of the frontal eye field and its effects on covert spatial attention. *Journal of Neurophysiology*, *91*(1), 152–162.

Moran, J., & Desimone, R. (1985). Selective attention gates visual processing in the extrastriate cortex. *Science*, *229*(4715), 782–784.

Mruczek, R. E., & Sheinberg, D. L. (2005). Distractor familiarity leads to more efficient visual search for complex stimuli. *Perception & Psychophysics*, *67*(6), 1016–1031.

Mruczek, R. E., & Sheinberg, D. L. (2007). Context familiarity enhances target processing by inferior temporal cortex neurons. *Journal of Neuroscience*, *27*(32), 8533–8545.

Nassi, J. J., & Callaway, E. M. (2009). Parallel processing strategies of the primate visual system. *Nature Reviews. Neuroscience*, *10*(5), 360–372.

Peters, A., & Payne, B. R. (1993). Numerical relationships between geniculocortical afferents and pyramidal cell modules in cat primary visual cortex. *Cerebral Cortex*, *3*(1), 69–78.

Ratliff, F., & Riggs, L. A. (1950). Involuntary motions of the eye during monocular fixation. *Journal of Experimental Psychology*, *40*(6), 687.

Reynolds, J. H., Chelazzi, L., & Desimone, R. (1999). Competitive mechanisms subserve attention in macaque areas V2 and V4. *Journal of Neuroscience*, *19*(5), 1736–1753.

Rolls, E. T., Aggelopoulos, N. C., & Zheng, F. (2003). The receptive fields of inferior temporal cortex neurons in natural scenes. *Journal of Neuroscience*, *23*(1), 339–348.

Rucci, M. (2008). Fixational eye movements, natural image statistics, and fine spatial vision. *Network (Bristol, England)*, *19*(4), 253–285.

Schall, J. D., Morel, A., King, D. J., & Bullier, J. (1995). Topography of visual cortex connections with frontal eye field in macaque: Convergence and segregation of processing streams. *Journal of Neuroscience*, *15*(6), 4464–4487.

Schmolesky, M. T., Wang, Y., Hanes, D. P., Thompson, K. G., Leutgeb, S., Schall, J. D., et al. (1998). Signal timing across the macaque visual system. *Journal of Neurophysiology*, *79*(6), 3272–3278.

Sheinberg, D. L., & Logothetis, N. K. (2001). Noticing familiar objects in real world scenes: The role of temporal cortical neurons in natural vision. *Journal of Neuroscience*, *21*(4), 1340–1350.

Sillito, A. M., Grieve, K. L., Jones, H. E., Cudeiro, J., & Davis, J. (1995). Visual cortical mechanisms detecting focal orientation discontinuities. *Nature*, *378*(6556), 492–496.

Simoncelli, E. P., & Olshausen, B. A. (2001). Natural image statistics and neural representation. *Annual Review of Neuroscience*, *24*, 1193–1216.

Stanton, G. B., Bruce, C. J., & Goldberg, M. E. (1995). Topography of projections to posterior cortical areas from the macaque frontal eye fields. *Journal of Comparative Neurology*, *353*(2), 291–305.

Tamura, H., & Tanaka, K. (2001). Visual response properties of cells in the ventral and dorsal parts of the macaque inferotemporal cortex. *Cerebral Cortex*, *11*(5), 384–399.

Tanaka, T., Nishida, S., Aso, T., & Ogawa, T. (2013). Visual response of neurons in the lateral intraparietal area and saccadic reaction time during a visual detection task. *European Journal of Neuroscience*, *37*(6), 942–956.

Thomas, N. W., & Paré, M. (2007). Temporal processing of saccade targets in parietal cortex area LIP during visual search. *Journal of Neurophysiology*, *97*(1), 942–947.

Thompson, K. G., & Bichot, N. P. (2005). A visual salience map in the primate frontal eye field. *Progress in Brain Research*, *147*, 251–262.

Treisman, A. M., & Gelade, G. (1980). A feature-integration theory of attention. *Cognitive Psychology, 12*(1), 97–136.

Tsotsos, J. K. (2008). *What roles can attention play in recognition?* Paper presented at the Development and Learning: 7th IEEE International Conference.

Ungerleider, L. G., & Haxby, J. V. (1994). "What" and "where" in the human brain. *Current Opinion in Neurobiology, 4*(2), 157–165.

Ungerleider, L. G., & Mishkin, M. (1982). Two cortical visual systems. In D. J. Ingle, M. A. Goodale, & R. J. W. Mansfield (Eds.), *Analysis of visual behavior* (pp. 549–586). Cambridge, MA: MIT Press.

Van Essen, D. C., Anderson, C. H., & Felleman, D. J. (1992). Information processing in the primate visual system: An integrated systems perspective. *Science, 255*(5043), 419–423.

Vinje, W. E., & Gallant, J. L. (2000). Sparse coding and decorrelation in primary visual cortex during natural vision. *Science, 287*(5456), 1273–1276.

Webster, M. J., Bachevalier, J., & Ungerleider, L. G. (1994). Connections of inferior temporal areas TEO and TE with parietal and frontal cortex in macaque monkeys. *Cerebral Cortex, 4*(5), 470–483.

Wolfe, J. M. (1994). Guided search 2.0: A revised model of visual search. *Psychonomic Bulletin & Review, 1*, 202–238.

Wolfe, J. M. (2007). Guided search 4.0: Current progress with a model of visual search. In W. Gray (Ed.), *Integrated models of cognitive systems* (pp. 99–119). Oxford: Oxford University Press.

Wolfe, J. M., & Horowitz, T. S. (2004). What attributes guide the deployment of visual attention and how do they do it? *Nature Reviews. Neuroscience, 5*(6), 495–501.

Wolfe, J. M., Vo, M. L., Evans, K. K., & Greene, M. R. (2011). Visual search in scenes involves selective and nonselective pathways. *Trends in Cognitive Sciences, 15*(2), 77–84.

Woloszyn, L., & Sheinberg, D. L. (2012). Effects of long-term visual experience on responses of distinct classes of single units in inferior temporal cortex. *Neuron, 74*(1), 193–205.

Zhou, H., & Desimone, R. (2011). Feature-based attention in the frontal eye field and area V4 during visual search. *Neuron, 70*(6), 1205–1217.

6

Neural Systems for Visual Scene Recognition

Russell A. Epstein

What Is a Scene?

If you were to look out my office window at this moment, you would see a campus vista that includes a number of trees, a few academic buildings, and a small green pond. Turning your gaze the other direction, you would see a room with a desk, a bookshelf, a rug, and a couch. Although the objects are of interest in both cases, what you see in each view is more than just a collection of disconnected objects—it is a coherent entity that we colloquially label a "scene." In this chapter I describe the neural systems involved in the perception and recognition of scenes. I focus in particular on the parahippocampal place area (PPA), a brain region that plays a central role in scene processing, emphasizing the many new studies that have expanded our understanding of its function in recent years.

Let me first take a moment to define some terms. By a "scene" I mean a section of a real-world environment (or an artificial equivalent) that typically includes both foreground objects and fixed background elements (such as walls and a ground plane) and that can be ascertained in a single view (Epstein & MacEvoy, 2011). For example, a photograph of a room, a landscape, or a city street is a scene—or, more precisely, an image of a scene. In this conceptualization "scenes" are contrasted with "objects," such as shoes and bottles, hawks and hacksaws, which are discrete, potentially movable entities without background elements that are bounded by a single contour. This definition follows closely on the one offered by Henderson and Hollingworth (1999), who emphasized the same distinction between scenes and objects and who made the point that scenes are often semantically coherent and even nameable. As a simple heuristic, one could say that objects are spatially compact entities one acts upon, whereas scenes are spatially distributed entities one acts within (Epstein, 2005).

Why should the visual system care about scenes? First and foremost, because scenes are places in the world. The fact that I can glance out at a scene and quickly identify it as "Locust Walk" or "Rittenhouse Square" means that I have an easy way to determine my current location, for example, if I were to become lost while taking a walk around Philadelphia. Of course, I could also figure this out by identifying individual objects, but the scene as a whole provides a much more stable and discriminative constellation of place-related cues. Second, because scenes provide important information about the objects that are likely to occur in a place and the actions that one should perform there (Bar, 2004). If I am hungry, for example, it makes more sense to look for something to eat in a kitchen than in a classroom. For this function it may be more important to recognize the scene as a member of a general scene category rather than as a specific unique place as one typically wants to do during spatial navigation. Finally, one might want to evaluate qualities of the scene that are independent of its category or identity, for example, whether a city street looks safe or dangerous, or whether travel along a path in the woods seems likely to bring one to food or shelter. I use the term *scene recognition* to encompass all three of these tasks (identification as a specific exemplar, classification as a member of a general category, evaluation of reward-related or aesthetic properties).

Previous behavioral work has shown that human observers have an impressive ability to recognize even briefly presented real-world scenes. In a classic series of studies Potter and colleagues (Potter, 1975, 1976; Potter & Levy, 1969) reported that subjects could detect a target scene within a sequence of scene distracters with 75% accuracy when the visual system was blitzed by scenes at a rate of 8 per second. The phenomenology of this effect is quite striking: although the scenes go by so quickly that most seem little more than a blur, the target scene jumps into awareness—even when it is cued by nothing more than a verbal label that provides almost no information about its exact appearance (e.g., “picnic”). The fact that we can select the target from the distracters implies that every scene in the sequence must have been processed up to the level of meaning (or *gist*). Related results were obtained by Biederman (1972), who reported that recognition of a single object within a briefly flashed (300–700 ms) scene was more accurate when the scene was coherent than when it was jumbled up into pieces. This result indicates that the human visual system can extract the meaning of a complex visual scene within a few hundred milliseconds and can use it to facilitate object recognition (for similar results, see Antes, Penland, & Metzger, 1981; Biederman, Rabinowitz, Glass, & Stacy, 1974; Fei-Fei, Iyer, Koch, & Perona, 2007; Thorpe, Fize, & Marlot, 1996).

Although one might argue that scene recognition in these earlier studies reduces simply to recognition of one or two critical objects, subsequent work has provided evidence that this is not the complete story. Scenes can also be identified based on their whole-scene characteristics, such as their overall spatial layout. Schyns and Oliva (1994) demonstrated that subjects could classify briefly flashed (30 ms) scenes into categories even if the images were filtered to remove all high-spatial-frequency information, leaving only the overall layout of coarse blobs, which conveyed little information about individual objects. More recently Greene and Oliva (2009b) developed a scene recognition model that operated on seven global properties: openness, expansion, mean depth, temperature, transience, concealment, and navigability. These properties predicted the performance of human observers insofar as scenes that were more similar in the property space were more often misclassified by the observers. Indeed, observers ascertained these global properties of outdoor scenes *prior to* identifying their basic-level category, suggesting that categorization may rely on these global properties (Greene & Oliva, 2009a). Computational modeling work has given further credence to the idea that scenes can be categorized based on whole-scene information by showing that human recognition of briefly presented scenes can be simulated by machine recognition systems that operate solely on the texture statistics of the image (Renninger & Malik, 2004).

In sum, both theoretical considerations and experimental data suggest that the visual system contains dedicated systems for scene recognition. In the following sections I describe the neuroscientific evidence that supports this proposition.

Scene-Responsive Brain Areas

Functional magnetic resonance imaging (fMRI) studies have identified several brain regions that respond preferentially to scenes (figure 6.1, plate 12). Of these, the first discovered, and the most studied, is the parahippocampal place area (PPA). This ventral occipitotemporal region responds much more strongly when people view scenes (landscapes, cityscapes, rooms) than when they view isolated single objects, and does not respond at all when people view faces (Epstein & Kanwisher, 1998). The scene-related response in the PPA is extremely reliable (Julian, Fedorenko, Webster, & Kanwisher, 2012): in my lab we have scanned hundreds of subjects, and we almost never encounter a person without a PPA.

The PPA is a functionally defined, rather than an anatomically defined region. This can lead to some confusion, as there is a tendency to conflate the PPA with parahippocampal cortex (PHC), its anatomical namesake. Although they are partially overlapping, these two regions are not the same. The PPA includes the posterior end of the PHC but extends beyond it posteriorly into the lingual gyrus and laterally across the collateral sulcus into the fusiform gyrus. Indeed, a recent study suggested that the most reliable locus of PPA activity may be on the fusiform (lateral) rather than the parahippocampal (medial) side of the collateral sulcus (Nasr et al., 2011). Some earlier studies reported activation in the lingual gyrus in response to houses and buildings (Aguirre, Zarahn, & D'Esposito, 1998). It seems likely that this "lingual landmark area" is equivalent to the PPA or at least the posterior portion of it. As we will see, the notion that the PPA is a "landmark" area turns out to be fairly accurate.

The PPA is not the only brain region that responds more strongly to scenes than to other visual stimuli. A large locus of scene-evoked activation is commonly observed

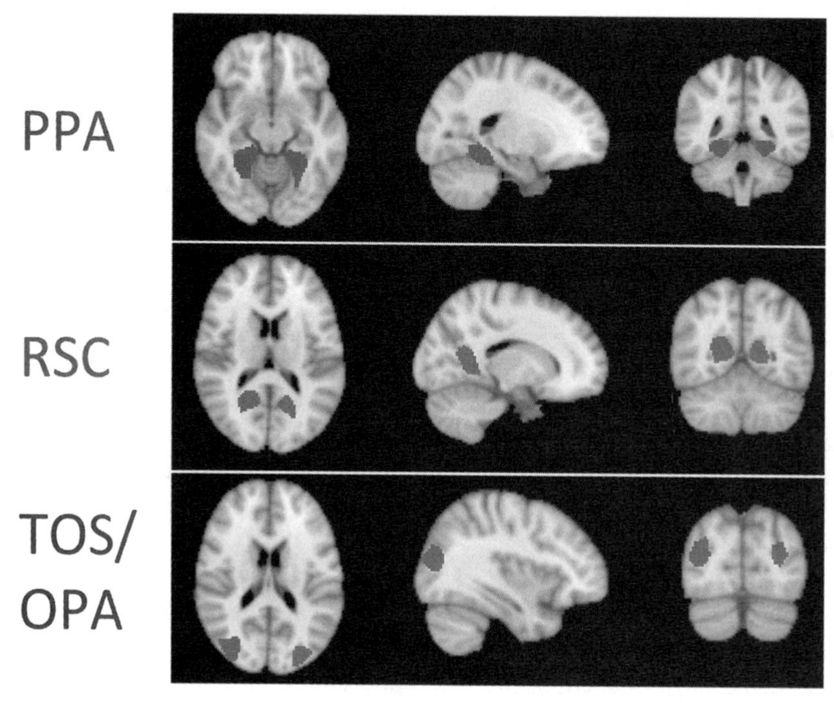

Figure 6.1 (plate 12)
Three regions of the human brain—the PPA, RSC, and TOS/OPA—respond preferentially to visual scenes. Shown here are voxels for which >80% of subjects (n = 42) have significant scenes > objects activation. Regions were defined using the algorithmic group-constrained subject-specific (GSS) method (Julian et al., 2012).

in the retrosplenial region extending posteriorly into the parietal-occipital sulcus. This has been labeled the retrosplenial complex (RSC) (Bar & Aminoff, 2003). Once again there is a possibility of confusion here because the functionally defined RSC is not equivalent to the retrosplenial cortex, which is defined based on cytoarchitecture and anatomy rather than fMRI response (Vann, Aggleton, & Maguire, 2009). A third region of scene-evoked activity is frequently observed near the transverse occipital sulcus (TOS) (Hasson, Levy, Behrmann, Hendler, & Malach, 2002); this has also been labeled the occipital place area (OPA) (Dilks, Julian, Paunov, & Kanwisher, 2013). Scene-responsive "patches" have been observed in similar areas in macaque monkeys, although the precise homologies still need to be established (Kornblith, Cheng, Ohayon, & Tsao, 2013; Nasr et al., 2011).

The PPA and RSC appear to play distinct but complementary roles in scene recognition. Whereas the PPA seems to be primarily involved in perceptual analysis of the scene, the RSC seems to play a more mnemonic role, which is best characterized as connecting the local scene to the broader spatial environment (Epstein & Higgins, 2007; Park & Chun, 2009). This putative division of labor is supported by several lines of evidence. The RSC is more sensitive than the PPA to place familiarity (Epstein, Higgins, Jablonski, & Feiler, 2007). That is, response in the RSC to photographs of familiar places is much greater than response to unfamiliar places. In contrast the PPA responds about equally to both (with a slight but significant advantage for the familiar places). The familiarity effect in the RSC suggests that it may be involved in situating the scene relative to other locations, since this operation can be performed only for places that are known. The minimal familiarity effect in the PPA, on the other hand, suggests that it supports perceptual analysis of the local (i.e., currently visible) scene that does not depend on long-term knowledge about the depicted place. When subjects are explicitly asked to retrieve spatial information about a scene, such as where the scene was located within the larger environment or which compass direction the camera was facing when the photograph was taken, activity is increased in the RSC but not in the PPA (Epstein, Parker, & Feiler, 2007). This suggests that the RSC (but not the PPA) supports spatial operations that can extend beyond the scene's boundaries (Park, Intraub, Yi, Widders, & Chun, 2007). A number of other studies have demonstrated RSC involvement in spatial memory, which I do not review here (Epstein, 2008; Vann et al., 2009; Wolbers & Buchel, 2005).

This division of labor between the PPA and the RSC is also supported by neuropsychological data (Aguirre & D'Esposito, 1999; Epstein, 2008). When the PPA is damaged due to stroke, patients have difficulty identifying places and landmarks, and they report that their sense of a scene as a coherent whole has been lost. Their ability to identify discrete objects within the scene, on the other hand, is largely unimpaired. This can lead to some remarkable behavior during navigation, such as attempting to recognize a house based on a small detail such as a mailbox or a door knocker rather than its overall appearance. Notably, some of these patients still retain long-term spatial knowledge—for example, they can sometimes draw maps showing the spatial relationships between the places that they cannot visually recognize. Patients with damage to RSC, on the other hand, display a very different problem. They can visually recognize places and buildings without difficulty, but they cannot use these landmarks to orient themselves in large-scale space. For example, they can look at a building and name it without hesitation, but they cannot tell from this whether they are facing north, south, east, or west, and they cannot point to any other location that is not immediately visible. It is as if they can perceive the scenes around them normally, but these scenes are "lost in space"—unmoored from their broader spatial context.

The idea that the PPA is involved in perceptual analysis of the currently visible scene gains further support from the discovery of retinotopic organization in this region (Arcaro, McMains, Singer, & Kastner, 2009). Somewhat unexpectedly, the PPA appears to contain not one but two retinotopic maps, both of which respond more strongly to scenes than to objects. This finding suggests that the PPA might in fact be a compound of two visual subregions whose individual functions are yet to be differentiated. Both of these subregions respond especially strongly to stimulation in the periphery (Arcaro et al., 2009; Levy, Hasson, Avidan, Hendler, & Malach, 2001; Levy, Hasson, Harel, & Malach, 2004), a pattern that contrasts with object-preferring regions such as the lateral occipital complex, which respond more strongly to visual stimulation near the fovea. This relative bias for peripheral information makes sense given that information about scene identity is likely to be obtainable from across the visual field. In contrast, objects are more visually compact and are usually foveated when they are of interest. Other studies have further confirmed the existence of retinotopic organization in the PPA by showing that its response is affected by the location of the stimulus relative to the fixation point but not by the location of the stimulus on the screen when these two quantities are dissociated by varying the fixation position (Golomb & Kanwisher, 2012; Ward, MacEvoy, & Epstein, 2010).

Thus, the overall organization of the PPA appears to be retinotopic. However, retinotopic organization does not preclude the possibility that the region might encode information about stimulus identity that is invariant to retinal position (DiCarlo, Zoccolan, & Rust, 2012; Schwarzlose, Swisher, Dang, & Kanwisher, 2008). Indeed, Sean MacEvoy and I observed that fMRI adaptation when scenes were repeated at different retinal locations was almost as great as adaptation when scenes were repeated at the same retinal location, consistent with position invariance (MacEvoy & Epstein, 2007). Golomb and colleagues (2011) similarly observed adaptation when subjects moved their eyes over a stationary scene image, thus varying retinotopic input. Interestingly, adaptation was also observed in this study when subjects moved their eyes in tandem with a moving scene, a manipulation that kept retinal input constant. Thus, PPA appears to represent scenes in an intermediate format that is neither fully dependent on the exact retinal image nor fully independent of it.

The observation that the PPA appears to act as a visual region, with retinotopic organization, seems at first glance to conflict with the traditional view that the PHC is part of the medial temporal lobe memory system. However, once again, we must be careful to distinguish between the PPA and the PHC. Although the PHC in monkeys is usually divided into two subregions, TF and TH, some neuroanatomical studies have indicated the existence of a posterior subregion that has been labeled TFO (Saleem, Price, & Hashikawa, 2007). Notably, the TFO has a prominent layer IV, making it cytoarchitechtonically more similar to the adjoining visual region V4 than to either TF or TH. This suggests that the TFO may be a visually responsive region. The PPA may be an amalgam of the TFO and other visually responsive territory. TF and TH, on the other hand, may be more directly involved in spatial memory. As we will see, a key function of the PPA may be extracting spatial information from visual scenes.

Beyond the PPA, RSC, and TOS, a fourth region that has been implicated in scene processing is the hippocampus. Although this region does not typically activate above baseline during scene perception or during mental imagery of familiar places (O'Craven & Kanwisher, 2000), it does activate when subjects are asked to construct detailed imaginations of novel scenes (Hassabis, Kumaran, & Maguire, 2007). Furthermore, patients with damage to the hippocampus are impaired on this scene construction task, insofar as their imaginations have fewer details and far less spatial coherence than those offered by normal subjects (Hassabis et al., 2007; but see Squire et al., 2010). Other neuropsychological studies have found that hippocampally damaged patients are impaired at remembering the spatial relationships between scene elements when these relationships must be accessed from multiple points of view (Hartley et al., 2007; King, Burgess, Hartley, Vargha-Khadem, & O'Keefe, 2002). These results suggest that the hippocampus forms an allocentric "map" of a scene that allows different scene elements to be assigned to different within-scene locations. However, it is unclear how important this map is for scene recognition under normal circumstances, as perceptual deficits after hippocampal damage are subtle (Lee et al., 2005; Lee, Yeung, & Barense, 2012). Thus, I focus here on the role of occipitotemporal visual regions in scene recognition, especially the PPA.

What Does the PPA Do?

We turn now to the central question of this chapter: *how* does the PPA represent scenes in order to recognize them? I first consider the representational level encoded by the PPA—that is, whether the PPA represents scene categories, individual scene/place exemplars, or specific views. Then I discuss the content of the information encoded in the PPA—whether it encodes the geometric structure of scenes, nongeometric visual quantities, or information about objects. As we will see, recent investigations of these issues lead us to a more nuanced view of the PPA under which it represents more than just scenes.

Categories versus Places versus Views

As noted in the beginning of this chapter, scenes can be recognized in several different ways. They can be identified as a member of a general scene category ("kitchen"), as a specific place ("the kitchen on the fifth floor of the Center for Cognitive Neuroscience"), or even as a distinct view of a place ("the CCN kitchen, observed from the south"). Which of these representational distinctions, if any, are made by the PPA?

The ideal way to answer this question would be to insert electrodes into the PPA and record from individual neurons. This would allow us to determine whether individual units—or multiunit activity patterns—code categories, places, or views (DiCarlo et al., 2012). Although some single-unit recordings have been made from medial temporal lobe regions—including PHC—in presurgical epilepsy patients (Ekstrom et al., 2003; Kreiman, Koch, & Fried, 2000; Mormann et al., 2008), no study has explicitly targeted the tuning of neurons in the PPA. Thus, we turn instead to neuroimaging data.

There are two neuroimaging techniques that can be used to probe the representational distinctions made by a brain region: multivoxel pattern analysis (MVPA) and fMRI adaptation (fMRIa). In MVPA one examines the multivoxel activity patterns elicited by different stimuli to determine which stimuli elicit patterns that are similar and which stimuli elicit patterns that are distinct (Cox & Savoy, 2003; Haxby et al., 2001). In fMRIa one examines the response to items presented sequentially under the hypothesis that response to an item will be reduced if it is preceded by an identical or representationally similar item (Grill-Spector, Henson, & Martin, 2006; Grill-Spector & Malach, 2001).

MVPA studies have shown that activity patterns in the PPA contain information about the scene category being viewed. These patterns can be used to reliably distinguish among beaches, forests, highways, and the like (Walther, Caddigan, Fei-Fei, & Beck, 2009). The information in these patterns does not appear to be epiphenomenal—when scenes are presented very briefly and masked to make recognition difficult, the categories confused by the subjects are the ones that are "confused" by the PPA. Thus, the representational distinctions made by the PPA seem to be closely related to the representational distinctions made by human observers. It is also possible to classify scene category based on activity patterns in a number of other brain regions, including RSC, early visual cortex, and (in some studies but not others) object-sensitive regions such as lateral occipital complex (LOC). However, activation patterns in these regions are not as tightly coupled to behavioral performance as activation patterns in the PPA (Walther et al., 2009; Walther, Chai, Caddigan, Beck, & Fei-Fei, 2011).

These MVPA results suggest that the PPA might represent scenes in terms of categorical distinctions—or at least, in such a way that categories are easily distinguishable. But what about more fine-grained distinctions? In the studies described above, each image of a "beach" or a "forest" depicted a different place, yet they were grouped together for analysis to find a common pattern for each category. To determine if the PPA represents individual places, we scanned University of Pennsylvania students while they viewed many different images of familiar landmarks (buildings and statues) that signify unique locations on the Penn campus (Morgan, MacEvoy, Aguirre, & Epstein, 2011). We were able to decode landmark identity from PPA activity patterns with high accuracy (figure 6.2). Moreover, accuracy for decoding of Penn landmarks was equivalent to accuracy for decoding of scene categories, as revealed by results from a contemporaneous experiment performed on the same subjects in the same scan session (Epstein & Morgan, 2012). Thus, the PPA appears to encode information that allows both scene categories and individual scenes (or at least, individual familiar landmarks) to be distinguished. But—as we discuss in the next section—the precise nature of that information, and how it differs from the information that allows such discriminations to be made in other areas such as early visual cortex and RSC, still needs to be determined.

Findings from fMRI adaptation studies are only partially consistent with these MVPA results. On one hand, fMRIa studies support the idea that the PPA distinguishes between different scenes (Ewbank, Schluppeck, & Andrews, 2005; Xu, Turk-Browne, & Chun, 2007). For example, in the Morgan et al. (2011) study described above, we observed reduced PPA response (i.e., adaptation) when images of the same landmark were shown sequentially, indicating that it considered the two images of the same landmark to be representationally similar (figure 6.2). However, we did *not* observe adaptation when scene category was repeated—for example, when images of two different beaches were shown in succession (Epstein & Morgan, 2012). Thus, if one only had the adaptation results, one would conclude that the PPA represents individual landmarks or scenes but does not group those scenes into categories.

Indeed, other fMRIa studies from my laboratory have suggested that scene representations in the PPA can be even more stimulus specific. When we present two views of the same scene in sequence—for example, an image of a building viewed from the southeast followed by an image of the same building viewed from the southwest—the two images only partially cross-adapt each other (Epstein, Graham, & Downing, 2003; Epstein, Higgins, et al., 2007; Epstein, Parker, & Feiler, 2008). This indicates that the PPA treats these different views as representationally distinct items, even though they may depict many of the same details (e.g., the same front door, the same building facade, the same statue in front of the building). Strikingly, even overlapping images that are cut out from a larger scene panorama are treated as distinct items by the PPA (Park & Chun, 2009).

What are we to make of this apparent discrepancy between the fMRIa and MVPA results? The most likely explanation is that MVPA and fMRIa interrogate different aspects of the PPA neural code (Epstein & Morgan, 2012). For example, fMRIa may reflect processes that operate at the level of the single unit (Drucker & Aguirre, 2009), whereas MVPA might reveal coarser topographical organization along the cortical surface (Freeman, Brouwer, Heeger, & Merriam, 2011; Sasaki et al., 2006). In this scenario PPA neurons would encode viewpoint-specific representations of individual scenes, which are then grouped together on the cortical surface according to place and category. Insofar as fMRIa indexes the neurons, it would reveal coding of views, with some degree of cross-adaptation between similar views. MVPA, on the other hand, would reveal the coarser coding by places and categories. However, other scenarios are also possible.[1] A full resolution of this topic will require a more thorough understanding of the neural mechanisms underlying MVPA and fMRIa. Nevertheless, we can make the preliminary conclusion that the PPA encodes information that

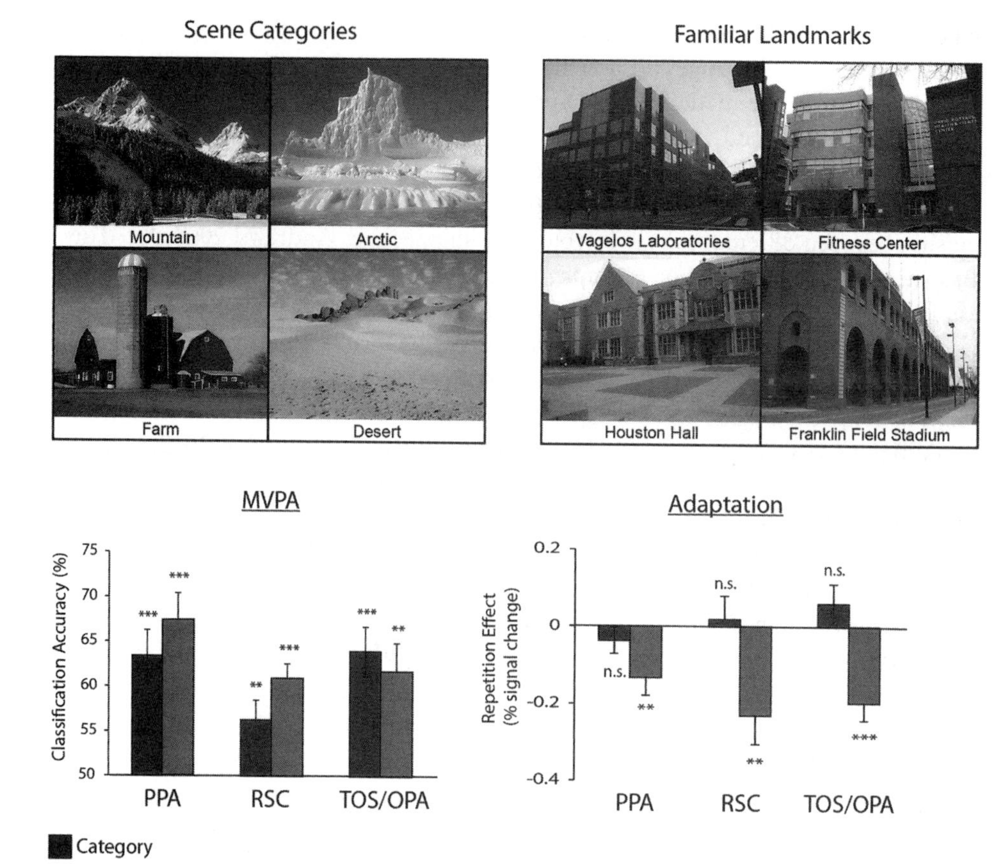

Figure 6.2
Coding of scene categories and landmarks in the PPA, RSC, and TOS/OPA. Subjects were scanned with fMRI while viewing a variety of images of 10 scene categories and 10 familiar campus landmarks (four examples shown). Multivoxel pattern analysis (MVPA) revealed coding of both category and landmark identity in all three regions (bottom left). In contrast, adaptation effects were observed only when landmarks were repeated—repetition of scene category had no effect (bottom right). One interpretation is that fine-grain organization within scene regions reflects coding of features that are specific to individual landmarks and scenes, whereas coarse-grain organization reflects grouping by category. However, other interpretations are possible. Adapted from Epstein and Morgan (2012).

allows it to make distinctions at all three representational levels: category, scene/place identity, and view.

But how does the PPA do it? What kind of information about the scene does the PPA extract in order to distinguish among scene categories, scene exemplars, and views? It is this question we turn to next.

Coding of Scene Geometry

Scenes contain fixed background elements such as walls, building facades, streets, and natural topography. These elements constrain movement within a scene and thus are relevant for navigation. Moreover, because these elements are fixed and durable, they are likely to be very useful cues for scene recognition. In fact behavioral studies suggest that both humans and animals preferentially use information about the geometric layout of the local environment to reorient themselves after disorientation (Cheng, 1986; Cheng & Newcombe, 2005; Gallistel, 1990; Hermer & Spelke, 1994). Thus, an appealing hypothesis is that the PPA represents information about the geometric structure of the local scene as defined by the spatial layout of these fixed background elements.

Consistent with this view, the original report on the PPA obtained evidence that the region was especially sensitive to these fixed background cues (Epstein & Kanwisher, 1998). The response of the PPA to scenes was not appreciably reduced when all the movable objects were removed from the scene—specifically, when all the furniture was removed from a room, leaving just bare walls. In contrast the PPA responded only weakly to the objects alone when the background elements were not present. When scene images were fractured into surface elements that were then rearranged so that they no longer depicted a three-dimensional space, response in the PPA was significantly reduced. In a follow-up study the PPA was shown to respond strongly even to "scenes" made out of Lego blocks, which were clearly not real-world places but had a similar geometric organization (Epstein, Harris, Stanley, & Kanwisher, 1999). From these results, we concluded that the PPA responds to stimuli that have a scene-like but not an object-like geometry.

Two recent contemporaneous studies have taken these findings a step further by showing that multivoxel activity patterns in the PPA distinguish between scenes based on their geometry. The first study, by Park and colleagues (2011), looked at activity patterns elicited during viewing of scenes that were grouped according to either spatial expanse (open vs. closed) or content (urban vs. natural). These "supercategories" were distinguishable from each other; furthermore, when patterns were misclassified by the PPA, it was more likely that the content than the spatial expanse was classified wrong, suggesting that the representation of spatial expanse was more salient than the representation of scene content. The second study, by Kravitz and colleagues (2011), looked at multivoxel patterns elicited by 96 scenes drawn from 16 categories, this time grouped by three factors: expanse (open vs. closed), content (natural vs. man-made), and distance to scene elements (near vs. far). These PPA activity patterns were distinguishable on the basis of expanse and distance but not on the basis of content. Moreover, scene categories could not be reliably distinguished when the two spatial factors (expanse and distance) were controlled for, suggesting that previous demonstrations of category decoding may have been leveraging the spatial differences between categories—for example, the fact that highway scenes tend to be open whereas forest scenes tend to be closed.

Thus, the PPA does seem to encode information about scene geometry. Moreover, it seems unlikely that this geometric coding can be explained by low-level visual differences that tend to correlate with geometry. When the images in the Park et al. (2011) experiment were phase scrambled so that spatial frequency differences were retained but geometric and content information was eliminated, PPA classification fell to chance levels. Furthermore, Walther and colleagues (2011) demonstrated cross categorization between photographs and line drawings, a manipulation that preserves geometry and content while changing many low-level visual features.

Can this be taken a step further by showing that the PPA encodes something more detailed than whether a scene is open or closed? In a fascinating study Dilks and colleagues (2011) used fMRI adaptation to test whether the PPA was sensitive to mirror-reversal of a scene. Strikingly, the PPA showed almost as much adaptation to a mirror-reversed version of a scene as it did to the original version. In contrast, the RSC and TOS treated mirror-reversed scenes as new items. This result could indicate that the PPA primarily encodes nonspatial aspects of the scene such as its spatial frequency distribution, color, and objects, all of which are unchanged by mirror reversal. Indeed, as we see in the next two sections, the PPA is in fact sensitive to these properties. However, an equally good account of the Dilks result is that the PPA represents spatial information, but in a way that is invariant to left-right reversal. For example, the PPA could encode distances and angles between scene elements in an unsigned manner—mirror reversal leaves the magnitudes of these quantities unchanged while changing the direction of angles (clockwise becomes counterclockwise) and the x-coordinate (left becomes right).

In any case the Dilks et al. (2011) results suggest that the PPA may encode quantities that are useful for identifying a scene but are less useful for calculating one's orientation relative to the scene. To see this, imagine the simplest case, a scene consisting of an array of discrete identifiable points in the frontoparallel plane. Mirror-reversal changes the implied viewing direction 180° (i.e., if the original image depicts the array viewed from the south, so that one sees the points A-B-C in order from left to right, the mirror-reversed image depicts the array viewed from the north, so that one sees the points C-B-A from left to right). A brain region involved in calculating one's orientation relative to the scene (e.g., RSC) should be sensitive to this manipulation; a brain region involved in identifying the scene (e.g., PPA) should not be. This observation is consistent with the neuropsychological evidence reviewed earlier that suggests that the PPA is more involved in place recognition, whereas the RSC is more involved in using scene information to orient oneself within the world.

Perhaps the strongest evidence that the PPA encodes geometric information comes from a study that showed PPA activation during haptic exploration of "scenes" made out of Lego blocks (Wolbers, Klatzky, Loomis, Wutte, & Giudice, 2011). As noted above, we previously observed that the PPA responds more strongly when subjects view Lego scenes than when they view "objects" made out of the same materials. Wolbers and colleagues observed the same scene advantage during haptic exploration. Moreover, they observed this scene-versus-object difference both in normal sighted subjects and also in subjects who were blind from an early age. This is an important control because it shows that PPA activity during haptic exploration cannot be explained by visual imagery. These results suggest that the PPA extracts geometric representations of scenes that can be accessed through either vision or touch.

Coding of Visual Properties

The strongest version of the spatial layout hypothesis is that the PPA *only* represents geometric information—a "shrink-wrapped" representation of scene surfaces that eschews any information about the color, texture, or material properties of these surfaces. However, recent studies have shown that the story is more complicated: in addition to coding geometry, the PPA also seems to encode purely visual (i.e., nongeometric) qualities of a scene.

A series of studies from Tootell and colleagues has shown that the PPA is sensitive to low-level visual properties of an image. The first study in the series showed that the PPA responds more strongly to high-spatial-frequency (HSF) images than to low-spatial-frequency (LSF) images (Rajimehr, Devaney, Bilenko, Young, & Tootell, 2011). This HSF preference is found not only for scenes but also for simpler stimuli such as checkerboards. The second study found that the PPA exhibits a cardinal orientation bias, responding more strongly to stimuli that have the majority of their edges oriented vertically/horizontally than to stimuli that have the majority of their edges oriented obliquely (Nasr & Tootell, 2012). As with the HSF preference, this cardinal-orientation bias can be observed both for natural scenes (by tilting them to different degrees) and for simpler stimuli such as arrays of squares and line segments. As the authors of these studies note, these biases might reflect PPA tuning for low-level visual features that are typically found in scenes. For example, scene images usually contain more HSF information than images of faces or objects; the ability to process this HSF information would be useful for localizing spatial discontinuities caused by boundaries between scene surfaces. The cardinal orientation bias might relate to the fact that scenes typically contain a large number of vertical and horizontal edges, both in natural and man-made environments, because surfaces in scenes are typically oriented by reference to gravity.

The PPA has also been shown to be sensitive to higher-level visual properties. Cant and Goodale (2007) found that it responded more strongly to objects when subjects attend to the material properties of the objects (e.g., whether it is made out of metal or wood, whether it is hard or soft) than when they attend to the shape of the objects. Although the strongest differential activation in the studies is in a collateral sulcus region posterior to the PPA, the preference for material properties extends anteriorly into the PPA (Cant, Arnott, & Goodale, 2009; Cant & Goodale, 2007). This may indicate sensitivity in the collateral sulcus generally and the PPA in particular to color and texture information, the processing of which can be a first step toward scene recognition (Gegenfurtner & Rieger, 2000; Goffaux et al., 2005; Oliva & Schyns, 2000). In addition material properties might provide important cues for scene recognition (Arnott, Cant, Dutton, & Goodale, 2008): buildings can be distinguished based on whether they are made of brick or wood; forests are "soft," whereas urban scenes are "hard."

In a recent study Cant and Xu (2012) took this line of inquiry a step further by showing that the PPA is sensitive not just to texture and material properties but also to the visual summary statistics of images (Ariely, 2001; Chong & Treisman, 2003). To show this they used an fMRI adaptation paradigm in which subjects viewed images of object ensembles—for example, an array of strawberries or baseballs viewed from above. Adaptation was observed in the PPA (and in other collateral sulcus regions) when ensemble statistics were repeated—for example, when one image of a pile of baseballs was followed by another image of a similar pile. Adaptation was also observed for repetition of surface textures that were not decomposable into individual objects. In both cases the stimulus might be considered a type of scene, but viewed from close-up, so that only the pattern created by the surface or repeated objects is visible, without background elements or depth. The fact that the PPA adapts to repetitions of these "scenes" without geometry strongly suggests that it codes nongeometric properties in addition to geometry.

Coding of Objects

Now we turn to the final kind of information that the PPA might extract from visual scenes: information about individual objects. At first glance the idea that the PPA is concerned with individual objects may seem like a bit of a contradiction. After all, the PPA is typically defined based on greater response to scenes than to objects. Furthermore, as discussed above, the magnitude of the PPA response to scenes does not seem to be affected by the presence or absence of individual objects within the scene (Epstein & Kanwisher, 1998). Nevertheless, a number of recent studies have shown that the PPA is sensitive to spatial qualities of objects when the objects are presented not as part of a scene, but in isolation. As we will see, this suggests that the division between scene and object is a bit less than absolute, at least as far as the PPA is concerned.

Indeed, there is evidence for a graded boundary between scenes and objects in the original paper on the PPA, which examined response to four stimulus categories: scenes, houses, common everyday objects, and faces (Epstein & Kanwisher, 1998). The response to scenes in the PPA was significantly greater than the response to the next-best stimulus, which was houses (see also Mur et al., 2012). However, the response to houses (shown without background) was numerically greater than the response to objects, and the response to objects was numerically greater than the response to faces. Low-level visual differences between the categories might explain some of these effects—for example, the fact that face images tend to have less power in the high spatial frequencies, or the fact that images of houses tend to have more horizontal and vertical edges than images of objects and faces. However, it is also possible that the PPA really does care about the categorical differences between houses, objects, and faces. One way of interpreting this ordering of responses is to posit that the PPA responds more strongly to stimuli that are more useful as landmarks. A building is a good landmark because it is never going to move, whereas faces are terrible landmarks because people almost always change their positions.

Even within the catchall category of common everyday objects, we can observe reliable differences in PPA responses that may relate to landmark suitability. Konkle and Oliva (2012) showed that a region of posterior parahippocampal cortex that partially overlaps with the PPA responds more strongly to large objects (e.g., car, piano) than to small objects (e.g., strawberry, calculator), even when the stimuli have equivalent retinal size. Similarly, Amit and colleagues (2012) and Cate and colleagues (2011) observed greater PPA activity to objects that were perceived as being larger or more distant, where size and distance were implied by the presence of Ponzo lines defining a minimal scene.

The response in the PPA to objects can even be modulated by their navigational history. Janzen and Van Turennout (2004) familiarized subjects with a large number of objects during navigation through a virtual museum. Some of the objects were placed at navigational decision points (intersections), and others were placed at less navigationally relevant locations (simple turns). The subjects later viewed the same objects in the scanner along with previously unseen foils, and were asked to judge whether each item had been in the museum or not. Objects that were previously encountered at navigational decision points elicited greater response in the PPA than objects previously encountered at other locations within the maze. Interestingly, this decision point advantage was found even for objects that subjects did not explicitly remember seeing. A later study found that this decision-point advantage was reduced for objects appearing at two different decision points (Janzen & Jansen, 2010), consistent with the idea that the PPA responds to the decision-point objects because they uniquely specify a navigationally relevant location. In other words the decision-point objects have become landmarks. We subsequently replicated these results in an experiment that examined response to buildings at decision points and nondecision points along a real-world route (Schinazi & Epstein, 2010).

Observations such as these suggest that the PPA is in fact sensitive to the spatial qualities of objects. Two groups have advanced theories about the functions of the PPA under which scene-based and object-based responses are explained by a single mechanism. First, Bar and colleagues have proposed that the PPA is a subcomponent of a parahippocampal mechanism for processing contextual associations, by which they mean associations between items that typically occur together in the same place or situation. For example, a toaster and a coffee maker are contextually associated because they typically co-occur in a kitchen, and a picnic basket and a blanket are contextually associated because they typically co-occur at a picnic. According to the theory, the PPA represents spatial contextual associations whereas the portion of parahippocampal cortex anterior to the PPA represents nonspatial contextual associations (Aminoff, Gronau, & Bar, 2007). Because scenes are filled with spatial relationships, the PPA responds strongly to scenes. Evidence for this idea comes from a series of studies that observed greater parahippocampal activity when subjects were viewing objects that are strongly associated with a given context (for example, a beach ball or a stove) than when viewing objects that are not strongly associated to any context (for example, an apple or a Rubik's cube) (Bar, 2004; Bar & Aminoff, 2003; Bar, Aminoff, & Schacter, 2008). A second theory has been advanced by Mullally and Maguire (2011), who suggest that the PPA responds strongly to stimuli that convey a sense of surrounding space. Evidence in support of this theory comes from the fact that the PPA activates more strongly when subjects imagine objects that convey a strong sense of surrounding space than when they imagine objects that have weak "spatial definition." Objects with high spatial definition tend to be large and fixed whereas low-spatial-definition objects tend to be small and movable. In this view, a scene is merely the kind of object with the highest spatial definition of all.

Is either of these theories correct? It has been difficult to determine which object property is the essential driver of PPA response, in part because the properties of interest tend to covary with each other: large objects tend to be fixed in space, have strong contextual associations, and define the space around them and are typically viewed at greater distances. Furthermore, the aforementioned studies did not directly compare the categorical advantage for scenes over objects to the effect of object-based properties. Finally, the robustness of object-based effects has been unclear. The context effect, for example, is quite fragile: it can be eliminated by simply changing the presentation rate and controlling for low-level differences (Epstein & Ward, 2010),

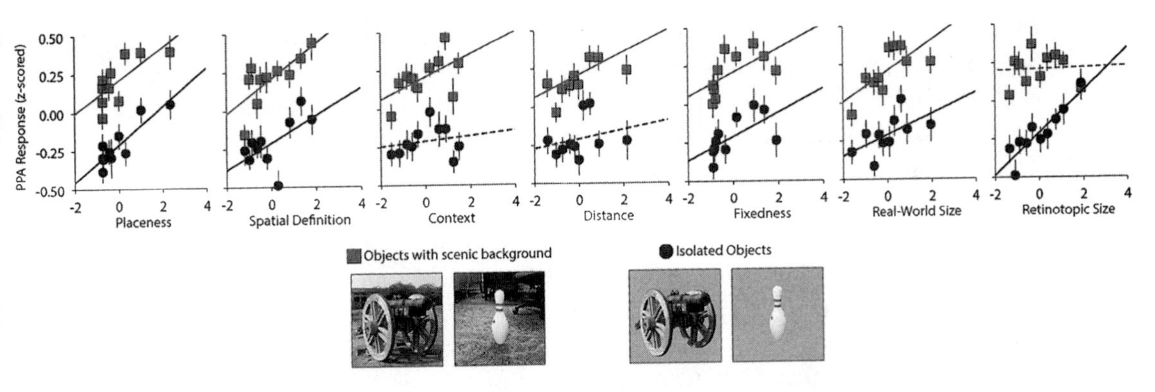

Figure 6.3
Sensitivity of the PPA to object characteristics. Subjects were scanned with fMRI while viewing 200 objects, shown either on a scenic background or in isolation. Response in the PPA depended on object properties that reflect the landmark suitability of the item; however, there was also a categorical offset for objects within scenes (squares) compared to isolated objects (circles). For purposes of display, items are grouped into sets of 20 based on their property scores. Solid trend lines indicate a significant effect; dashed lines are nonsignificant. Adapted from Troiani, Stigliani, Smith, and Epstein (2014).

and it has failed to replicate under other conditions as well (Mullally & Maguire, 2011; Yue, Vessel, & Biederman, 2007).

To clarify these issues we ran a study in which subjects viewed 200 different objects, each of which had been previously rated along six different stimulus dimensions: physical size, distance, fixedness, spatial definition, contextual associations, and placeness (i.e., the extent to which the object was "a place" instead of "a thing") (Troiani, Stigliani, Smith, & Epstein, 2014; see figure 6.3). The objects were either shown in isolation or immersed in a scene with background elements. The results indicated that the PPA was sensitive to all six object properties (and, in addition, to retinotopic extent); however, we could not identify a unique contribution from any one of them. In other words all of the properties seemed to relate to a single underlying factor that drives the PPA, which we labeled the "landmark suitability" of the object. Notably, this object-based factor was not sufficient to explain all of the PPA response on its own because there was an additional categorical difference between scenes and objects: response was greater when the objects were shown as part of a scene than when they were shown in isolation, over and above the response to the spatial properties of the objects. This "categorical" difference between scenes and objects might reflect difference in visual properties—for example, the fact that the scenes afford statistical summary information over a wider portion of the visual field.

Thus, the PPA does seem to be sensitive to spatial properties of objects, responding more strongly to objects that are more suitable as landmarks. The fact that the PPA encodes this information might explain the fact that previous multivoxel pattern analysis (MVPA) studies have found it possible to decode object identity within the

PPA. Interestingly, the studies that have done this successfully have generally used large fixed objects as stimuli (Diana, Yonelinas, & Ranganath, 2008; Harel, Kravitz, & Baker, 2013; MacEvoy & Epstein, 2011), whereas a study that failed to find this decoding used objects that were small and manipulable (Spiridon & Kanwisher, 2002). This is consistent with the idea that the PPA does not encode object identity per se but rather encodes spatial information that inheres to some objects but not others. Also, it is of note that all of the studies that have examined object coding in the PPA have either looked at the response to these objects in isolation or when shown as the central, clearly dominant object within a scene (Bar et al., 2008; Harel et al., 2013; Troiani et al., 2012). Thus, it remains unclear whether the PPA encodes information about objects when they form just a small part of a larger scene. Indeed, as we see below, recent evidence tends to argue against this idea.

Putting It All Together

The research reviewed above suggests that the PPA represents geometric information from scenes, nonspatial visual information from scenes, and spatial information that can be extracted from both scenes and objects. How do we put this all together in order to understand the function of the PPA? My current view is that it is not possible to explain all of these results using a single cognitive mechanism. In particular, the fact that the PPA represents both spatial and nonspatial information suggests the existence of two mechanisms within the PPA: one for processing spatial information and one for processing the visual properties of the stimulus.

One possibility is that these two mechanisms are anatomically separated. Recall that Arcaro and colleagues (2009) found two distinct visual maps in the PPA. Recent work examining the anatomical connectivity within the PPA has found an anterior-posterior gradient whereby the posterior PPA connects more strongly to visual cortices and the anterior PPA connects more strongly to the RSC and the parietal lobe (Baldassano, Beck, & Fei-Fei, 2013). In other words the posterior PPA gets more visual input, and the anterior PPA gets more spatial input. This gradient is reminiscent of a division reported in the neuropsychological literature: patients with damage to the posterior portion of the lingual-parahippocampal region have a deficit in landmark recognition that is observed in both familiar and unfamiliar environments, whereas patients with damage located more anteriorly in the parahippocampal cortex proper have a deficit in topographical learning that mostly impacts navigation in novel environments (Aguirre & D'Esposito, 1999). Thus, it is possible that the posterior PPA processes the visual properties of scenes, whereas the anterior PPA incorporates spatial information about scene geometry (and also objects, if they have such spatial information associated with them). The two parts of the PPA might work together to allow recognition of scenes (and other landmarks) based on both visual and spatial properties. Interestingly, a recent fMRI study in the macaque found two distinct scene-responsive regions in the general vicinity of the PPA, which were labeled the medial place patch (MPP) and the lateral place patch (LPP) (Kornblith et al., 2013). These might correspond to the anterior and posterior PPA in humans (Epstein & Julian, 2013).

Another possibility is that the PPA supports two recognition mechanisms that are temporally rather than spatially separated. In this scenario, the PPA first encodes the visual properties of the scene and then later extracts information about scene geometry. Some evidence for this idea comes from two intracranial EEG (i.e., electrocorticography) studies that recorded from the parahippocampal region in presurgical epilepsy patients. The first study (Bastin, Committeri, et al., 2013) was motivated by earlier fMRI work examining response in the PPA when subjects make different kinds of spatial judgments. In these earlier studies the PPA and RSC responded more strongly when subjects reported which of two small objects was closer to the wing of a building than when they reported which was closer to a third small object or to themselves. That is, the PPA and RSC were more active when the task required the use of an environment-centered rather than an object- or viewer-centered reference frame (Committeri et al., 2004; Galati, Pelle, Berthoz, & Committeri, 2010). When presurgical epilepsy patients were run on this paradigm, increased power in the gamma oscillation band was observed at parahippocampal contacts for landmark-centered compared to the viewer-centered judgments, consistent with the previous fMRI results. Notably, this increased power occurred at 600–800 ms poststimulus, suggesting that information about the environmental reference frame was activated quite late, after perceptual processing of the scene had been completed. The second study (Bastin, Vidal, et al., 2013) was motivated by previous fMRI results indicating that the PPA responds more strongly to buildings than to other kinds of objects (Aguirre et al., 1998). Buildings have an interesting intermediate status halfway between objects and scenes. In terms of visual properties they are more similar to objects (i.e., discrete convex entities with a definite boundary), but in terms of spatial properties, they are more similar to scenes (i.e., large, fixed entities that define the space around them). If the PPA responds to visual properties early but spatial properties late, then it should treat buildings as objects initially but as scenes later on. Indeed, this was exactly what was found: in the earliest components of the response, scenes were distinguishable from buildings and objects, but buildings and objects were not distinguishable from each other. A differential response to buildings versus nonbuilding objects was not observed until significantly later.

These results suggest the existence of two stages of processing in the PPA. The earlier stage may involve processing of purely visual information—for example, the analysis of visual features that are unique to scenes or the calculation of statistical summaries across the image, which would require more processing and hence more activity for scenes than for objects. The later stage may involve processing of spatial information and possibly also conceptual information about the meaning of the stimulus as a place. In this scenario the early stage processes the appearance of the scene from the current point of view, whereas the later stage abstracts geometric information about the scene, which allows it to be represented in either egocentric or allocentric coordinates. The viewpoint-specific snapshot extracted in the first stage may suffice for scene recognition, whereas the spatial information extracted in the second stage may facilitate cross talk between the PPA representation of the local scene and spatial representations in the RSC and hippocampus (Kuipers, Modayil, Beeson, MacMahon, & Savelli, 2004). This dual role for the PPA could explain its involvement in both scene recognition and spatial learning (Aguirre & D'Esposito, 1999; Bohbot et al., 1998; Epstein, DeYoe, Press, Rosen, & Kanwisher, 2001; Ploner et al., 2000).

Object-Based Scene Recognition

A central theme of the preceding section is that the PPA represents scenes in terms of whole-scene characteristics, such as geometric layout or visual summary statistics. Even when the PPA responds to objects, it is typically because the object is acting as a landmark or potential landmark—in other words, because the object is a signifier for a place and thus has become a kind of "scene" in its own right. There is little evidence that the PPA uses information about the objects within a scene for scene recognition. This neuroscientific observation dovetails nicely with behavioral and computational work that suggest that such whole-scene characteristics are used for scene recognition (Fei-Fei & Perona, 2005; Greene & Oliva, 2009b; Oliva & Torralba, 2001; Renninger & Malik, 2004).

However, there are certain circumstances in which the objects within a scene might provide important information about its identity or category. For example, a living room and a bedroom are primarily distinguishable on the basis of their furniture—a living room contains a sofa whereas a bedroom contains a bed—rather than on the basis of their overall geometry (Quattoni & Torralba, 2009). This observation suggests that there might be a second, object-based route to scene recognition, which might exploit information about the identities of the objects with a scene or their spatial relationships (Biederman, 1981; Davenport & Potter, 2004).

MacEvoy and I obtained evidence for such an object-based scene recognition mechanism in an fMRI study (MacEvoy & Epstein, 2011; see figure 6.4, plate 13). We reasoned that a brain region involved in object-based scene recognition should encode information about within-scene objects when subjects view scenes. To test this we examined the multivoxel activity patterns elicited by four different scene categories (kitchens, bathrooms, intersections, and playgrounds) and eight different objects that were present in these scenes (stoves and refrigerators; bathtubs and toilets; cars and

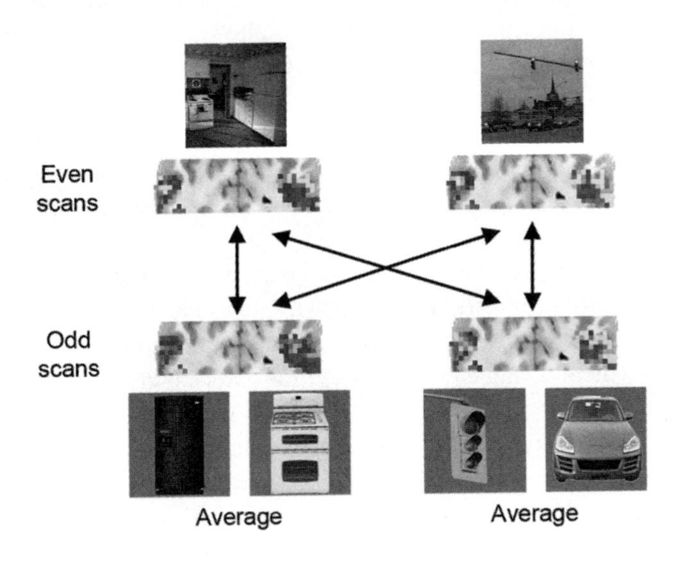

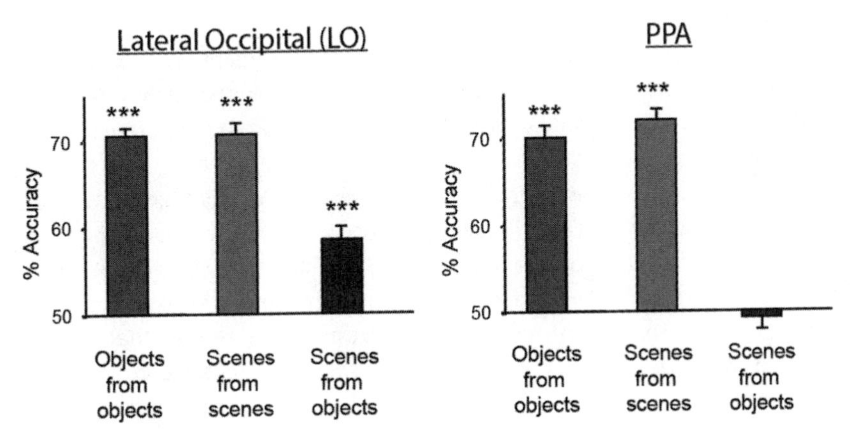

Figure 6.4 (plate 13)
Evidence for an object-based scene recognition mechanism in the lateral occipital (LO) cortex. Multivoxel activity patterns elicited during scene viewing (four categories: kitchen, bathroom, intersection, playground) were classified based on activity patterns elicited by two objects characteristic of the scenes (e.g., stove and refrigerator for kitchen). Although objects could be classified from object patterns and scenes from scene patterns in both the LO and the PPA, only LO showed above-chance scene-from-object classification. This suggests that scenes are represented in LO (but not in the PPA) in terms of their constituent objects. Adapted from MacEvoy and Epstein (2011).

traffic lights; slides and swing sets). We then looked for similarities between the scene-evoked and object-evoked patterns. Strikingly, we found that scene patterns were predictable on the basis of the object-evoked patterns; however, this relationship was not observed in the PPA but in the object-sensitive lateral occipital (LO) cortex (Grill-Spector, Kourtzi, & Kanwisher, 2001; Malach et al., 1995). More specifically, the patterns evoked by the scenes in this region were close to the averages of the patterns evoked by the objects characteristic of the scenes. Simply put, LO represents kitchens as the average of stoves and refrigerators, bathrooms as the average of toilets and bathtubs.

We hypothesized that by averaging the object-evoked patterns, LO might be creating a code that allows scene identity (or *gist*) to be extracted when subjects attend broadly to the scene as a whole but still retains information about the individual objects that can be used if any one of them is singled out for attention. Indeed, in a related study, when subjects looked at briefly presented scenes with the goal of finding a target object (in this case, a person or an automobile), LO activity patterns reflected the target object but not the nontarget object, even when the nontarget object was present (Peelen, Fei-Fei, & Kastner, 2009). Thus, LO can represent either multiple objects within the scene or just a single object, depending on how attention is allocated as a consequence of the behavioral task (Treisman, 2006).

A very different finding was observed in the PPA in our experiment. The multivoxel patterns in this region contained information about the scenes and also about the objects when the objects were presented in isolation. That is, the scene patterns were distinguishable from each other, as were the object patterns. However, in contrast to LO, where the scene patterns were well predicted by the average of the object patterns, here there was no relationship between the scene and object patterns. That is, the PPA had a pattern for kitchen and a pattern for refrigerator, but there was no similarity between these two patterns. (Nor, for that matter, was there similarity between contextually related patterns: stoves and refrigerators were no more similar than stoves and traffic lights.) Whereas LO seems to construct scenes from their constituent objects, the PPA considers scenes and their constituent objects to be unrelated items. Although at first this may seem surprising, it makes sense if the PPA represents global properties of the stimulus. The spatial layout of a kitchen is unlikely to be strongly related to the spatial axes defined by a stove that constitutes only a small part of a whole. Similarly, the visual properties of individual objects are likely to be swamped when they are seen as part of a real-world scene.

Thus, it is feasible that LO might support a second pathway for scene recognition based on the objects within the scene. But is this object-based information used to guide recognition behavior? The evidence on this point is unclear. In a behavioral version of our fMRI experiment, we asked subjects to make category judgments on briefly presented and masked versions of the kitchen, bathroom, intersection, and playground scenes. To determine the influence of the objects on recognition, images were either presented in their original versions, or with one or both of the objects obscured by a noise mask. Recognition performance was impaired by obscuring the objects, with greater decrement when both objects were obscured than when only one object was obscured. Furthermore, the effect of obscuring the objects could not be entirely explained by the fact that this manipulation degraded the image as a whole. Rather, the results suggested the parallel operation of object-based and image-based pathways for scene recognition.

Additional evidence on this point comes from studies that have examined scene recognition after LO is damaged, or interrupted with transcranial magnetic stimulation (TMS). Steeves and colleagues (2004) looked at the scene recognition abilities of patient D.F., who sustained bilateral damage to her LO subsequent to carbon monoxide poisoning. Although this patient was almost completely unable to recognize objects on the basis of their shape, she was able to classify scenes into six different categories when they were presented in color (although performance was abnormal for grayscale images). Furthermore, her PPA was active when performing this task. A TMS study on normal subjects found a similar result (Mullin & Steeves, 2011): stimulation to LO disrupted classification of objects into natural and manmade but actually *increased* performance on the same task for scenes. Another study found no impairment on two scene discrimination tasks after TMS stimulation to LO but significant impairment after stimulation to the TOS (Dilks et al., 2013). In sum, the evidence thus far suggests that LO might not be necessary for scene recognition under many circumstances. This does not necessarily contradict the two-pathways view, but it does suggest that the whole-scene pathway through the PPA is primary. Future experiments should attempt to determine what scene recognition tasks, if any, require LO.

Conclusions

The evidence reviewed above suggests that our brains contain specialized neural machinery for visual scene recognition, with the PPA in particular playing a central role. Recent neuroimaging studies have significantly expanded our understanding of the function of the PPA. Not only does the PPA encode the spatial layout of scenes, it also encodes visual properties of scenes and spatial information that can potentially be extracted from both scenes and objects. This work leads us to a more nuanced understanding of the PPA's function under which it represents scenes but also other stimuli that can act as navigational landmarks. It also suggests the possibility that the PPA may not be a unified entity but might be fractionable into two functionally or anatomically distinct parts. Complementing this PPA work are studies indicating that there might be a second pathway for scene recognition that passes through the lateral occipital cortex. Whereas the PPA represents scenes based on whole-scene characteristics, LO represents scenes based on the identities of within-scene objects.

The study of scene perception is a rapidly advancing field, and it is likely that new discoveries will require us to further refine our understanding of its neural basis. In particular, as noted above, very recent reports have identified scene-responsive regions in the macaque monkey (Nasr et al., 2011), and neuronal recordings from these regions have already begun to expand on the results obtained by fMRI studies (Kornblith et al., 2013; see Epstein & Julian, 2013, for discussion). Thus, we must be cautious about drawing conclusions that are too definitive. Nevertheless, these caveats aside, it is remarkable how well the different strands of research into the neural basis of scene recognition have converged into a common story. A central goal of cognitive neuroscience is to understand the neural systems that underlie different cognitive abilities. Within the realm of scene recognition, I believe the field can claim some modicum of success.

Acknowledgments

I thank Joshua Julian and Steve Marchette for helpful comments. Supported by the National Science Foundation Spatial Intelligence and Learning Center (SBE-0541957) and National Institutes of Health (EY-022350 and EY-022751).

Note

1. In Epstein and Morgan (2012) we consider two other possible scenarios. Under the first scenario fMRIa operates at the synaptic input to each unit (Epstein et al., 2008; Sawamura, Orban, & Vogels, 2006), whereas MVPA indexes neuronal or columnar tuning (Kamitani & Tong, 2005; Swisher et al., 2010). If this scenario is correct, the PPA might be conceptualized as taking viewpoint-specific inputs and converting them into representations of place identity and scene category. Under the second scenario, fMRIa reflects the operation of a dynamic mechanism that incorporates information about moment-to-moment expectations (Summerfield, Trittschuh, Monti, Mesulam, & Egner, 2008), whereas MVPA reflects more stable representational distinctions, coded at the level of the neuron, column, or cortical map (Kriegeskorte, Goebel, & Bandettini, 2006).

References

Aguirre, G. K., & D'Esposito, M. (1999). Topographical disorientation: A synthesis and taxonomy. *Brain*, *122*, 1613–1628.

Aguirre, G. K., Zarahn, E., & D'Esposito, M. (1998). An area within human ventral cortex sensitive to "building" stimuli: Evidence and implications. *Neuron*, *21*, 373–383.

Aminoff, E., Gronau, N., & Bar, M. (2007). The parahippocampal cortex mediates spatial and nonspatial associations. *Cerebral Cortex*, *17*(7), 1493–1503.

Amit, E., Mehoudar, E., Trope, Y., & Yovel, G. (2012). Do object-category selective regions in the ventral visual stream represent perceived distance information? *Brain and Cognition*, *80*(2), 201–213.

Antes, J. R., Penland, J. G., & Metzger, R. L. (1981). Processing global information in briefly presented pictures. *Psychological Research*, *43*(3), 277–292.

Arcaro, M. J., McMains, S. A., Singer, B. D., & Kastner, S. (2009). Retinotopic organization of human ventral visual cortex. *Journal of Neuroscience*, *29*(34), 10638–10652.

Ariely, D. (2001). Seeing sets: Representation by statistical properties. *Psychological Science*, *12*(2), 157–162.

Arnott, S. R., Cant, J. S., Dutton, G. N., & Goodale, M. A. (2008). Crinkling and crumpling: An auditory fMRI study of material properties. *NeuroImage*, *43*(2), 368–378.

Baldassano, C., Beck, D. M., & Fei-Fei, L. (2013). Differential connectivity within the parahippocampal place area. *NeuroImage*, *75*, 228–237.

Bar, M. (2004). Visual objects in context. *Nature Reviews Neuroscience*, *5*(8), 617–629.

Bar, M., & Aminoff, E. M. (2003). Cortical analysis of visual context. *Neuron*, *38*, 347–358.

Bar, M., Aminoff, E. M., & Schacter, D. L. (2008). Scenes unseen: The parahippocampal cortex intrinsically subserves contextual associations, not scenes or places per se. *Journal of Neuroscience*, *28*(34), 8539–8544.

Bastin, J., Committeri, G., Kahane, P., Galati, G., Minotti, L., Lachaux, J. P., et al. (2013). Timing of posterior parahippocampal gyrus activity reveals multiple scene processing stages. *Human Brain Mapping*, *34*(6), 1357–1370.

Bastin, J., Vidal, J. R., Bouvier, S., Perrone-Bertolotti, M., Benis, D., Kahane, P., et al. (2013). Temporal components in the parahippocampal place area revealed by human intracerebral recordings. *Journal of Neuroscience*, *33*(24), 10123–10131.

Biederman, I. (1972). Perceiving real-world scenes. *Science*, *177*(4043), 77–80.

Biederman, I. (1981). On the semantics of a glance at a scene. In M. Kubovy & J. R. Pomerantz (Eds.), *Perceptual organization* (pp. 213–263). Hillsdale, NJ: Lawrence Erlbaum Associates.

Biederman, I., Rabinowitz, J. C., Glass, A. L., & Stacy, E. W. J. (1974). On the information extracted from a glance at a scene. *Journal of Experimental Psychology*, *103*(3), 597–600.

Bohbot, V. D., Kalina, M., Stepankova, K., Spackova, N., Petrides, M., & Nadel, L. (1998). Spatial memory deficits in patients with lesions to the right hippocampus and to the right parahippocampal cortex. *Neuropsychologia*, *36*(11), 1217–1238.

Cant, J. S., Arnott, S. R., & Goodale, M. A. (2009). fMR-adaptation reveals separate processing regions for the perception of form and texture in the human ventral stream. *Experimental Brain Research*, *192*(3), 391–405.

Cant, J. S., & Goodale, M. A. (2007). Attention to form or surface properties modulates different regions of human occipitotemporal cortex. *Cerebral Cortex*, *17*(3), 713–731.

Cant, J. S., & Xu, Y. (2012). Object ensemble processing in human anterior-medial ventral visual cortex. *Journal of Neuroscience*, *32*(22), 7685–7700.

Cate, A. D., Goodale, M. A., & Kohler, S. (2011). The role of apparent size in building- and object-specific regions of ventral visual cortex. *Brain Research*, *1388*, 109–122.

Cheng, K. (1986). A purely geometric module in the rat's spatial representation. *Cognition*, *23*(2), 149–178.

Cheng, K., & Newcombe, N. S. (2005). Is there a geometric module for spatial orientation? Squaring theory and evidence. *Psychonomic Bulletin & Review*, *12*(1), 1–23.

Chong, S. C., & Treisman, A. (2003). Representation of statistical properties. *Vision Research*, *43*(4), 393–404.

Committeri, G., Galati, G., Paradis, A. L., Pizzamiglio, L., Berthoz, A., & LeBihan, D. (2004). Reference frames for spatial cognition: Different brain areas are involved in viewer-, object-, and landmark-centered judgments about object location. *Journal of Cognitive Neuroscience*, *16*(9), 1517–1535.

Cox, D. D., & Savoy, R. L. (2003). Functional magnetic resonance imaging (fMRI) "brain reading": Detecting and classifying distributed patterns of fMRI activity in human visual cortex. *NeuroImage*, *19*, 261–270.

Davenport, J. L., & Potter, M. C. (2004). Scene consistency in object and background perception. *Psychological Science*, *15*(8), 559–564.

Diana, R. A., Yonelinas, A. P., & Ranganath, C. (2008). High-resolution multi-voxel pattern analysis of category selectivity in the medial temporal lobes. *Hippocampus*, *18*(6), 536–541.

DiCarlo, J. J., Zoccolan, D., & Rust, N. C. (2012). How does the brain solve visual object recognition? *Neuron*, *73*(3), 415–434.

Dilks, D., Julian, J. B., Kubilius, J., Spelke, E. S., & Kanwisher, N. (2011). Mirror-image sensitivity and invariance in object and scene processing pathways. *Journal of Neuroscience*, *33*(31), 11305–11312.

Dilks, D. D., Julian, J. B., Paunov, A. M., & Kanwisher, N. (2013). The occipital place area (OPA) is causally and selectively involved in scene perception. *Journal of Neuroscience*, *33*(4), 1331–1336.

Drucker, D. M., & Aguirre, G. K. (2009). Different spatial scales of shape similarity representation in lateral and ventral LOC. *Cerebral Cortex*, *19*(10), 2269–2280.

Ekstrom, A. D., Kahana, M. J., Caplan, J. B., Fields, T. A., Isham, E. A., Newman, E. L., et al. (2003). Cellular networks underlying human spatial navigation. *Nature*, *425*(6954), 184–188.

Epstein, R. A. (2005). The cortical basis of visual scene processing. *Visual Cognition*, *12*(6), 954–978.

Epstein, R. A. (2008). Parahippocampal and retrosplenial contributions to human spatial navigation. *Trends in Cognitive Sciences*, *12*(10), 388–396.

Epstein, R. A., DeYoe, E. A., Press, D. Z., Rosen, A. C., & Kanwisher, N. (2001). Neuropsychological evidence for a topographical learning mechanism in parahippocampal cortex. *Cognitive Neuropsychology*, *18*(6), 481–508.

Epstein, R. A., Graham, K. S., & Downing, P. E. (2003). Viewpoint specific scene representations in human parahippocampal cortex. *Neuron*, *37*, 865–876.

Epstein, R. A., Harris, A., Stanley, D., & Kanwisher, N. (1999). The parahippocampal place area: Recognition, navigation, or encoding? *Neuron*, *23*(1), 115–125.

Epstein, R. A., & Higgins, J. S. (2007). Differential parahippocampal and retrosplenial involvement in three types of visual scene recognition. *Cerebral Cortex*, *17*(7), 1680–1693.

Epstein, R. A., Higgins, J. S., Jablonski, K., & Feiler, A. M. (2007). Visual scene processing in familiar and unfamiliar environments. *Journal of Neurophysiology*, *97*(5), 3670–3683.

Epstein, R. A., & Julian, J. B. (2013). Scene areas in humans and macaques. *Neuron*, *79*(4), 615–617.

Epstein, R. A., & Kanwisher, N. (1998). A cortical representation of the local visual environment. *Nature*, *392*(6676), 598–601.

Epstein, R. A., & MacEvoy, S. P. (2011). Making a scene in the brain. In L. Harris & M. Jenkin (Eds.), *Vision in 3D environments* (pp. 255–279). Cambridge: Cambridge University Press.

Epstein, R. A., & Morgan, L. K. (2012). Neural responses to visual scenes reveals inconsistencies between fMRI adaptation and multivoxel pattern analysis. *Neuropsychologia*, *50*(4), 530–543.

Epstein, R. A., Parker, W. E., & Feiler, A. M. (2007). Where am I now? Distinct roles for parahippocampal and retrosplenial cortices in place recognition. *Journal of Neuroscience*, *27*(23), 6141–6149.

Epstein, R. A., Parker, W. E., & Feiler, A. M. (2008). Two kinds of fMRI repetition suppression? Evidence for dissociable neural mechanisms. *Journal of Neurophysiology*, *99*, 2877–2886.

Epstein, R. A., & Ward, E. J. (2010). How reliable are visual context effects in the parahippocampal place area? *Cerebral Cortex*, *20*(2), 294–303.

Ewbank, M. P., Schluppeck, D., & Andrews, T. J. (2005). fMR-adaptation reveals a distributed representation of inanimate objects and places in human visual cortex. *NeuroImage*, *28*(1), 268–279.

Fei-Fei, L., Iyer, A., Koch, C., & Perona, P. (2007). What do we perceive in a glance of a real-world scene? *Journal of Vision*, *7*(1), 1–29.

Fei-Fei, L., & Perona, P. (2005). A Bayesian hierarchical model for learning natural scene categories. *Computer Vision and Pattern Recognition*, *2*, 524–531.

Freeman, J., Brouwer, G. J., Heeger, D. J., & Merriam, E. P. (2011). Orientation decoding depends on maps, not columns. *Journal of Neuroscience*, *31*(13), 4792–4804.

Galati, G., Pelle, G., Berthoz, A., & Committeri, G. (2010). Multiple reference frames used by the human brain for spatial perception and memory. *Experimental Brain Research*, *206*(2), 109–120.

Gallistel, C. R. (1990). *The organization of learning.* Cambridge, MA: MIT Press.

Gegenfurtner, K. R., & Rieger, J. (2000). Sensory and cognitive contributions of color to the recognition of natural scenes. *Current Biology*, *10*(13), 805–808.

Goffaux, V., Jacques, C., Mouraux, A., Oliva, A., Schyns, P. G., & Rossion, B. (2005). Diagnostic colours contribute to the early stages of scene categorization: Behavioural and neurophysiological evidence. *Visual Cognition*, *12*(6), 878–892.

Golomb, J. D., Albrecht, A., Park, S., & Chun, M. M. (2011). Eye movements help link different views in scene-selective cortex. *Cerebral Cortex*, *21*(9), 2094–2102.

Golomb, J. D., & Kanwisher, N. (2012). Higher level visual cortex represents retinotopic, not spatiotopic, object location. *Cerebral Cortex*, *22*(12), 2794–2810.

Greene, M. R., & Oliva, A. (2009a). The briefest of glances: The time course of natural scene understanding. *Psychological Science*, *20*(4), 464–472.

Greene, M. R., & Oliva, A. (2009b). Recognition of natural scenes from global properties: Seeing the forest without representing the trees. *Cognitive Psychology*, *58*(2), 137–176.

Grill-Spector, K., Henson, R., & Martin, A. (2006). Repetition and the brain: Neural models of stimulus-specific effects. *Trends in Cognitive Sciences*, *10*(1), 14–23.

Grill-Spector, K., Kourtzi, Z., & Kanwisher, N. (2001). The lateral occipital complex and its role in object recognition. *Vision Research*, *41*(10–11), 1409–1422.

Grill-Spector, K., & Malach, R. (2001). fMR-adaptation: A tool for studying the functional properties of human cortical neurons. *Acta Psychologica*, *107*(1–3), 293–321.

Harel, A., Kravitz, D. J., & Baker, C. I. (2013). Deconstructing visual scenes in cortex: Gradients of object and spatial layout information. *Cerebral Cortex*, *23*(4), 947–957.

Hartley, T., Bird, C. M., Chan, D., Cipolotti, L., Husain, M., Vargha-Khadem, F., et al. (2007). The hippocampus is required for short-term topographical memory in humans. *Hippocampus*, *17*(1), 34–48.

Hassabis, D., Kumaran, D., & Maguire, E. A. (2007). Using imagination to understand the neural basis of episodic memory. *Journal of Neuroscience*, *27*(52), 14365–14374.

Hasson, U., Levy, I., Behrmann, M., Hendler, T., & Malach, R. (2002). Eccentricity bias as an organizing principle for human high-order object areas. *Neuron*, *34*(3), 479–490.

Haxby, J. V., Gobbini, M. I., Furey, M. L., Ishai, A., Schouten, J. L., & Pietrini, P. (2001). Distributed and overlapping representations of faces and objects in ventral temporal cortex. *Science*, *293*(5539), 2425–2430.

Henderson, J. M., & Hollingworth, A. (1999). High-level scene perception. *Annual Review of Psychology*, *50*, 243–271.

Hermer, L., & Spelke, E. S. (1994). A geometric process for spatial reorientation in young children. *Nature*, *370*(6484), 57–59.

Janzen, G., & Jansen, C. (2010). A neural wayfinding mechanism adjusts for ambiguous landmark information. *NeuroImage*, *52*(1), 364–370.

Janzen, G., & Van Turennout, M. (2004). Selective neural representation of objects relevant for navigation. *Nature Neuroscience*, *7*(6), 673–677.

Julian, J. B., Fedorenko, E., Webster, J., & Kanwisher, N. (2012). An algorithmic method for functionally defining regions of interest in the ventral visual pathway. *NeuroImage*, *60*(4), 2357–2364.

Kamitani, Y., & Tong, F. (2005). Decoding the visual and subjective contents of the human brain. *Nature Neuroscience*, *8*(5), 679–685.

King, J. A., Burgess, N., Hartley, T., Vargha-Khadem, F., & O'Keefe, J. (2002). Human hippocampus and viewpoint dependence in spatial memory. *Hippocampus*, *12*(6), 811–820.

Konkle, T., & Oliva, A. (2012). A real-world size organization of object responses in occipito-temporal cortex. *Neuron*, *74*(6), 1114–1124.

Kornblith, S., Cheng, X., Ohayon, S., & Tsao, D. Y. (2013). A network for scene processing in the macaque temporal lobe. *Neuron*, *79*(4), 766–781.

Kravitz, D. J., Peng, C. S., & Baker, C. I. (2011). Real-world scene representations in high-level visual cortex: It's the spaces more than the places. *Journal of Neuroscience*, *31*(20), 7322–7333.

Kreiman, G., Koch, C., & Fried, I. (2000). Imagery neurons in the human brain. *Nature*, *408*(6810), 357–361.

Kriegeskorte, N., Goebel, R., & Bandettini, P. (2006). Information-based functional brain mapping. *Proceedings of the National Academy of Sciences of the United States of America*, *103*(10), 3863–3868.

Kuipers, B., Modayil, J., Beeson, P., MacMahon, M., & Savelli, F. (2004). *Local metrical and global topological maps in the hybrid spatial semantic hierarchy*. Paper presented at the IEEE International Conference on Robotics and Automation.

Lee, A. C., Bussey, T. J., Murray, E. A., Saksida, L. M., Epstein, R. A., Kapur, N., et al. (2005). Perceptual deficits in amnesia: Challenging the medial temporal lobe "mnemonic" view. *Neuropsychologia*, *43*(1), 1–11.

Lee, A. C., Yeung, L. K., & Barense, M. D. (2012). The hippocampus and visual perception. *Frontiers in Human Neuroscience*, *6*, 91.

Levy, I., Hasson, U., Avidan, G., Hendler, T., & Malach, R. (2001). Center-periphery organization of human object areas. *Nature Neuroscience*, *4*(5), 533–539.

Levy, I., Hasson, U., Harel, M., & Malach, R. (2004). Functional analysis of the periphery effect in human building related areas. *Human Brain Mapping*, *22*(1), 15–26.

MacEvoy, S. P., & Epstein, R. A. (2007). Position selectivity in scene and object responsive occipitotemporal regions. *Journal of Neurophysiology*, *98*, 2089–2098.

MacEvoy, S. P., & Epstein, R. A. (2011). Constructing scenes from objects in human occipitotemporal cortex. *Nature Neuroscience*, *14*(10), 1323–1329.

Malach, R., Reppas, J. B., Benson, R. R., Kwong, K. K., Jiang, H., Kennedy, W. A., et al. (1995). Object-related activity revealed by functional magnetic resonance imaging in human occipital cortex. *Proceedings of the National Academy of Sciences of the United States of America*, *92*, 8135–8139.

Morgan, L. K., MacEvoy, S. P., Aguirre, G. K., & Epstein, R. A. (2011). Distances between real-world locations are represented in the human hippocampus. *Journal of Neuroscience*, *31*(4), 1238–1245.

Mormann, F., Kornblith, S., Quiroga, R. Q., Kraskov, A., Cerf, M., Fried, I., et al. (2008). Latency and selectivity of single neurons indicate hierarchical processing in the human medial temporal lobe. *Journal of Neuroscience*, *28*(36), 8865–8872.

Mullally, S. L., & Maguire, E. A. (2011). A new role for the parahippocampal cortex in representing space. *Journal of Neuroscience*, *31*(20), 7441–7449.

Mullin, C. R., & Steeves, J. K. (2011). TMS to the lateral occipital cortex disrupts object processing but facilitates scene processing. *Journal of Cognitive Neuroscience*, *23*(12), 4174–4184.

Mur, M., Ruff, D. A., Bodurka, J., De Weerd, P., Bandettini, P. A., & Kriegeskorte, N. (2012). Categorical, yet graded—single-image activation profiles of human category-selective cortical regions. *Journal of Neuroscience*, *32*(25), 8649–8662.

Nasr, S., Liu, N., Devaney, K. J., Yue, X., Rajimehr, R., Ungerleider, L. G., et al. (2011). Scene-selective cortical regions in human and non-human primates. *Journal of Neuroscience*, *31*(39), 13771–13785.

Nasr, S., & Tootell, R. B. H. (2012). A cardinal orientation bias in scene-selective visual cortex. *Journal of Neuroscience*, *32*(43), 14921–14926.

O'Craven, K. M., & Kanwisher, N. (2000). Mental imagery of faces and places activates corresponding stimulus-specific brain regions. *Journal of Cognitive Neuroscience*, *12*(6), 1013–1023.

Oliva, A., & Schyns, P. G. (2000). Diagnostic colors mediate scene recognition. *Cognitive Psychology*, *41*(2), 176–210.

Oliva, A., & Torralba, A. (2001). Modeling the shape of the scene: A holistic representation of the spatial envelope. *International Journal of Computer Vision*, *42*(3), 145–175.

Park, S., Brady, T. F., Greene, M. R., & Oliva, A. (2011). Disentangling scene content from spatial boundary: Complementary roles for the PPA and LOC in representing real-world scenes. *Journal of Neuroscience*, *31*(4), 1333–1340.

Park, S., & Chun, M. M. (2009). Different roles of the parahippocampal place area (PPA) and retrosplenial cortex (RSC) in scene perception. *NeuroImage*, *47*(4), 1747–1756.

Park, S., Intraub, H., Yi, D. J., Widders, D., & Chun, M. M. (2007). Beyond the edges of a view: Boundary extension in human scene-selective visual cortex. *Neuron*, *54*(2), 335–342.

Peelen, M. V., Fei-Fei, L., & Kastner, S. (2009). Neural mechanisms of rapid natural scene categorization in human visual cortex. *Nature*, *460*(7251), 94–97.

Ploner, C. J., Gaymard, B. M., Rivaud-Pechoux, S., Baulac, M., Clemenceau, S., Samson, S., et al. (2000). Lesions affecting the parahippocampal cortex yield spatial memory deficits in humans. *Cerebral Cortex*, *10*(12), 1211–1216.

Potter, M. C. (1975). Meaning in visual scenes. *Science*, *187*, 965–966.

Potter, M. C. (1976). Short-term conceptual memory for pictures. *Journal of Experimental Psychology. Human Learning and Memory*, *2*(5), 509–522.

Potter, M. C., & Levy, E. I. (1969). Recognition memory for a rapid sequence of pictures. *Journal of Experimental Psychology*, *81*(1), 10–15.

Quattoni, A., & Torralba, A. (2009). Recognizing indoor scenes. *CVPR: 2009 IEEE Conference on Computer Vision and Pattern Recognition, 1–4*, 413–420.

Rajimehr, R., Devaney, K. J., Bilenko, N. Y., Young, J. C., & Tootell, R. B. H. (2011). The parahippocampal place area responds preferentially to high spatial frequencies in humans and monkeys. *PLoS Biology*, *9*(4), e1000608.

Renninger, L. W., & Malik, J. (2004). When is scene identification just texture recognition? *Vision Research*, *44*, 2301–2311.

Saleem, K. S., Price, J. L., & Hashikawa, T. (2007). Cytoarchitectonic and chemoarchitectonic subdivisions of the perirhinal and parahippocampal cortices in macaque monkeys. *Journal of Comparative Neurology*, *500*(6), 937–1006.

Sasaki, Y., Rajimehr, R., Kim, B. W., Ekstrom, L. B., Vanduffel, W., & Tootell, R. B. H. (2006). The radial bias: A different slant on visual orientation sensitivity in human and nonhuman primates. *Neuron*, *51*(5), 661–670.

Sawamura, H., Orban, G. A., & Vogels, R. (2006). Selectivity of neuronal adaptation does not match response selectivity: A single-cell study of the fMRI adaptation paradigm. *Neuron*, *49*(2), 307–318.

Schinazi, V. R., & Epstein, R. A. (2010). Neural correlates of real-world route learning. *NeuroImage*, *53*(2), 725–735.

Schwarzlose, R. F., Swisher, J. D., Dang, S., & Kanwisher, N. (2008). The distribution of category and location information across object-selective regions of visual cortex. *Proceedings of the National Academy of Sciences of the United States of America*, *105*(11), 4447–4452.

Schyns, P. G., & Oliva, A. (1994). From blobs to boundary edges: Evidence for time- and spatial-scale-dependent scene recognition. *Psychological Science*, *5*(4), 195–200.

Spiridon, M., & Kanwisher, N. (2002). How distributed is visual category information in human occipito-temporal cortex? An fMRI study. *Neuron*, *35*, 1157–1165.

Squire, L. R., van der Horst, A. S., McDuff, S. G., Frascino, J. C., Hopkins, R. O., & Mauldin, K. N. (2010). Role of the hippocampus in remembering the past and imagining the future. *Proceedings of the National Academy of Sciences of the United States of America*, *107*(44), 19044–19048.

Steeves, J. K., Humphrey, G. K., Culham, J. C., Menon, R. S., Milner, A. D., & Goodale, M. A. (2004). Behavioral and neuroimaging evidence for a contribution of color and texture information to scene classification in a patient with visual form agnosia. *Journal of Cognitive Neuroscience*, *16*(6), 955–965.

Summerfield, C., Trittschuh, E. H., Monti, J. M., Mesulam, M. M., & Egner, T. (2008). Neural repetition suppression reflects fulfilled perceptual expectations. *Nature Neuroscience*, *11*(9), 1004–1006.

Swisher, J. D., Gatenby, J. C., Gore, J. C., Wolfe, B. A., Moon, C. H., Kim, S. G., et al. (2010). Multiscale pattern analysis of orientation-selective activity in the primary visual cortex. *Journal of Neuroscience*, *30*(1), 325–330.

Thorpe, S., Fize, D., & Marlot, C. (1996). Speed of processing in the human visual system. *Nature*, *381*(6582), 520–522.

Treisman, A. (2006). How the deployment of attention determines what we see. *Visual Cognition*, *14*(4–8), 411–443.

Troiani, V., Stigliani, A., Smith, M. E., & Epstein, R. A. (2014). Multiple object properties drive scene-selective regions. *Cerebral Cortex*, *24*(4), 883–897.

Vann, S. D., Aggleton, J. P., & Maguire, E. A. (2009). What does the retrosplenial cortex do? *Nature Reviews Neuroscience*, *10*(11), 792–802.

Walther, D. B., Caddigan, E., Fei-Fei, L., & Beck, D. M. (2009). Natural scene categories revealed in distributed patterns of activity in the human brain. *Journal of Neuroscience*, *29*(34), 10573–10581.

Walther, D. B., Chai, B., Caddigan, E., Beck, D. M., & Fei-Fei, L. (2011). Simple line drawings suffice for functional MRI decoding of natural scene categories. *Proceedings of the National Academy of Sciences of the United States of America*, *108*(23), 9661–9666.

Ward, E. J., MacEvoy, S. P., & Epstein, R. A. (2010). Eye-centered encoding of visual space in scene-selective regions. *Journal of Vision*, *10*(14), 1–12.

Wolbers, T., & Buchel, C. (2005). Dissociable retrosplenial and hippocampal contributions to successful formation of survey representations. *Journal of Neuroscience*, *25*(13), 3333–3340.

Wolbers, T., Klatzky, R. L., Loomis, J. M., Wutte, M. G., & Giudice, N. A. (2011). Modality-independent coding of spatial layout in the human brain. *Current Biology*, *21*(11), 984–989.

Xu, Y., Turk-Browne, N. B., & Chun, M. M. (2007). Dissociating task performance from fMRI repetition attenuation in ventral visual cortex. *Journal of Neuroscience*, *27*(22), 5981–5985.

Yue, X., Vessel, E. A., & Biederman, I. (2007). The neural basis of scene preferences. *Neuroreport*, *18*(6), 525–529.

7

Putting Scenes in Context

Elissa M. Aminoff

Human vision can understand an image of a scene extremely quickly and effortlessly. However, the mechanisms mediating scene understanding are still being explored. This chapter proposes that scene understanding is not derived from a unique, isolated cognitive process but rather is part of a more general mechanism of associative processing. When a person is understanding a scene, it is the collection of associations, meaning the co-occurrence of objects, the spatial relations among these objects, and other statistical regularities associated with scene categories that are processed. The object-to-object relations and spatial relations define the scene and signify a scene category. Framing scene understanding as associative processing provides a framework for not only a bottom-up flow of information in which the visual stimulus is analyzed but also a top-down flow of information. Associative processing is used to generate predications and expectations about what is likely to appear in the environment (Bar, 2007), which facilitates additional scene processing. Without this facilitation, scene understanding would be extremely slow. This chapter discusses an associative processing approach to scene understanding that provides a testable model for future research.

Scenes are rich stimuli that carry vast amounts of information in a wide variety of domains. For example, if you look at figure 7.1A, you can quickly identify the scene as a pier and predict the water likely extends beyond the boundaries of the picture presented; if you were there, you would be able to walk out on the pier for some distance; and given the pier lights are on, it is likely close to evening time. Most theories explaining the mechanisms underlying scene understanding concentrate within a single dimension of the spatial domain (e.g., three-dimentionality, or geometric layout, or spatial boundary), which may be plausible given figure 7.1A. However, in regard to figure 7.1B, it quickly becomes apparent that scene understanding involves more than spatial processing of a single dimension because it is quickly understood that this scene is bizarre given that a giant size rubber duck is in a city river. In the case of this image there are many associative violations: a rubber duck in a river and the size of the duck compared with the size of the people. This exemplifies how the relations within the scene create a meaningful

Figure 7.1
Examples of scenes.

whole stimulus. Along these same lines, the description of figure 7.1A included characteristics derived from associative processing: predicting the water extended beyond the boundaries of the scene and that the picture was taken close to evening. The scene was also characterized by the joint presence of the boardwalk and the water underneath, which are strongly associated together. Moreover, if scene understanding could be reduced to a single spatial dimension, then understanding the differences between figure 7.1C, a dining room, and figure 7.1D, a conference room, would be very difficult given the similar spatial structure of these two scenes. These provide clear examples of how mechanisms mediating scene understanding should include processing the associations among the elements of a scene, both the object-to-object relations and the spatial relations.

This chapter proposes a mechanism of scene understanding by which properties of a scene that have high associative strength (e.g., objects, such as a place setting, and the spatial layout, a large central surface surrounded by items) are extracted first and used to access long-term representations of contexts and scene categories. These long-term representations are then used to generate predictions and expectations about the environment, facilitating information processing. This theory provides a parsimonious explanation of scene understanding that incorporates the elegant theories previously proposed in scene understanding into a more comprehensive mechanism.

Existing Theories of Scene Understanding

Previous theories of scene understanding, many inspired by fMRI results, concentrate on trying to describe a single dimension of a scene that provides the most meaningful information. This earlier research revealed valuable steps in our understanding of scene perception and of the type of scene information that is represented in the cortex. For example, scene categories can be characterized by global properties that take into account overall spatial layout (e.g., "spatial envelope") (Oliva & Torralba, 2001; Torralba & Oliva, 2003), which has been shown to predict behavioral performance in scene categorization (Greene & Oliva, 2009). Moreover, the neural pattern of scene representation carries meaningful information in this regard as well. Spatial boundary (e.g., an open or closed expanse) (Kravitz, Peng, & Baker, 2011; Park, Brady, Greene, & Oliva, 2011), geometric layout (Epstein & Ward, 2010), and three-dimensionality (Mullally & Maguire, 2011) are all encoded within the activity related to processing scenes. However, there is a problem with each of these theories. First, the different theories are mutually exclusive. Studies providing evidence for each theory independently demonstrate that their proposed spatial dimension is the property reflected in the neural activity related to processing scenes, without including a role for the other dimensions to play a role as well. *Each* of the dimensions listed above is an important aspect of a scene, and all should be reflected in the accompanying neural signal, which is not accounted for in these theoretical accounts. Second, the previous discussion of scene understanding accounts exclusively for spatial dimensions of a scene, which misses both the spatial relations within the scene and information on the scene within nonspatial domains. For example, each of the theories mentioned above treats both figure 7.1A and figure 7.1B with the same validity, which is obviously missing an important dimension of differences between these scenes.

A third problem with existing theories of the mechanisms underlying scene understanding is that scene perception engages regions of the brain that are not exclusive to scenes or even to spatial processing. To find "scene-selective" regions of the brain, typically fMRI signal is measured when the participant is viewing scenes compared with viewing objects, faces, and phase-scrambled images. This contrast reveals three main sites of activity: the parahippocampal cortex (which may also extend into more posterior lingual regions; PHC; also known as the parahippocampal place area; PPA), the retrosplenial complex (which includes the retrosplenial cortex, and regions of the posterior cingulate and precuneus; RSC), and a lateral region near the parietal-occipital junction (LPC) (e.g., Aguirre, Detre, Alsop, & D'Esposito, 1996; Dilks, Julian, Paunov, & Kanwisher, 2013; Epstein & Kanwisher, 1998; Henderson, Larson, & Zhu, 2008). Both the PHC and the RSC are also implicated in episodic memory (e.g., Addis, Wong, & Schacter, 2007; Davachi, Mitchell, & Wagner, 2003; Diana, Yonelinas, & Ranganath, 2010; Ranganath, 2010), even when nonspatial stimuli are used (Kirwan & Stark, 2004). And the PHC and RSC are also part of the default mode (Buckner, Andrews-Hanna, & Schacter, 2008; Raichle et al., 2001), which is typically more active at rest than while on task. How can these rather disparate functions all activate the same regions of the brain? If the processing had been only spatial, than episodic memory regarding nonspatial content should not engage these regions. And moreover, why would it be part of the default network? This chapter proposes the reason for this overlap of functional activity is related to a fundamental cognitive process involved in scene understanding, episodic memory, and default mode: associative processing (Aminoff, Kveraga, & Bar, 2013; Bar, Aminoff, Mason, & Fenske, 2007). Associative processing is critical in scene understanding in order to examine the relations among the units of a scene and to understand the scene with respect to a larger context. Along these lines, a scene can be thought of as a collection of associative elements that are repeatedly encountered together. This provides a mechanism for defining the differences in the validity between figure 7.1A and B. Associative processing is also critical in episodic memory by forming a memory of an event by creating associations among the different elements of the episode and using long-term contextual associations to fill in details. Last, associative processing can explain what is occurring during the "rest" periods that define the default network. These are times in which an individual is not devoid of thought but rather is engaged in free associative thought or mind wandering, which engages associative processing (Mason et al., 2007). Thus, associative processing can explain why each of these processes engages overlapping regions of the cortex.

Associative Processing Explanation of Scene Understanding

An associative processing approach to scene understanding addresses each of the concerns listed above. First, it includes processing of information along different dimensions—the only requirement is to have associative relevance. For example, spatial boundary, such as expanse, is strongly associated with outdoor landscapes, and thus, activity related to scene understanding may reflect the processing of spatial boundary because it is highly associative to a type of scene. Second, associative processing designates the relations among elements in a scene in both spatial (e.g., the size of the duck compared to the size of the people) and nonspatial domains (e.g., a duck in a river and not in a bathtub) as mediating scene understanding. Third, scene understanding is not explained through an isolated process, which should have special neural regions dedicated only to scene understanding, but rather as part of a more general mechanism of associative processing, which is involved in other cognitive domains such as episodic memory and mind wandering.

If associative processing is the underlying mechanism involved in scene understanding, then the associations within a scene should modulate neural activity related to scene processing. To test this, scene processing in the brain was compared for scenes with a foreground object that had strong contextual associations (e.g., a scene with a tennis racket in the center) versus scenes with a foreground object that had weak contextual associations (e.g., a scene with a vase in the center; figure 7.2A, plate 14) (Bar, Aminoff, & Schacter, 2008). More associative processing would occur if the foreground object were strongly associated with a specific context (e.g., a tennis racket—tennis court) because the stimulus would elicit activation of related associations (e.g., tennis net, the lines of the court, sounds of tennis balls being hit). In comparison, a scene with a foreground object that had weak contextual associations (e.g., a vase, which can be found anywhere) would not elicit any additional associations. Results demonstrated, more fMRI signal was associated with processing the scenes with the strong contextual foreground object than with the scenes with a weak contextual foreground object. This differential activity was found within the "scene-selective" regions, the RSC, PHC, and LPC, and an additional region within the medial prefrontal cortex (MPFC). These two conditions of scenes were matched for spatial information; for example, both sets of scenes are three-dimensional and contain geometric layout, and thus, the differential activity could not be attributed to processing of the spatial properties of the scene. In contrast, this differential activity supported the proposal that associative processing was mediating scene understanding because the stimuli with more associative information activated scene-selective regions to a greater extent.

If associative processing was modulating the activity within these scene-selective regions, then these regions should be active for nonscene stimuli that are also highly associative. This was tested in a number of ways. First, activity in these regions elicited for single objects was investigated. As mentioned above, some objects are strongly associated with a context (e.g., shower curtain—bathroom), and some objects are weakly associated with many contexts and have no specific context strongly associated with them (e.g., folding chair). Moreover, some objects are strongly associated with contexts that are not tied to a specific place but are still strongly associated with a clustering of objects, such as a cake at birthday party, referred to as a nonspatial context. When the fMRI activity elicited for viewing objects that were strongly associated with either a spatial context or a nonspatial context was compared with that evoked by viewing objects that were weakly associated with many contexts, differential activity was found within the scene-selective regions—the RSC, PHC, and LPC—and again an additional region within the MPFC (figure 7.2B, plate 14) (Aminoff, Schacter, & Bar, 2008; Bar & Aminoff, 2003; Kveraga et al., 2011; Shenhav, Barrett, & Bar, 2012). This supported the proposal that these regions are processing associations at large and that the mechanism mediating scene understanding is part of this more general mechanism and is not unique to scenes. However, scenes typically activate these regions to a much larger extent than individual objects due to the richness of

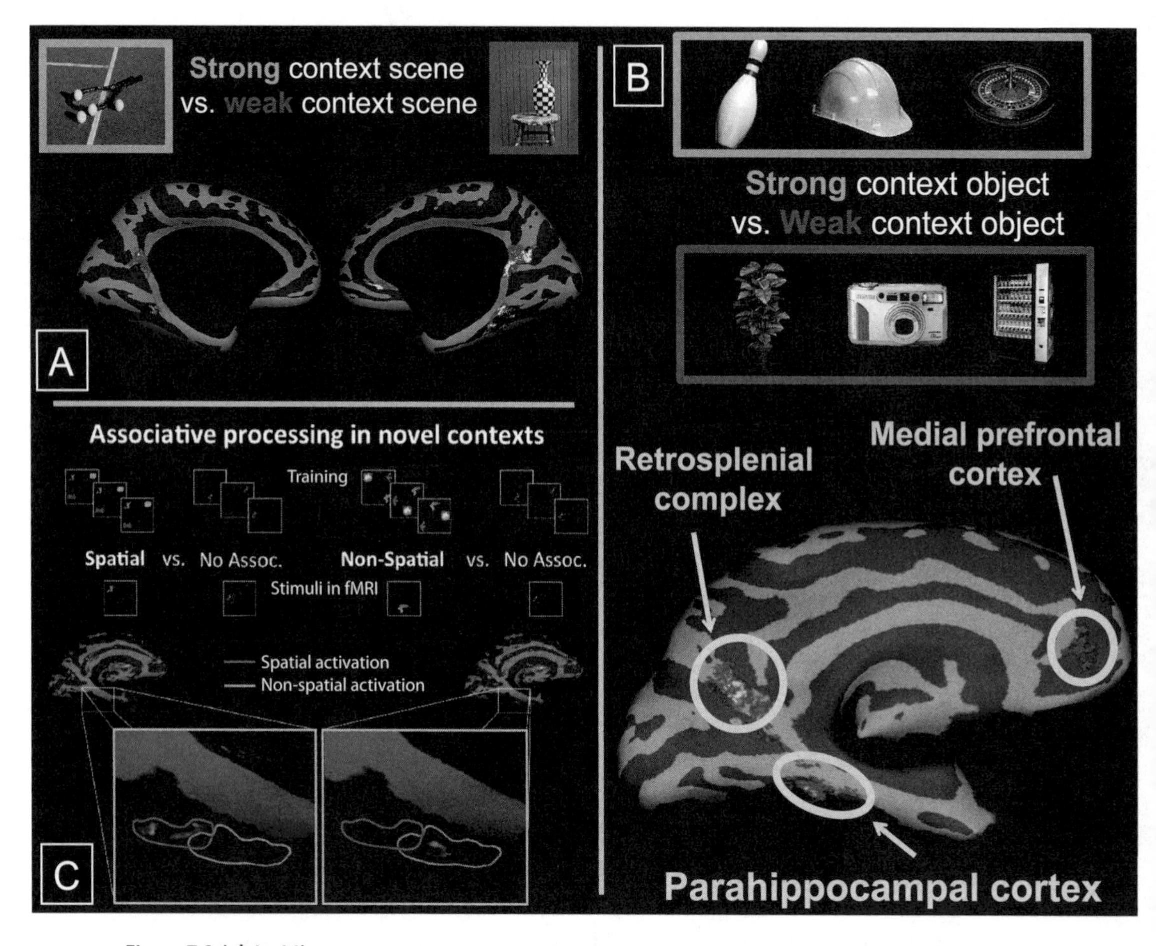

Figure 7.2 (plate 14)
Associative processing in the brain. (A) Differential activity related to processing scenes with a foreground object with strong contextual associations compared with processing scenes with a foreground object with weak contextual associations. Three main sites of activity are the scene-selective regions of the parahippocampal cortex and retrosplenial complex and the medial prefrontal cortex. (B) Differential activity related to processing single objects in isolation, comparing activity elicited for objects with strong contextual associations (e.g., roulette wheel) with objects that were weakly associated with many contexts (e.g., plants). Results were seen in the retrosplenial complex, the parahippocampal cortex, and the medial prefrontal cortex. (C) Associative processing of meaningless shapes. Participants were trained to learn associations between the shapes and locations, and some shapes that were trained had no associations. Using fMRI we compared activity elicited for just single shapes that only differed based on the associations learned during training. Comparing activity elicited for shapes with strong associations with shapes with no associations yielded differential activity within the parahippocampal cortex (zoomed-in view). This activation was divided with anterior regions of the PHC processing nonspatial associations and posterior regions of the PHC processing spatial associations.

the associations within a scene. A scene reveals many more associations to process than a single object and thus elicits these regions to a greater amount.

Objects are always encountered embedded within a scene, and therefore, an alternative interpretation of the differential activity related to object processing was that instead of associative processing it was an indirect activation of scene processing in which spatial processing along a single dimension may not be disputed (e.g., Mullally & Maguire, 2011). To address this issue, associative processing in extremely minimal stimuli (associations between meaningless shapes within a grid trained over a 2-week period) was examined in comparison to scene processing (figure 7.2C, plate 14; Aminoff, Gronau, & Bar, 2007). Thus, this was the second method for examining the neural signal related to associative processing in nonscene stimuli. In this study the specific hypothesis addressed the parahippocampal region alone, targeting whether the "parahippocampal place area" may not be specific for places but rather for stimuli with strong associations. If overlapping activity within the PHC were found for processing scenes and for processing the associations for the meaningless shapes, this would provide strong evidence that associative processing is the mechanism that underlies scene understanding. Indeed, this is what was found. When the fMRI signal elicited for shapes that had strong contextual associations was compared with that for shapes that had no associations, differential activity was found within the PHC, overlapping with activity elicited for scenes. Even with such minimal stimuli (meaningless shapes), these highly controlled associations elicited activity within the scene-selective parahippocampal region, supporting an associative processing interpretation of the functional role of this region.

This collection of studies provides evidence supporting the proposal that associative processing should be attributed to these regions previously thought to be selective for scene processing, and thus, scene understanding should be considered a product of associative processing. So far, in this discussion all brain regions (RSC, PHC, LPC, and MPFC) have been investigated together without consideration of the unique contribution of each region to scene understanding and to associative processing in general. To elaborate on a potential mechanism mediating scene understanding and associative processing, it requires an explanation that accounts for the *network* of brain regions sensitive to scenes rather than focusing on just one region or a single function. This chapter proposes a mechanism (figure 7.3) that attributes the RSC as storing and processing a long-term representation of a context, or scene category, that is built up over a lifetime in which regularities are extracted and represented within a prototypical representation of the context—a "context frame" (Aminoff et al., 2008; Bar & Ullman, 1996). The PHC, on the other hand, acts as a liaison between the context frames stored in the RSC and the current environment. And finally, the MPFC uses the associations stored in the context frames of the RSC to make predictions about what will next occur in the environment, facilitating subsequent cognition (Bar,

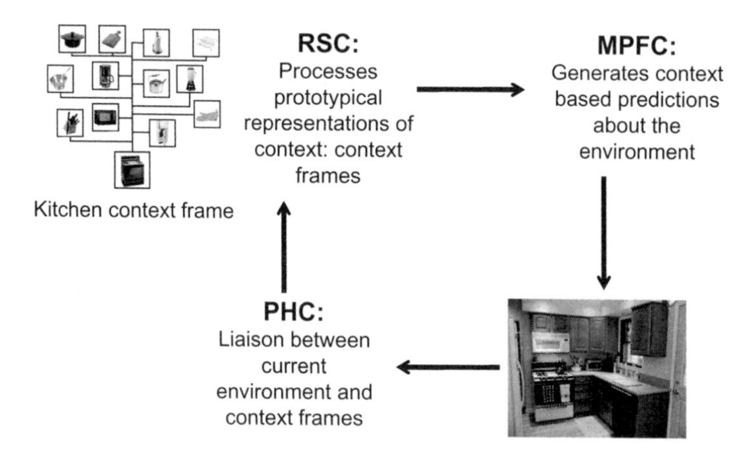

Figure 7.3
A proposed psychological and neural mechanism underlying scene understanding. Information with contextual significance is first extracted and processed in the PHC. This contextual information is then used to activate the relevant context frame in the RSC. The associations within the context frame are then used to generate expectations and predictions about future interactions in the environment within the MPFC.

2007). Not as much work has been dedicated to exploring the role of the LPC in associative processing, and thus, no concrete proposal of the functional role of the LPC is being made at this time.

Context Frames in the Retrosplenial Cortex

Throughout a lifetime we repeatedly experience certain contexts and categories of scenes. This allows us to step into a kitchen that we have never been in before and still easily identify the room as a kitchen, know that the sink is for washing dishes and not brushing teeth, and expect to find a refrigerator with food inside. Matching the perception of a scene with a long-term memory representation of the related scene category or context is a critical aspect to scene understanding. It provides heuristics and shortcuts to facilitate and guide perception of the scene. These long-term memory representations, termed *context frames* (Bar & Ullman, 1996), contain prototypical information about a scene category or a context defined by the regularities extracted over repeated exposure to different exemplars, which is similar to concepts such as *schema* (Bartlett, 1932). For example, a context frame of a bathroom would include a shower, toilet, sink, toothbrush, and so on. It would also include spatial relations such as the toilet paper being next to the toilet and the sink placed in front of a mirror. However, the information within a context frame is general enough to apply to any exemplar (e.g., a bathtub would be represented, but whether the tub was a claw-foot bathtub would not necessarily be included). The prototypical nature of this representation allows it to be applicable to all different exemplars of the context and to be applied to new instances of the context to help guide behavior, expectations, and interaction with the environment (Bar, 2007). To ascribe a role to the RSC in processing context frames, the RSC should exhibit the ability to (1) integrate across multiple stimuli to create a network of associations inherent to a context frame; (2) abstract across multiple exemplars to a combined, prototypical representation; and, last, (3) relate specific instances to the greater context. Previous research on the RSC provides strong evidence to delegate this region as storing and processing these context frames.

1. *Integrating across multiple stimuli* The RSC is located in a highly integrative region of the brain. It has reciprocal connections with the hippocampal formation, the parahippocampal region, thalamic nuclei, prefrontal cortex, superior temporal sulcus, areas V4, and other cingulate regions (Kobayashi & Amaral, 2003, 2007). Therefore, it receives many inputs both in modal and amodal regions of the cortex. This provides a key environment to integrate over multiple signals to develop a context frame. Studies of RSC-lesioned rats have shown a direct link between integrating across multiple stimuli and RSC function. RSC-lesioned rats are impaired in learning associations among multiple stimuli when they are presented simultaneously or serially (Keene & Bucci, 2008; Robinson, Keene, Iaccarino, Duan, & Bucci, 2011). This impairment was specific to learning the associations among multiple stimuli, not in learning single stimulus-stimulus associations. This impairment in RSC-lesioned rats suggests a direct link between the ability to integrate across many associations and RSC function. Thus, providing evidence that the RSC is involved in integrating across multiple stimuli—a critical condition necessary for processing and storing a context frame, which contains a network of associations.

2. *Integrating across multiple exemplars* Within spatial memory, the RSC has been shown to process a representation that integrates across multiple exemplars. For example, the RSC processes an allocentric perspective of space (Committeri et al., 2004; Vann & Aggleton, 2002). This allocentric perspective has also been shown to correlate with map learning (Wolbers & Buchel, 2005). In order to develop an allocentric map it is necessary to integrate across multiple egocentric maps. Once multiple egocentric maps of different heading directions can be integrated, the allocentric map can be extracted. An analogous process can be involved in developing and accessing context frames in contextual associative processing. The multiple exemplars of a dining room (e.g., your dining room, your parents' dining room) are integrated together to provide a representation of the dining room context (i.e., analogous to the allocentric map) not associated with any specific exemplar (i.e., analogous to separate egocentric maps of the same area).

3. *Relate specific instances to a broader context* Another requirement in processing context frames is to extrapolate a stimulus to its broader context. This is critical for understanding a scene; for example, in figure 7.1A it is understood that the entrance of the pier is on land and extends out to the portion we see over the water. Without linking a scene to the broader context, processes related to scene understanding, such as navigation, would be nearly impossible. Previous research supports a role for the RSC in this type of extrapolation. Park and colleagues (Park & Chun, 2009; Park, Intraub, Yi, Widders, & Chun, 2007) have shown an integral role of the RSC in relating a single picture of a scene to a larger context. For example, panoramic scenes were divided into three different sections. Using a repetition paradigm, the RSC treated three different, but continuous, views of a panoramic scene in the same way it would respond to three identical scene representations (Park & Chun, 2009). By treating each scene as the same, the RSC appears to have extrapolated all three views into the same broader context. This extraction of the stimulus to a larger context was also shown when fMRI was used to examine the phenomenon of boundary extension—a false memory of a scene with wider boundaries then was actually presented (see chapter 1 by Intraub). In this study, Park et al. found that the RSC responded to a picture of a scene and a picture of the same scene with a wide angle view in the same way, suggesting that on first presentation of a scene the RSC extracts it to the larger view (i.e., extending the boundaries) (Park et al., 2007). Epstein and colleagues have also shown that the RSC is involved in processing a single scene (e.g., a specific location on campus) within a larger context (e.g., the whole campus) (Epstein, 2008; Epstein & Higgins, 2007). These sets of studies reveal a role of the RSC in taking a single stimulus and relating it to the broader context with which it is associated. This is an analogous process used in contextual associative processing in which a specific exemplar of a context (e.g., *my* bedroom) is related to the broader, prototypical, representation of context (i.e., context frame of a bedroom).

If context frames are the representations that mediate contextual processing, and if the RSC is involved in processing context frames, one could infer that the neural signal elicited in the RSC reflects a process in which a highly contextual object (e.g., an oven) activates a relevant context frame (e.g., kitchen), which leads to the activation of related associations (e.g., refrigerator, pots, the smell of freshly baked cookies) within the context frame. By this mechanism seeing an oven can activate related associations such as a refrigerator. This leads to the hypothesis that seeing one object will lead to activation of other objects related to the context but not necessarily present in the scene, and this will be reflected in the neural signal of the RSC. If these related objects are activated at the same time the original object was first perceived, it may subsequently be difficult to remember which object was actually present in the scene and lead to false recognition of related objects. To test this, a false recognition paradigm was used to investigate whether, on seeing a strong contextual object (e.g., a bed), the processing related to accessing the related context frame, reflected in the activity elicited from the RSC, would predict subsequent false recognition for a related object (e.g., thinking one also saw a dresser) (Aminoff et al., 2008). The prediction was that the more the context frame was processed, the more likely related objects were processed, which increases the likelihood of subsequent false memory of the related object. The less a context frame was processed, the less likely related objects were processed, and thus, a smaller chance of false recognition of the related object. This false recognition would be predicted by increased BOLD signal in the RSC during encoding at study. We tested this hypothesis using fMRI to look at activity within the context brain network at encoding and found that increased BOLD signal in the RSC predicted subsequent false recognition of the related item (Aminoff et al., 2008). This study supported the proposal that context frames are stored and processed in the RSC and can be activated by the process of relating a presented object into its greater context, which in turn activates related objects.

Although this may be considered a fault of cognition (Schacter, 1999), it illustrates a critical aspect to scene understanding: that a scene is not processed as an indivisible unit in isolation but rather is processed by being related to the context most strongly associated with it, and this is the function of the RSC. When processing a scene the perception does not just include the bottom-up analysis of the stimulus but also includes top-down influences generated from the activation of related associations in a context frame. This false recognition result exposes this mechanism, which, instead of resulting in weaknesses, typically results in facilitation of cognition. For example, quickly activating the related context frame can prime objects, yielding faster recognition of related objects. Scenes are rich stimuli that carry a lot of information, and activation of the related context frame can reduce the load of information processed based on expectations built from previous experiences within that scene category or context. But a critical aspect in this mechanism is finding the aspects of a scene that carry strong contextual associations, to access the related context frame. For this step in the mechanism, we turn to the functional role of the PHC.

The Parahippocampal Cortex: Liaison from Context Frames to the Current Environment

Linking the current environment to stored representations of context is critical for fluid cognition. For example, walking into a room and seeing a desk chair on wheels we can quickly identify the room as an office based on our previous experience. This is the result of the current stimulus, the desk chair, linking to the stored context frame of an office. Similarly, in navigation, a current location is relayed to the stored representation of the town in order to create a path to the goal destination. Research suggests that this functional role can be attributed to the PHC. To provide evidence, two assumptions need to be verified: (1) sensitivity to contextually relevant information and (2) sensitivity to the current environment.

1. *Sensitivity to contextual information* As previously discussed, the PHC was sensitive to processing the contextual associations of scenes (Bar et al., 2008), objects (Aminoff et al., 2008; Bar & Aminoff, 2003; Kveraga et al., 2011; Shenhav et al., 2012), and learned contextual associations with novel shapes (Aminoff, Bar, & Schacter, 2007). This has been shown through a number of methods and paradigms (Diana, Yonelinas, & Ranganath, 2008; Peters, Daum, Gizewski, Forsting, & Suchan, 2009). The PHC also has been shown to selectively process properties of an environment that have strong associations. For example, Janzen and Van Turennout (2004) have demonstrated that the PHC elicits more activation for objects that are relevant for navigation (i.e., at decision points) compared with objects that are not (Janzen & Van Turennout, 2004). Moreover, familiarity, which provides a rich context, also modulated activity within the PHC during navigation (well-learned vs. novel routes) and landmark processing (famous vs. unfamiliar buildings) (Brown, Ross, Keller, Hasselmo, & Stern, 2010; Leveroni et al., 2000; Rauchs et al., 2008). In addition, chess boards presented with a configuration of pieces in legitimate positions compared with chess boards presented with a random configuration elicited activity in the parahippocampal cortex (Bilalić, Langner, Erb, & Grodd, 2010). The PHC was also sensitive to contextual information in a social-cognitive domain. Contextually relevant information, such as social hierarchical information (e.g., superior and inferior ranking), was found to modulate PHC activity (Zink et al., 2008). In addition, action perception of meaningful movement compared with meaningless movement yielded differential activity within the PHC (Rumiati et al., 2005; Schubotz, Korb, Schiffer, Stadler, & von Cramon, 2012). And within the auditory domain the PHC responded to highly contextual sounds (e.g., the sound of a fax machine) over weak contextual (e.g., rain) or meaningless sounds (Arnott, Cant, Dutton, & Goodale, 2008; Engel, Frum, Puce, Walker, & Lewis, 2009; Engelien et al., 2006). Thus, through a wide array of stimuli, in many different domains of information, the PHC has a clear role in processing highly contextual stimuli. The PHC is not just processing any stimuli but in fact processes only stimuli that have contextual relevance, which presumably are used to activate the relevant context frames. Therefore, when it is involved in scene understanding, it does not process the entire scene but processes parts of a scene that have contextual relevance.

2. *Sensitivity to the current environment* To assume a functional role for the PHC as the liaison between the current environment and the stored context frame, it is important to show that the PHC not only is sensitive to contextual information (as is the RSC) but is modulated by and interacts with the current environment. The PHC activity reflects the current environment by exhibiting sensitivity to the current stimulus specifically rather than extracting just the gist or abstracting it from present

physical form. For example, the PHC has been sensitive to specific exemplars. The PHC is typically characterized as viewpoint specific and showing individuation of exemplars from the same category (Epstein & Higgins, 2007; Park & Chun, 2009; Sung, Kamba, & Ogawa, 2008), but see Stevens, Kahn, Wig, and Schacter (2012) for important hemispheric differences. In addition, the PHC is sensitive to whether a strong contextual object is presented within a background and elicits more activity when the object is embedded within a background; in contrast, the RSC does not show this sensitivity and elicits similar activity for a strong contextual object in isolation or embedded within a background (Bar & Aminoff, 2003), presumably because both stimuli activate the same context frame. Moreover, the number of contextual associations can modulate the signal in the PHC. This was demonstrated by showing increased PHC activity for the more complex scenes (Bar et al., 2008; Chai, Ofen, Jacobs, & Gabrieli, 2010; Qin, van Marle, Hermans, & Fernández, 2011). Complexity can be reinterpreted as an increase in the number of associations present. It was also demonstrated in the number of contextual associations in memory encoding and retrieval (Tendolkar et al., 2007, 2008). Thus, when more associations are visible in the present stimulus, more contextual information is processed by the PHC, which then presumably activates the relevant context frame in the RSC. The RSC thus shows activity relating to abstracting the stimulus from the specific episode, whereas the PHC shows sensitivity to the current experience of the stimulus.

Altogether, the PHC is found to be sensitive to the current environment and selectively processes information that has strong contextual associations. Therefore, in this mechanism, a scene is first perceived within the PHC, with the intention to extract meaningful contextual associations. This extracted information is then used to access the relevant context frame, which is stored in the RSC. This suggests that in scene understanding, the visual system is constantly trying to link the current environment with our prior experience, which can help facilitate subsequent cognition. Facilitating cognition describes the last stage in this mechanism, which is suggested to occur in the MPFC.

Contextual Predictions and the Medial Prefrontal Cortex

Contextual processing can facilitate cognition by generating expectations and predictions about the next events to occur in the environment. This is the role attributed to the MPFC. Associations of a context frame can be used to anticipate what is not directly perceived in the current environment. This is critical for understanding pictures of scenes, such as anticipating that the pier in figure 7.1A eventually leads to land, and the water extends beyond the boundaries of the photograph presented. This is also critical for navigation; for example, when I open the door to my office, I know what to expect and know how to navigate to reach the elevators. Contextual processing is the proposed mechanism that yields these expectations.

To attribute this role to the MPFC there should be evidence that the MPFC is involved in generating expectations and predictions. To directly test this we examined a memory phenomenon, boundary extension, introduced earlier in the chapter. This phenomenon is a distortion of memory for scenes that are remembered with extended boundaries. In order to extend the boundaries of the scene, it is necessary to create expectations of what occurs beyond the boundaries presented, and those expectations are necessarily derived through contextual associations. We found activity in the MPFC, which overlaps the contextual processing regions, predicted subsequent boundary extension for scenes (Aminoff, Bar, & Schacter, 2007). Thus, the scenes that were remembered with extended boundaries were encoded in conjunction with increased MPFC activity. We interpret this result to indicate that the MPFC created context-based expectations of what occurred beyond the boundaries of the scene presented, confirming the role of the MPFC.

A core process of these contextually related predictions is to use an already established knowledge base (i.e., a context frame) to make predictions about the future. The MPFC has been linked to various cognitive processes, which at first may seem confusing; however, they all share this common core process. For example, the MPFC is strongly implicated in social cognition and understanding the minds of others (Mitchell, Macrae, & Banaji, 2006). In order to do this one uses past experiences to predict the thoughts of another person's mind. The MPFC, especially within the ventral regions, has also been strongly linked to reward and outcome processing in humans and in nonhuman primates (Buckley et al., 2009; Luk & Wallis, 2009; Noonan et al., 2010; Ridderinkhof, van den Wildenberg, Segalowitz, & Carter, 2004). This processing of expected outcome and reward can be derived from past experiences and established context frames, which help guide decision making. The MPFC has also been implicated in a number of forms of memory-related processing. Autobiographical memory relies heavily on this region, both in recalling the past and in the ability to simulate future events, which relies on the library of autobiographical past to construct the future (Addis et al., 2007). Confidence in memory and the act of judging whether a memory is retrievable are also found to activate this region (Schnyer, Nicholls, & Verfaellie, 2005). Similarly, memory formation and retrieval related to schema congruency, and thus expectation congruency, have been linked to the MPFC (Maguire, Frith, & Morris, 1999; van Kesteren, Fernández, Norris, & Hermans, 2010; van Kesteren, Ruiter, Fernández, & Henson, 2012). In addition, Summerfield and Koechlin (2008) demonstrated that the MPFC was involved in matching outcomes with perceptual expectations. As seen in this literature, the MPFC is related to many different cognitive processes; however, a common link across them is the accessing of memory to generate predictions. Thus, a natural extension of this is to demonstrate a role for the MPFC in generating contextual predictions to facilitate scene understanding.

Putting the Pieces Together: A Neural Mechanism Underlying Contextual Associative Processing and Scene Understanding

Throughout this chapter, scenes are not discussed as a unique class of visual stimuli that are processed in isolation; rather, scenes are found to be processed via their associations in both bottom-up and top-down information streams. It is proposed that when a scene is processed, its information including contextual relevance is extracted via processing within the PHC. This information is then used to access the relevant context frame to place the scene into a larger and more meaningful context via processing by the RSC, which provides input to generate expectations and predictions about what is next to occur in the environment or what is occurring beyond the currently visible scene via processing in the MPFC (figure 7.3).

In the sections above evidence has linked a functional role to each region; however, critical evidence for this neural mechanism requires a demonstration of structural and functional communication among these regions.

Structural anatomical connections among the MPFC, RSC, and PHC have been documented in both humans and monkeys. By use of tracer methods, extensive bidirectional anatomical connections have been shown between the PHC and RSC, between the RSC and MPFC, and to a lesser extent between the PHC and MPFC (Barbas, Ghashghaei, Dombrowski, & Rempel-Clower, 1999; Kobayashi & Amaral, 2003, 2007; Kondo, Saleem, & Price, 2005; Lavenex, Suzuki, & Amaral, 2004). In humans, to investigate anatomical connections, diffusion tensor imaging (DTI) is used to infer white matter tracts between different regions. DTI enabled anatomical connections to be found between the PHC and the RSC (focused in the caudal regions) and between the MPFC and RSC (focused in rostral regions) (Greicius, Supekar, Menon, & Dougherty, 2009; Qi et al., 2012). Thus, the neural architecture does indeed support a network of communication among these three regions (Kravitz, Saleem, Baker, & Mishkin, 2011).

Although there is neural architecture supporting a network linking these three different regions, it does not elucidate how the MPFC, RSC, and PHC functionally communicate while processing the contextual associations that are fundamental to scene understanding. To examine this, the spatiotemporal dynamics and cross communication of these regions were investigated using magnetoencephalography (MEG) (Kveraga et al., 2011). In this investigation, examining the phase synchrony among the MPFC, RSC, PHC, and early visual areas assessed functional connectivity. Participants were asked to recognize pictures of objects, which, unknown to the participants, included stimuli with strong or weak contextual associations. MEG phase synchrony analyses demonstrated that there was significantly more phase locking in the neural signal across the context cortical network (PHC, RSC, and MPFC) for the strong contextual objects than for the weak contextual objects. The onset of the synchronous activity among the regions revealed the time course of contextual processing in the visual domain. Differential neural signals related to contextual processing began early (~150–220 ms from stimulus onset) between the early visual areas and the PHC and was followed by significant phase locking between the PHC and RSC (~170–240 ms). The RSC continued contextually related processing with synchronous activity with the early visual areas (~310–360 ms), which was followed by a period of synchronous activity between the RSC and MPFC (~370–400 ms) (Kveraga et al., 2011). These results elucidate the time course of information flow within the cortical network processing contextual associations.

The time course of contextual processing revealed by this MEG study provides evidence for the proposed psychological and neural mechanism underlying contextual processing and scene understanding (figure 7.3). Early in visual processing the PHC extracts the relevant contextual information for the stimuli, demonstrated by the significant phase locking with the early visual regions. This information is then fed to the RSC, activating the relevant context frame. The associations within the context frame are proposed to provide important feedback information to the early visual areas. And finally, the MPFC responds to the context frame selection in the RSC by generating contextually relevant expectations.

Conclusions

The goal of this chapter was to elucidate a mechanism of scene understanding that takes into account that scenes are not simply a class of visual stimuli with a particular geometric layout but rather are laden with rich contextual associations. The overlap between regions of the brain that process context, scenes, episodic memory, and stimulus-independent thought is not a coincidence and indicates an inherent commonality of relying on experience-based associations. The roles of the PHC, RSC, and MPFC in scene understanding were investigated and discussed within a framework of contextual associative processing. This mechanism includes the extracting of contextual associative information from the environment in the PHC, which activates the relevant context frame in the RSC. The associations within the context frame provide the source for the MPFC to generate predictions, which facilitate cognition and our interaction with the environment. Associative processing provides an account of scene understanding that explains the complexity and diversity of the information processed in a scene. Scene processing is not just a bottom-up perception but rather is extrapolated into the context it is associated with that has been learned over a lifetime, providing top-down feedback to guide further processing.

Acknowledgment

This work was supported by the Office of Naval Research MURI contract N000141010934

References

Addis, D. R., Wong, A. T., & Schacter, D. L. (2007). Remembering the past and imagining the future: Common and distinct neural substrates during event construction and elaboration. *Neuropsychologia*, *45*(7), 1363–1377.

Aguirre, G. K., Detre, J. A., Alsop, D. C., & D'Esposito, M. (1996). The parahippocampus subserves topographical learning in man. *Cerebral Cortex*, *6*(6), 823–829.

Aminoff, E. M., Bar, M., & Schacter, D. L. (2007). *How the brain extends the boundaries of a scene in memory*. Paper presented at the Society of Neuroscience, San Diego, CA.

Aminoff, E. M., Gronau, N., & Bar, M. (2007). The parahippocampal cortex mediates spatial and nonspatial associations. *Cerebral Cortex*, *17*(7), 1493–1503.

Aminoff, E. M., Kveraga, K., & Bar, M. (2013). The role of parahippocampal cortex in cognition. *Trends in Cognitive Sciences*, *17*(8), 379–390.

Aminoff, E. M., Schacter, D. L., & Bar, M. (2008). The cortical underpinnings of context-based memory distortion. *Journal of Cognitive Neuroscience*, *20*(12), 2226–2237.

Arnott, S. R., Cant, J. S., Dutton, G. N., & Goodale, M. A. (2008). Crinkling and crumpling: An auditory fMRI study of material properties. *NeuroImage*, *43*(2), 368–378.

Bar, M. (2007). The proactive brain: Using analogies and associations to generate predictions. *Trends in Cognitive Sciences*, *11*(7), 280–289.

Bar, M., Aminoff, E. M., Mason, M., & Fenske, M. (2007). The units of thought. *Hippocampus*, *17*(6), 420–428.

Bar, M., & Aminoff, E. M. (2003). Cortical analysis of visual context. *Neuron*, *38*, 347–358.

Bar, M., Aminoff, E. M., & Schacter, D. L. (2008). Scenes unseen: The parahippocampal cortex intrinsically subserves contextual associations, not scenes or places per se. *Journal of Neuroscience*, *28*(34), 8539–8544.

Bar, M., & Ullman, S. (1996). Spatial context in recognition. *Perception*, *25*(3), 343–352.

Barbas, H., Ghashghaei, H., Dombrowski, S. M., & Rempel-Clower, N. L. (1999). Medial prefrontal cortices are unified by common connections with superior temporal cortices and distinguished by input from memory-related areas in the rhesus monkey. *Journal of Comparative Neurology*, *410*(3), 343–367.

Bartlett, F. (1932). *Remembering: A study in experimental and social psychology*. Cambridge: Cambridge University Press.

Bilalić, M., Langner, R., Erb, M., & Grodd, W. (2010). Mechanisms and neural basis of object and pattern recognition: A study with chess experts. *Journal of Experimental Psychology. General*, *139*(4), 728–742.

Brown, T. I., Ross, R. S., Keller, J. B., Hasselmo, M. E., & Stern, C. E. (2010). Which way was I going? Contextual retrieval supports the disambiguation of well learned overlapping navigational routes. *Journal of Neuroscience*, *30*(21), 7414–7422.

Buckley, M. J., Mansouri, F. A., Hoda, H., Mahboubi, M., Browning, P. G. F., Kwok, S. C., et al. (2009). Dissociable components of rule-guided behavior depend on distinct medial and prefrontal regions. *Science*, *325*(5936), 52–58.

Buckner, R. L., Andrews-Hanna, J. R., & Schacter, D. L. (2008). The brain's default network: Anatomy, function, and relevance to disease. *Annals of the New York Academy of Sciences*, *1124*, 1–38.

Chai, X. J., Ofen, N., Jacobs, L. F., & Gabrieli, J. D. E. (2010). Scene complexity: Influence on perception, memory, and development in the medial temporal lobe. *Frontiers in Human Neuroscience*, *4*, 21.

Committeri, G., Galati, G., Paradis, A. L., Pizzamiglio, L., Berthoz, A., & LeBihan, D. (2004). Reference frames for spatial cognition: different brain areas are involved in viewer-, object-, and landmark-centered judgments about object location. *Journal of Cognitive Neuroscience*, *16*(9), 1517–1535.

Davachi, L., Mitchell, J. P., & Wagner, A. D. (2003). Multiple routes to memory: Distinct medial temporal lobe processes build item and source memories. *Proceedings of the National Academy of Sciences of the United States of America*, *100*(4), 2157–2162.

Diana, R. A., Yonelinas, A. P., & Ranganath, C. (2008). High-resolution multi-voxel pattern analysis of category selectivity in the medial temporal lobes. *Hippocampus*, *18*(6), 536–541.

Diana, R. A., Yonelinas, A. P., & Ranganath, C. (2010). Medial temporal lobe activity during source retrieval reflects information type, not memory strength. *Journal of Cognitive Neuroscience*, *22*(8), 1808–1818.

Dilks, D. D., Julian, J. B., Paunov, A. M., & Kanwisher, N. (2013). The occipital place area (OPA) is causally and selectively involved in scene perception. *Journal of Neuroscience*, *33*(4), 1331–1336.

Engel, L. R., Frum, C., Puce, A., Walker, N. A., & Lewis, J. W. (2009). Different categories of living and non-living sound-sources activate distinct cortical networks. *NeuroImage*, *47*(4), 1778–1791.

Engelien, A., Tüscher, O., Hermans, W., Isenberg, N., Eidelberg, D., Frith, C., et al. (2006). Functional neuroanatomy of non-verbal semantic sound processing in humans. *Journal of Neural Transmission*, *113*(5), 599–608.

Epstein, R. A. (2008). Parahippocampal and retrosplenial contributions to human spatial navigation. *Trends in Cognitive Sciences*, *12*(10), 388–396.

Epstein, R. A., & Higgins, J. S. (2007). Differential parahippocampal and retrosplenial involvement in three types of visual scene recognition. *Cerebral Cortex*, *17*(7), 1680–1693.

Epstein, R. A., & Kanwisher, N. (1998). A cortical representation of the local visual environment. *Nature*, *392*(6676), 598–601.

Epstein, R. A., & Ward, E. J. (2010). How reliable are visual context effects in the parahippocampal place area? *Cerebral Cortex*, *20*(2), 294–303.

Greene, M. R., & Oliva, A. (2009). Recognition of natural scenes from global properties: Seeing the forest without representing the trees. *Cognitive Psychology*, *58*(2), 137–176.

Greicius, M. D., Supekar, K., Menon, V., & Dougherty, R. F. (2009). Resting-state functional connectivity reflects structural connectivity in the default mode network. *Cerebral Cortex*, *19*(1), 72–78.

Henderson, J. M., Larson, C. L., & Zhu, D. C. (2008). Full scenes produce more activation than close-up scenes and scene-diagnostic objects in parahippocampal and retrosplenial cortex: An fMRI study. *Brain and Cognition*, *66*(1), 40–49.

Janzen, G., & Van Turennout, M. (2004). Selective neural representation of objects relevant for navigation. *Nature Neuroscience*, *7*(6), 673–677.

Keene, C. S., & Bucci, D. J. (2008). Involvement of the retrosplenial cortex in processing multiple conditioned stimuli. *Behavioral Neuroscience*, *122*(3), 651–658.

Kirwan, C. B., & Stark, C. E. L. (2004). Medial temporal lobe activation during encoding and retrieval of novel face-name pairs. *Hippocampus*, *14*(7), 919–930.

Kobayashi, Y., & Amaral, D. G. (2003). Macaque monkey retrosplenial cortex: II. Cortical afferents. *Journal of Comparative Neurology*, *466*(1), 48–79.

Kobayashi, Y., & Amaral, D. G. (2007). Macaque monkey retrosplenial cortex: III. Cortical efferents. *Journal of Comparative Neurology*, *502*(5), 810–833.

Kondo, H., Saleem, K. S., & Price, J. L. (2005). Differential connections of the perirhinal and parahippocampal cortex with the orbital and medial prefrontal networks in macaque monkeys. *Journal of Comparative Neurology*, *493*(4), 479–509.

Kravitz, D. J., Peng, C. S., & Baker, C. I. (2011). Real-world scene representations in high-level visual cortex: It's the spaces more than the places. *Journal of Neuroscience*, *31*(20), 7322–7333.

Kravitz, D. J., Saleem, K. S., Baker, C. I., & Mishkin, M. (2011). A new neural framework for visuospatial processing. *Nature Reviews Neuroscience*, *12*(4), 217–230.

Kveraga, K., Ghuman, A. S., Kassam, K. S., Aminoff, E. A., Hämäläinen, M. S., Chaumon, M., et al. (2011). Early onset of neural synchronization in the contextual associations network. *Proceedings of the National Academy of Sciences of the United States of America*, *108*(8), 3389–3394.

Lavenex, P., Suzuki, W. A., & Amaral, D. G. (2004). Perirhinal and parahippocampal cortices of the macaque monkey: Intrinsic projections and interconnections. *Journal of Comparative Neurology*, *472*(3), 371–394.

Leveroni, C. L., Seidenberg, M., Mayer, A. R., Mead, L. A., Binder, J. R., & Rao, S. M. (2000). Neural systems underlying the recognition of familiar and newly learned faces. *Journal of Neuroscience*, *20*(2), 878–886.

Luk, C.-H., & Wallis, J. D. (2009). Dynamic encoding of responses and outcomes by neurons in medial prefrontal cortex. *Journal of Neuroscience*, *29*(23), 7526–7539.

Maguire, E. A., Frith, C. D., & Morris, R. G. (1999). The functional neuroanatomy of comprehension and memory: The importance of prior knowledge. *Brain*, *122*(10), 1839–1850.

Mason, M. F., Norton, M., Van Horn, J., Wegner, D., Grafton, S., & Macrae, C. (2007). Wandering minds: The default network and stimulus-independent thought. *Science*, *315*(5810), 393.

Mitchell, J. P., Macrae, C. N., & Banaji, M. R. (2006). Dissociable medial prefrontal contributions to judgments of similar and dissimilar others. *Neuron*, *50*(4), 655–663.

Mullally, S. L., & Maguire, E. A. (2011). A new role for the parahippocampal cortex in representing space. *Journal of Neuroscience*, *31*(20), 7441–7449.

Noonan, M. P., Walton, M. E., Behrens, T. E. J., Sallet, J., Buckley, M. J., & Rushworth, M. F. S. (2010). Separate value comparison and learning mechanisms in macaque medial and lateral orbitofrontal cortex. *Proceedings of the National Academy of Sciences of the United States of America*, *107*(47), 20547–20552.

Oliva, A., & Torralba, A. (2001). Modeling the shape of the scene: A holistic representation of the spatial envelope. *International Journal of Computer Vision*, *42*(3), 145–175.

Park, S., Brady, T. F., Greene, M. R., & Oliva, A. (2011). Disentangling scene content from spatial boundary: Complementary roles for the PPA and LOC in representing real-world scenes. *Journal of Neuroscience*, *31*(4), 1333–1340.

Park, S., & Chun, M. M. (2009). Different roles of the parahippocampal place area (PPA) and retrosplenial cortex (RSC) in panoramic scene perception. *NeuroImage*, *47*(4), 1747–1756.

Park, S., Intraub, H., Yi, D. J., Widders, D., & Chun, M. M. (2007). Beyond the edges of a view: Boundary extension in human scene-selective visual cortex. *Neuron*, *54*(2), 335–342.

Peters, J., Daum, I., Gizewski, E., Forsting, M., & Suchan, B. (2009). Associations evoked during memory encoding recruit the context-network. *Hippocampus*, *19*(2), 141–151.

Qi, R., Xu, Q., Zhang, L. J., Zhong, J., Zheng, G., Wu, S., et al. (2012). Structural and functional abnormalities of default mode network in minimal hepatic encephalopathy: A study combining DTI and fMRI. *PLoS ONE*, *7*(7), e41376.

Qin, S., van Marle, H. J. F., Hermans, E. J., & Fernández, G. (2011). Subjective sense of memory strength and the objective amount of information accurately remembered are related to distinct neural correlates at encoding. *Journal of Neuroscience*, *31*(24), 8920–8927.

Raichle, M. E., MacLeod, A. M., Snyder, A. Z., Powers, W. J., Gusnard, D. A., & Shulman, G. L. (2001). A default mode of brain function. *Proceedings of the National Academy of Sciences of the United States of America*, *98*(2), 676–682.

Ranganath, C. (2010). Binding items and contexts: The cognitive neuroscience of episodic memory. *Current Directions in Psychological Science*, *19*(3), 131–137.

Rauchs, G., Orban, P., Balteau, E., Schmidt, C., Degueldre, C., Luxen, A., et al. (2008). Partially segregated neural networks for spatial and contextual memory in virtual navigation. *Hippocampus*, *18*(5), 503–518.

Ridderinkhof, K. R., van den Wildenberg, W. P. M., Segalowitz, S. J., & Carter, C. S. (2004). Neurocognitive mechanisms of cognitive control: The role of prefrontal cortex in action selection, response inhibition, performance monitoring, and reward-based learning. *Brain and Cognition*, *56*(2), 129–140.

Robinson, S., Keene, C. S., Iaccarino, H. F., Duan, D., & Bucci, D. J. (2011). Involvement of retrosplenial cortex in forming associations between multiple sensory stimuli. *Behavioral Neuroscience*, *125*(4), 578–587.

Rumiati, R. I., Weiss, P. H., Tessari, A., Assmus, A., Zilles, K., Herzog, H., et al. (2005). Common and differential neural mechanisms supporting imitation of meaningful and meaningless actions. *Journal of Cognitive Neuroscience*, *17*(9), 1420–1431.

Schacter, D. L. (1999). The seven sins of memory: Insights from psychology and cognitive neuroscience. *American Psychologist*, *54*(3), 182.

Schnyer, D. M., Nicholls, L., & Verfaellie, M. (2005). The role of VMPC in metamemorial judgments of content retrievability. *Journal of Cognitive Neuroscience*, *17*(5), 832–846.

Schubotz, R. I., Korb, F. M., Schiffer, A.-M., Stadler, W., & von Cramon, D. Y. (2012). The fraction of an action is more than a movement: Neural signatures of event segmentation in fMRI. *NeuroImage*, *61*(4), 1195–1205.

Shenhav, A., Barrett, L. F., & Bar, M. (2012). Affective value and associative processing share a cortical substrate. *Cognitive, Affective & Behavioral Neuroscience*, *13*(1), 46–59.

Stevens, W. D., Kahn, I., Wig, G. S., & Schacter, D. L. (2012). Hemispheric asymmetry of visual scene processing in the human brain: Evidence from repetition priming and intrinsic activity. *Cerebral Cortex*, *22*(8), 1935–1949.

Summerfield, C., & Koechlin, E. (2008). A neural representation of prior information during perceptual inference. *Neuron*, *59*(2), 336–347.

Sung, Y.-W., Kamba, M., & Ogawa, S. (2008). Building-specific categorical processing in the retrosplenial cortex. *Brain Research*, *1234*, 87–93.

Tendolkar, I., Arnold, J., Petersson, K. M., Weis, S., Brockhaus-Dumke, A., van Eijndhoven, P., et al. (2007). Probing the neural correlates of associative memory formation: A parametrically analyzed event-related functional MRI study. *Brain Research*, *1142*, 159–168.

Tendolkar, I., Arnold, J., Petersson, K. M., Weis, S., Brockhaus-Dumke, A., van Eijndhoven, P., et al. (2008). Contributions of the medial temporal lobe to declarative memory retrieval: Manipulating the amount of contextual retrieval. *Learning & Memory (Cold Spring Harbor, N.Y.)*, *15*(9), 611–617.

Torralba, A., & Oliva, A. (2003). Statistics of natural image categories. *Network: Computation in Neural Systems*, *14*(3), 391–412.

van Kesteren, M. T. R., Fernández, G., Norris, D. G., & Hermans, E. J. (2010). Persistent schema-dependent hippocampal-neocortical connectivity during memory encoding and postencoding rest in humans. *Proceedings of the National Academy of Sciences of the United States of America*, *107*(16), 7550–7555.

van Kesteren, M. T. R., Ruiter, D. J., Fernández, G., & Henson, R. N. (2012). How schema and novelty augment memory formation. *Trends in Neurosciences*, *35*(4), 211–219.

Vann, S., & Aggleton, J. P. (2002). Extensive cytotoxic lesions of the rat retrosplenial cortex reveal consistent deficits on tasks that tax allocentric spatial memory. *Behavioral Neuroscience*, *116*(1), 85–94.

Wolbers, T., & Buchel, C. (2005). Dissociable retrosplenial and hippocampal contributions to successful formation of survey representations. *Journal of Neuroscience*, *25*(13), 3333–3340.

Zink, C. F., Tong, Y., Chen, Q., Bassett, D. S., Stein, J. L., & Meyer-Lindenberg, A. (2008). Know your place: Neural processing of social hierarchy in humans. *Neuron*, *58*(2), 273–283.

8

Fast Visual Processing of "In-Context" Objects

M. Fabre-Thorpe

Summary

Isolated objects do not exist in the natural world because they are virtually always embedded in contextual scenes. Despite such complexity, object recognition within scenes appears both effortless and virtually instantaneous for humans, whereas coping with natural scenes is still a major challenge for computer vision systems. Even in categorization tasks with briefly flashed (20 ms) natural scenes, humans can respond with short latencies (250–280 ms) to any exemplar from a wide range of categories (animal, human, vehicles, urban or natural scenes).

In daily life, online contextual information can facilitate the processing of all objects that can be expected in such a context. But the situation is different in fast visual categorization tasks in which subjects have to process a succession of unrelated briefly flashed scenes and therefore cannot make any predictions about the photograph that will be presented next. In such cases, context and embedded objects have to be processed at the same time in an ascending flow of visual processing. In the present chapter we discuss a series of experiments showing that even in rapid go/no-go manual categorization tasks such object/context interactions can be observed. Thus, even with briefly flashed stimuli, the context in which objects are presented can induce a performance benefit (or cost) depending on whether it is congruent or not with the object category assigned as target. This "contextual effect" is strengthened with increasing age and is not specific to humans in that monkeys' performance in the same tasks also shows the same effects.

A major point is that the "contextual effect" can be observed very early during processing, from the very first short-latency manual responses. To investigate the latency of the earliest object/context interactions we discuss a second series of experiments based on a forced saccadic choice task. In saccadic tasks, behavioral responses can be obtained at even shorter latencies, opening up a temporal window for studying contextual effects well before any manual responses have occurred. In such tasks some target categories (such as animals but not vehicles) can be detected before scene categorization occurs. This result rules out one hypothesis often made to account for fast categorization response speed, the claim that subjects would base their decision on a prediction made from global scene statistics without real processing of the target object. Moreover, the earliest influence of context on animal detection was seen at a latency that matches the earliest saccades performed to global contextual scene targets. Object/context interactions are discussed in terms of facilitatory or inhibitory network connections set up by experience between populations of selective neurons that are used (or not) to coactivate in daily life.

Introduction

When zapping from one TV channel to another, you can make sense of a scene and recognize objects in a fraction of a second; moreover, this complex processing appears both very fast and effortless.

This striking efficiency of the biological visual system is the result of innate predispositions combined with the shaping of the selectivity of the visual system by years of visual experience. When you are cooking a dinner in your kitchen or jogging across country on a Sunday morning, the continuous processing of the surrounding environment presets your sensory systems so that processing efficiency for different kinds of objects can be modulated. Indeed, in our world, objects can be strongly associated with particular environments or appear in redundant sequences so that you will be more likely to come across a frying pan in a kitchen and a cow in the fields than the reverse association. Because of such repetitive associations in daily life, visual processing of an object could be under strong modulations by the contextual environment in which the object appears.

The fact that the contextual information can facilitate or interfere in the domain of word or object recognition is not a new idea (e.g., Bruner, 1957; Morton, 1969; Norman, 1968). The recognition of a word is greatly facilitated or even influenced by the prior description of a context. In the sentence "I saw the bank," the word "bank" will have very different meanings in the context of navigating canal boats in locks and in the context of drawing money from a bank teller. In object perception the first evidence dates from the 1970s. Objects were recognized more efficiently when presented in coherent real-scene photographs than in jumbled scenes (Biederman, 1972), when primed by a semantically consistent (vs. inconsistent) scene (Palmer, 1975), or when presented in a consistent (vs. inconsistent) drawing of a scene (Boyce & Pollatsek, 1992; Boyce, Pollatsek, & Rayner, 1989). Not only were objects identified more accurately, they were also expected in specific locations and specific sizes. Indeed, in the early 1980s, Biederman, Mezzanotte, and Rabinowitz (1982) showed that when subjects were required to detect a target object specified by its name, their performance was impaired when the object embedded in the scene was violating a range of rules such as position, support, size, and probability of appearing in a given scene context. A sofa is not expected to be seen floating in the sky, and the artist Magritte, for example, has played with our expectations in many of his paintings.

The influence of scene frame on object processing was later challenged by Hollingworth and Henderson (Henderson & Hollingworth, 1999; Hollingworth & Henderson, 1998), who reported that after elimination of guesses and response biases, no advantage was found for the detection of consistent objects over inconsistent ones. They proposed a functional isolation model in which object identification processes are isolated from knowledge about the world. However, using a memory task, Hollingworth (2006) provided evidence that object representation and location were bound in memory to the scene context in which the object was seen. The author clearly restricted possible contextual effects to the exemplar-level recognition of objects in scenes (a toaster in a kitchen) and still challenged the possibility of contextual effect on the processing of object categories.

In most of these experiments the stimuli used were line drawings, but the influence of context on object processing has also been shown with manipulated color photographs that were shown only once to avoid any learning (Davenport & Potter, 2004). Object/background interactions were broadened to include object/object interactions that depended on whether objects were likely to appear (or not) in the same settings (Davenport, 2007); two objects embedded in a context are more accurately reported if they are related to each other.

Although there now seems to be a large consensus on the influence of contextual information on object perception (Bar, 2004; Bar & Aminoff, 2003; Goh et al., 2004; Gronau, Neta, & Bar, 2008; Joubert, Fize, Rousselet, & Fabre-Thorpe, 2008), the question of the temporal dynamics of such interactions is still wide open.

The influence of context is often seen as a top-down influence. When immersed in a given context, the representations of objects that are likely to appear might be pre-activated by expectation (Bar & Ullman, 1996). However, when subjects are required to process briefly flashed scenes depicting varied and unrelated scenes, no predictions can be made about the next photograph that will be presented. Context and embedded objects have to be processed from scratch in an ascending flow of visual processing.

In the priming task used by Palmer (1975), a scene prime was shown for 2 seconds, leaving ample time for analysis and contextual effects to develop and influence the processing of the following object, but in most other experiments scenes were presented for a brief duration. Drawings were shown for 150 ms and masked in the studies by Biederman et al. (1982) and Boyce et al. (1989). In Davenport and Potter (2004) photographs were displayed for only 80 ms and masked. With such short stimulus durations, the fact that object perception was shown to be influenced by contextual congruence demonstrated that the pertinent contextual information able to influence object processing could be picked up very fast, with only a glance at the scene.

Although it is clear that the pertinent contextual information can be extracted very quickly, little is known about the time needed to process this information and the minimum latency at which contextual information can interact with object processing. Reaction times are very rarely reported, and when reported, as in the study by Biederman et al. (1982), they were over 900 ms for object detection, leaving ample time for long-latency interactions. Along with others (Boyce & Pollatsek, 1992; Boyce et al., 1989), Biederman et al. (1982) made the assumption that the scene schema was computed first. Indeed, natural image statistics could be used in scene processing (Fiser & Aslin, 2001; Torralba & Oliva, 2003). The gist of a scene can be extracted on the basis of global image features, and this could provide early contextual information that could potentially influence object processing (Oliva & Torralba, 2006, 2007) or set constraints on possible object interpretations. In the model proposed by Bar and collaborators, a coarse "blurred" representation of a contextual frame might be sufficient to guide object processing by activating the most likely possible object(s) in such a contextual frame (Bar, 2004; Bar et al., 2006). In addition to the models based on the assumption that scenes are processed before objects in order to influence object perception, Davenport and Potter (2004) proposed an interactive model with mutual interactions between object and background. In agreement with such reciprocal influences, Joubert et al. (2007) found that scene categorization was impaired when a foreground object was present, especially when this object was incongruent.

In contrast with experiments suggesting very fast interactions between object- and context-processing streams, other data suggest a late effect that would depend on the activation of semantic information. Recording EEGs when subjects are processing objects embedded in a congruent or incongruent context, Ganis and Kutas (2003) reported the earliest signs of cerebral activity related to congruity versus incongruity in a late 300- to 500-ms window.

So far, the latency at which the earliest influences of context on object processing can be evidenced still remains to be determined. In the past few years we have investigated the temporal dynamics of object/context interactions with the animal/nonanimal rapid visual categorization task introduced by Thorpe, Fize, and Marlot (1996). Using a go/no-go manual response (go on animal, no-go otherwise), subjects are required to categorize natural photographs that are briefly flashed on the basis of whether or not they contain an animal. Human subjects are extremely efficient in such tasks despite the fact that they have no a priori knowledge about the next photograph they will have to process, and, if the photograph contains an animal, they have no prior information about its species, size, view angle or location. Moreover, the 20-ms stimulus presentation does not allow them to explore the photographs. Following the request to respond "as fast and as accurately as possible," they score 94% correct with a mean reaction time (RT) of about 380–400 ms. Disregarding anticipations, the earliest responses are observed at very short latencies (250–280 ms). Such fast responses might not even require conscious representations (Thorpe, Gegenfurtner, Fabre-Thorpe, & Bülthoff, 2001). This task appears very appropriate to test for contextual effects on object categories, and the RT distribution including very short RT response should allow the latency timing at which object processing is modulated by contextual information.

A Contextual Effect on Object Category

The temporal dynamics of contextual influence on object processing was investigated using the go/no-go rapid visual categorization task introduced by Thorpe et al. (1996). The stimuli were natural scenes that were manipulated so that animal and man-made distractor objects were cropped and pasted on new congruent or incongruent contexts (Joubert et al., 2008). Progressive transparency was used on the object contours to avoid sharp edges and to allow good integration of objects with their new background. Stimulus manipulation was done with great care: adjusting for scene and object luminance, checking for object saliency, and taking into account object location, orientation, and coherence (support, interposition, scale) (Biederman et al., 1982). But such manipulations are far from trivial. Despite careful manipulations, the comparison of rapid visual categorization using original versus manipulated but congruent scenes showed that, for most subjects (9 of 12), performance was slightly (2% in average) but significantly impaired in accuracy with an increase of about 15 ms in reaction time. Thus, manipulation of natural stimuli is not easy, and modulation of object processing by context was done using congruent and incongruent scenes that were all manipulated.

Scenes were considered congruent when an animal was presented in a natural context and a man-made object was presented in a man-made environment. Why did we choose the level of superordinate categories to test object/context congruency? It might be semantically very incongruent to see a polar bear in the jungle or a deer on an ice flow. However, in our experiments such manipulated scenes were considered "congruent." This choice was guided by preceding experiments (Joubert et al., 2007; Rousselet, Joubert, & Fabre-Thorpe, 2005) showing that processing scene gist at the superordinate level (natural vs. man-made) is faster than at more detailed categorization levels (sea or mountain scenes; indoor or outdoor scenes). In the same line, processing animals at the superordinate level is faster than processing them at the basic level, for example, categorizing animals as dogs or birds (Macé, Joubert, Nespoulous, & Fabre-Thorpe, 2009). To be very schematic, one knows it is an animal in a natural context before knowing that it is a bird flying over a lake. On average, processing at the superordinate level is faster by at least 50 ms, and thus, by using

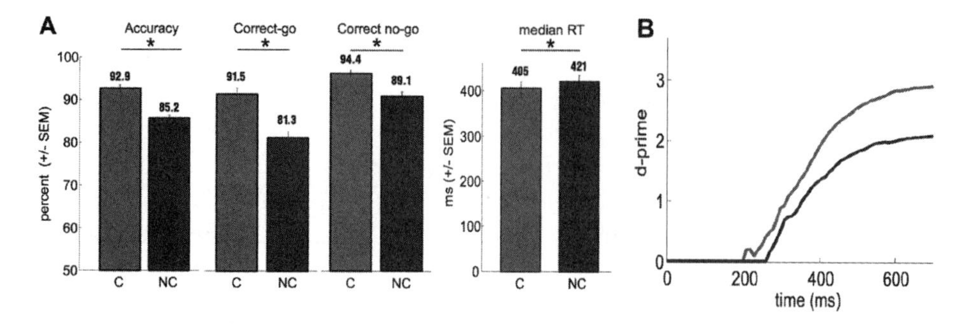

Figure 8.1 (plate 15)
Performance in the animal/nonanimal rapid visual categorization task. In A and B performance levels are shown in light gray for congruent stimuli and in dark gray for incongruent stimuli (see text). Note that for correct go responses congruent stimuli are animals in a natural context and that for correct no-go responses congruent photographs are man-made objects on man-made contexts. (A) Global accuracy, accuracy of correct go response on animal target, accuracy of correct no-go response on distractor, and median reaction time of correct go responses. (B) d-prime (*d'*), showing the shift toward longer latencies in the incongruent condition (in dark gray) relatively to the congruent condition (in light gray). Adapted from Joubert et al. (2008).

congruent object/context association at the superordinate level, we could investigate the modulation of object processing on responses with latencies ranging from 250 to 600 ms and hope to determine the earliest latency at which such modulation can be found.

The results (Joubert et al., 2008) showed a robust effect of contextual information on object processing with an overall significant drop of accuracy and increase of mean reaction time (RT) with incongruent stimuli (figure 8.1, plate 15). With a natural scene as background, subjects were more likely to produce a correct go response when an animal was present and less likely to be able to withhold their go response (correct no-go response) when the scene contained a man-made object. Thus, a natural background induced a bias toward responding that an animal was present. At the individual level the performance impairment was very consistent: the drop of accuracy induced by incongruent images was individually significant in all subjects, and the increase of mean reaction times was present in all subjects and reached significance in about half of them. This effect was not specific to the set of stimuli, as it was replicated using different object/background associations. It also did not depend on object saliency: when the most salient point of each stimulus was determined using the saliency toolbox (Walther & Koch, 2006) inspired by Itti and Koch (2001), the effect was always present regardless of whether the most salient point of the photograph was found on, close to, or far from the foreground object.

The most interesting observation concerned the temporal dynamics of object/context interactions. No minimum delay was necessary to observe a contextual influence on object categorization, as even the earliest responses produced by the subjects were modulated by scene context (as shown by the d' curve for the noncongruent condition that was shifted toward longer latencies by about 20–30 ms; see figure 8.1B, plate 15).

This experiment showed a clear effect of contextual information on object processing at the level of superordinate categories. Moreover, incongruent object/context associations affect the information accumulation rate even for responses with the shortest latencies, those around 250 ms, suggesting that such interactions happen very early in visual processing

Getting Rid of Biases Using Natural Images

Because the biological visual system has been optimized through years of experience to deal with our complex surrounding world, the use of natural photographs is essential to study its functioning. However, such stimuli are very difficult to manipulate and can induce many biases, as shown by the drop of human performance on original images compared with congruent photographs that have been manipulated (Joubert et al., 2008).

A main objective is to ensure that the effect of incongruent object/context associations is not the result of uncontrolled low-level features that would bias performance without any relation with stimulus congruence. Indeed, such low-level features could be processed much faster by the system and could bias results in terms of temporal dynamics of object/context interactions.

In order to get rid of such biases we built a set of 768 stimuli composed of 192 "quadrets." For each quadret (figure 8.2) four achromatic scenes were created associating an exemplar of the two superordinate contextual categories (man-made or natural) and an exemplar of the two superordinate object categories (animal or man-made object). For each object/background association the man-made and natural backgrounds had equal average luminance and RMS contrast, and the animal and man-made object vignettes had equal surface, center-of-mass location, luminance, and RMS contrast. Objects were pasted using progressive transparency as in the preceding experiment. A wide variety of animals (mammals, birds, insects, reptiles, amphibians, fish, and crustaceans) and man-made objects (means of transport, urban and house furniture, kitchen utensils, water containers, tools, toys, and various other objects) were used. Counterbalancing objects and backgrounds, removing colors, and controlling for object size and location, luminance, and contrast were employed to minimize all possible low-level features that could interfere and bias the effect of context on object processing.

With this carefully built set of stimuli we showed a similar modulation (in strength and timing) of object perception by the scene in which it is presented (Fize, Cauchoix,

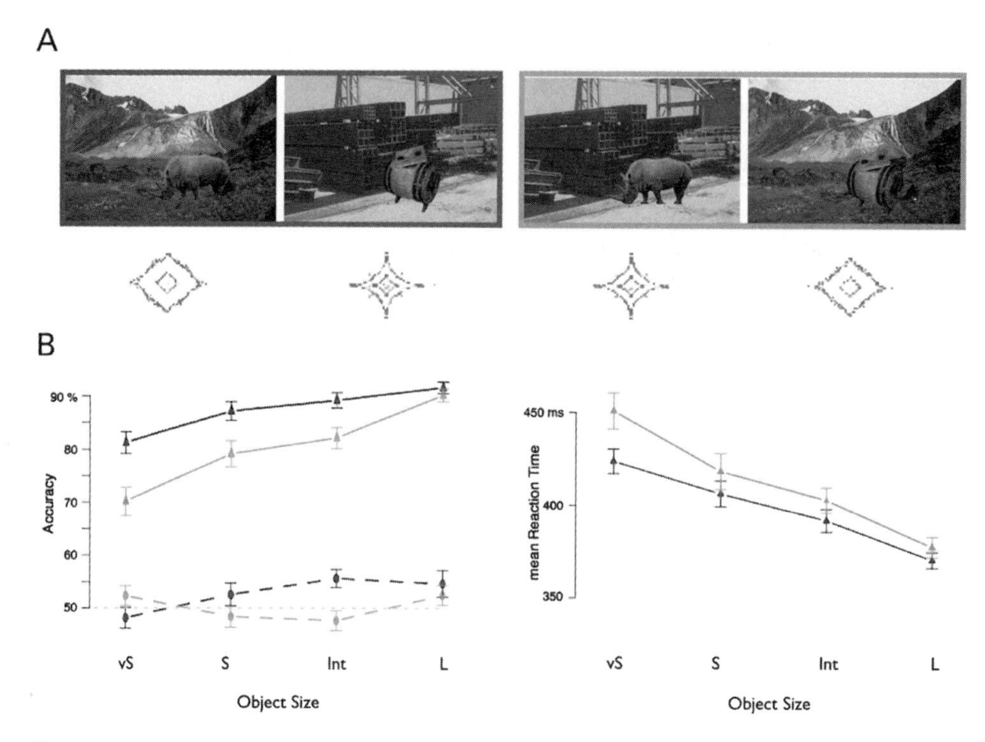

Figure 8.2
Top row, one example of the 192 sets of four stimuli used in Fize et al. (2011). With two backgrounds (natural and man-made) and two objects from different categories (animal targets and man-made object distracters), four black and white stimuli were built: two congruent (outlined in dark gray) and two incongruent stimuli (outlined in light gray). Below each stimulus example, the mean power spectrum averaged on the set of stimuli is illustrated, showing that the task cannot be solved using the stimuli's spectral signatures. The bottom row illustrates the accuracy performance and mean reaction times as a function of object/scene size ratio (from very small vS, >3%), to small (S), intermediate (Int), and large (L, >8%). Performance on congruent stimuli (in dark gray) and on incongruent stimuli (in light gray). Human performance is illustrated with full lines. The dotted lines (dark and light gray) illustrate the performance of computer simulations with congruent and incongruent stimuli, to which we refer later in the chapter.

& Fabre-Thorpe, 2011). Not unexpectedly, this effect is maximal when objects are small (<3% of the scene). But whereas one could think that the repetition of the same stimuli would make incongruence less striking, the very same impact on performance was observed between congruent and incongruent stimuli after three repetitions of the same stimuli.

The results clearly show that even when low-level cues are controlled and categories with a large variety of exemplars are used, there is a clear effect of context on the processing of object categories. This effect supports very early interactions between object- and context-processing flows (Biederman, 1972; Biederman et al.,

1982; Davenport & Potter, 2004). On the other hand, it goes against the claim made by Hollingworth (2006) that contextual information plays little or no role in object categorization. For Hollingworth, although contextual information might be retained in the image of an object, such contextual features would be lost when activation of large numbers of exemplar representations is combined, as needed in categorization. Our results clearly show that this is not the case.

A Contextual Effect Strengthened by Aging

In daily life, objects tend to appear in the same environment repetitively and in predictable locations. Over our lifetime some associations will slowly be lost (boats are not pulled by horses along canals any more), and we might have to learn new ones (computers in offices were not very common 30 years ago). But most object/context associations will be continuously reinforced, especially in terms of object categories: wild animals still appear in natural setting, whereas pieces of furniture will mainly be associated with indoor scenes. With aging, numerous repetitions of such associations should reinforce the memories of contextual rules and strengthen the learning of what objects are likely to occur in which environment. What would be the size of contextual modulations in healthy old adults performing this fast categorization task? Old adults have to deal with age-related cognitive decline, but, on the other hand, they could benefit from years of experience and strong object/context association rules that could help object perception. Such benefits have been shown recently in old adults involved in air traffic control (Nunes & Kramer, 2009) and in visual search using natural scenes and testing for contextual cuing (Neider & Kramer, 2011). Although subject to controversy (Smyth & Shanks, 2008), the implicit use of spatial configuration in performing visual search tasks has been reported in humans (Chun & Jiang, 1998, 2003). Chun and Jiang used a visual search task (T among Ls) in which certain displays were repeated so that the target systematically appeared in specific location within a given global configuration. Such targets were detected faster than others embedded in new configurations, but the global configuration was learned implicitly, as subjects were unable to explicitly discriminate new from repeated stimulus configurations (Chun & Jiang, 1998) or to predict the location of a missing target (Chun & Jiang, 2003). Contextual cuing was extended to natural scenes and shown to rely on global rather than local information (Brockmole, Castelhano, & Henderson, 2006; Brockmole & Henderson, 2006). Compared to the nonscene displays used by Chun and Jiang, the effect on search time is substantially larger, and the learning with natural scenes appears much faster. Moreover, memory of scene/target associations with natural scenes was shown to be explicit as subjects could recognize repeated scenes and recall target position. Comparing the effect of contextual cuing in old adults performing visual search in natural scenes where the target location is constrained or not in the scene, Neider and Kramer (2011) showed that target search in old adults (mean age of about 70 years) heavily relied on global context. Whereas young adults can shorten their mean RT by 320 ms when the location of the target is scene-constrained, the benefit for older adults is more than doubled! Spatial context information allows both young and old adults to restrict their search time in scene locations where the target is more likely to appear, but the benefit is much larger for old adults.

In a recent experiment (Rémy et al., 2013) we investigated the effect of incongruent versus congruent object/background associations through aging. We used a fast visual categorization task because the gathering of visual information is constrained by a brief presentation of the stimuli (100 ms) and RTs are constrained to <1 second to force fast responses. To test old and very old human subjects, a new set of quadret stimuli was built that used only color stimuli with uncluttered backgrounds and large foreground objects. The two object categories were strongly related to outdoor natural scenes (animals) or indoor urban scenes (pieces of furniture). An alternative forced-choice response (animal vs. furniture) was required. Four age groups were tested (young, mature, old, and very old subjects aged >75). In agreement with Fize et al. (2011) the use of large foreground objects necessary to test very old subjects induced only a minimal—but still present—contextual modulation in the group of young adults in terms of both accuracy and speed (figure 8.3, plate 16). With increasing age, the drop of accuracy and the RT increase induced by incongruent stimuli became progressively larger from about 2% and 13 ms in young adults to over 7% and 29 ms in very old adults. The effect was again shown to be very robust at the individual level, as 82 of the 92 subjects tested showed both a drop of accuracy and an increase in RT.

In this fast categorization task scene context thus has more influence on the responses produced by old than by young adults. Because aging is also associated with an increase of mean RT, the larger performance impairment for incongruent scene stimuli could simply be the result of an increased time window for context to influence object processing, but regardless of the group, the impairment size was not positively correlated with mean RT.

With aging, global scene processing would be preserved and could play an increased role in object perception, whereas object processing might be impaired. These results are in agreement with recent studies using an fMRI adaptation task (Goh et al., 2004) that showed a lack of adaptation to repeated objects in the context of a changing background, whereas adaptation could be evidenced with repeated backgrounds and changing foreground objects.

Given the temporal constraints of the fast visual categorization task—briefly flashed photographs (100 ms) and response times limited to 1s from stimulus onset—the large drop of performance with incongruent stimuli suggests that old adults are biased toward the processing of contextual information. However, very old adults

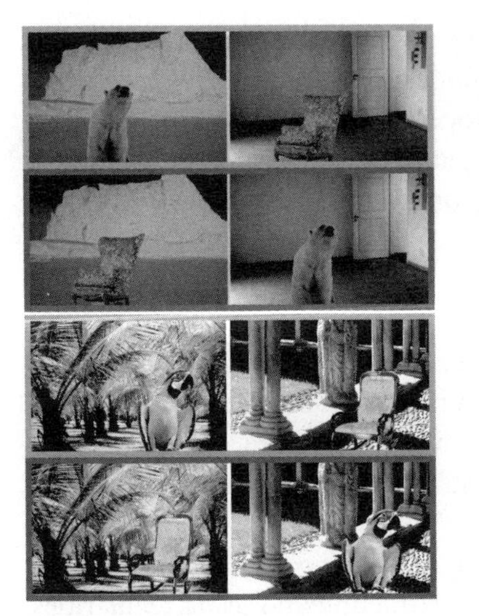

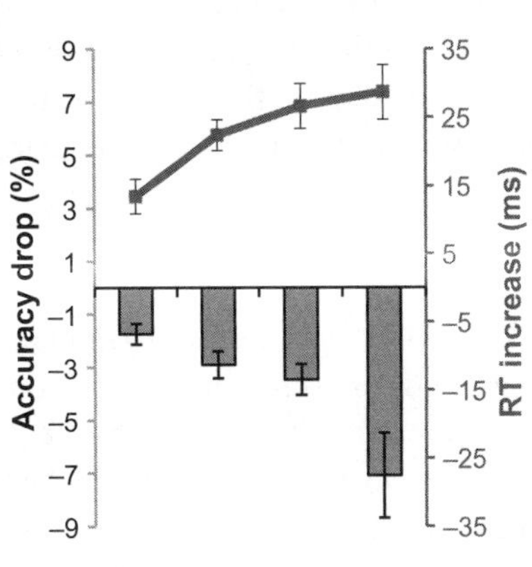

Figure 8.3 (plate 16)
On the left, two examples of the sets of four stimuli that were used by Rémy et al. (2013). With two backgrounds (natural and man-made) and two objects from different categories (animal and furniture), four stimuli were built: two congruent (first and third rows) and two incongruent stimuli (second and fourth rows). On the right, the drop of performance in accuracy and the increased response latencies observed when performance on incongruent stimuli is compared to performance on congruent stimuli. This "congruence" effect increases with age from young adult (Y) to mature adults (M), old (O), and very old adults (vO).

are still very good at performing the task despite the temporal constraints, scoring 94% correct with congruent scenes (95% correct for the young group), a robust performance that could depend on the strength of encoded object-context associations in the parahippocampal cortex (Aminoff, Gronau, & Bar, 2007; Bar & Aminoff, 2003; Goh et al., 2004).

The delayed response with incongruent object/context associations could also be explained in terms of accumulation of information without making any assumptions about the strength of high-level encoded associations. The visual system is very good at extracting and using regularities, and such regularities could be picked up from both objects and backgrounds. The speed at which an object representation will be accessed when presented in a given context would reflect the learned probability of occurrence of a set of features. In a congruent context the animal or furniture response would rely on a number of features provided by both the object and the context. In an incongruent context features from the object and from the context would be in conflict, and more information would be needed from object processing, resulting in a delayed response. All along the ventral visual pathway the weight of synaptic connections between selective populations of neurons to various features would be reinforced with their repetitive coactivation. Strengthening versus depressing these functional connections would induce behavioral facilitation versus interference. This alternative hypothesis (Fabre-Thorpe, 2011; Joubert et al., 2008) is developed later in this chapter and relies on the shaping of each step of the ventral visual pathway by repeated experience of object/object or object/context associations during ones' entire lifetime.

Contextual Effects in Nonhuman Primates

Fast visual categorization is not unique to human beings; nonhuman primates are also able to perform animal versus nonanimal or food versus nonfood fast visual categorizations of natural photographs (Fabre-Thorpe, Richard, & Thorpe, 1998), with surprisingly similar effects in both species even when stimuli are presented without color information and with extreme reductions in image contrast (Delorme, Richard, & Fabre-Thorpe, 2000; Macé, Delorme, Richard, & Fabre-Thorpe, 2010). However, as mentioned above, the visual system is very good at extracting and using regularities. Animals are much more common in natural environments, and the use of image statistics allows the introduction of contextual information in object detection. In fact, the processing of simple image statistics has been shown to predict the presence or absence of object categories in a photograph and different scene categories evidence very specific "spectral signatures" (Torralba & Oliva, 2003). Using 1000 training images and 1000 new images for testing, Torralba and Oliva showed that the presence/absence of animals, people, and vehicles in real-world images could be done with surprisingly good accuracy (around 80%). Note, however, that performance is close to chance when the animal size in the image is small relative to the surrounding context. The best results were obtained with images of large to very large animals. The performance of fast visual categorization could thus rely on the fast processing of global statistics rather than on abstract representations of a superordinate "animal" category. In such a case, the fact that monkeys can perform fast categorization of object is not enough to conclude that they have abstract representations of object categories but simply that their visual system is very good at using stimulus regularities.

Monkeys raised in the laboratory will have seen only a very small number of real animals. They have presumably learned to perform the task using the thousands of photographs (mainly from commercial databases) in which animals are presented in natural contexts. This means that they could use contextual regularities and implicit contextual learning as described above to perform the task. However, if monkeys base their response on global scene analysis rather than using a representation of "animal" objects, their performance should be totally disrupted by manipulation of object/context associations. To avoid a low-level bias as much as possible, monkeys were tested using the 768 stimuli organized in quadrets as shown in figure 8.2 with the corresponding spectral signatures calculated on each image set using the mean power spectrum (Torralba & Oliva, 2003). The results (Fize et al., 2011) clearly showed that monkeys base their responses on the presence or absence of the animal-object and not on the "spectral signature" of the stimuli. From the very first response toward these new unusual stimuli, their performance was around 70% correct—not just in terms of overall performance but also when separately analyzing performance with natural and urban contexts. The set of stimuli was especially difficult for the monkeys, as the 70% correct for monkeys has to be compared with a score of 80% correct in humans. The performance of computer simulations based on global statistics and run on the same stimuli (Oliva & Torralba, 2001) was at random (see figure 8.2).

However, as with humans, monkeys were biased by the nature of the context, and also as with humans, the effect of contextual incongruence on accuracy was much larger (around 10% drop of correct responses) with small objects (3% of the image) than with large objects (2% drop of accuracy for object >8% of the image).

Monkeys are much faster than humans in performing the task (Fabre-Thorpe et al., 1998), as their earliest responses (discarding anticipations) are observed at around 170 ms. Because the functional organization of their ventral visual pathway is much better known than that of humans, it was possible to evaluate the number of steps the visual information had to travel through. This evaluation (Fabre-Thorpe, 2011; Thorpe & Fabre-Thorpe, 2001) showed that there was effectively no time for time-consuming interactive loops and that it was difficult to avoid the conclusion that visual processing was mainly feedforward and largely parallel (figure 8.4).

Although faster than humans when performing on the manipulated stimuli, monkeys were not as fast as in preceding experiments using only original nonmanipulated photographs, and the latency of the earliest responses was observed at about 200 ms. However, for monkeys again, the interference effect due to incongruent object/background association was seen from the very first responses with a shift of 30 ms toward longer latencies when responses were produced with incongruent stimuli. Object/context interactions presumably have to be implemented in a feedforward model of visual processing. Scene background and object visual features are processed in parallel with early interactions and competition for accumulation of evidence to perform the task. Interestingly, this latency shift was identical (30 ms) for monkeys and humans, and we made the hypothesis (Fize et al., 2011) that the highly similar temporal dynamics of object/context interactions observed behaviorally are the signature of analogous fast visual mechanisms that locally process features for object and scene category.

Object/Context Interactions in a Feedforward Model of Visual Processing

In the model proposed by Bar and his group (Bar, 2004; Bar et al., 2006), a coarse processing of the context performed through the magnocellular dorsal visual pathway

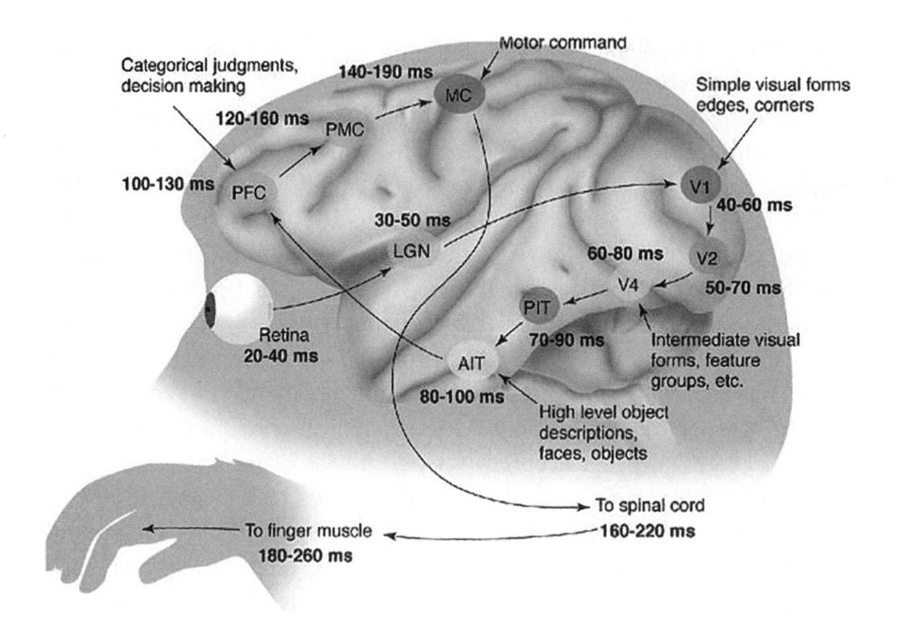

Figure 8.4
A feedforward progression of visual inputs along the ventral visual pathway in monkeys can explain fast response latencies. At each processing step (retina, lateral geniculate nucleus (LGN) up to posterior (PIT) and anterior (AIT) inferotemporal cortex), the shortest latency of recorded neuronal responses is indicated together with the usual response latency. From the anterior inferotemporal cortex (AIT) processing could continue through prefrontal (PFC), premotor (PMC) and motor cortex (MC) to trigger the hand movement; latencies are more speculative. From Thorpe and Fabre-Thorpe (2001), Seeking categories in the brain, *Science*, *291*, 260–263. Reprinted with permission from AAAS. Illustration: Carin Cain.

can influence object recognition (see also Bullier, 2001). Global low-spatial-frequency information about the context frame provides predictions about the set of objects that usually appear in such context. In parallel, the coarse processing of an object activates a set of possible shape-based candidate objects. Object interpretation is constrained by the interaction of the two parallel flows of information and is completed by high-spatial-frequency information.

This model can well account for the influence of context on object perception but would have more difficulty accounting for the influence of a foreground object on the processing of scene gist perception. Indeed, whereas the emphasis has been focused on the modulation of object processing by its surrounding background, a reciprocal influence has also been reported (Davenport & Potter, 2004; Joubert et al., 2007; Mack & Palmeri, 2010). Backgrounds were named more accurately when they contained a congruent, rather than an incongruent, object (Davenport & Potter, 2004), and in a fast visual categorization task using scene gists as targets, Joubert et al. (2007) showed that the presence of a foreground object in the scene delays gist categorization. This drop of performance is already significant when the object is congruent, but with an incongruent object, the temporal cost can reach 80 ms with an accuracy drop of about 14% relative to scenes that do not contain any foreground object.

In the model proposed by Bar the two low-spatial-frequency processing streams dealing with scene gist and objects are completely separated until they reach a high level of representation and then converge at the level of the inferotemporal cortex. An alternative possibility suggested by our group (Fabre-Thorpe, 2011; Macé, Thorpe, & Fabre-Thorpe, 2005) considers the low-spatial-frequency information processed within the ventral stream by the magnocellular pathway. At each processing stage of the ventral visual stream, the magnocellular pathway can feed back information to guide the processing of the slower parvocellular information in the preceding stages, allowing very early interactions between scene and object features. Interactions can take place between features of intermediate or even low complexity. According to this view, high-level representations of scenes or objects do not have to be accessed before interactions take place.

When performing a visual categorization task under strong temporal constraints, the visual system is presumably preset optimally by top-down influences, and pertinent familiar associations will be activated at each step of the visual pathways. The parallel processing of the scene-stimulus activates multiple populations of neurons selective for visual objects, object parts, or features. When the scene is congruent, such populations of selective neurons are used in coactivating together. On the other hand, a scene that contains an incongruent object will generate conflicting responses. Several populations of neurons that very seldom fire together would be active. The more incongruent the features in the scene, the greater the competition between these populations of neurons, and hence, the greater the competition between the go and the no-go motor output. The final perceptual decision would rely on an "accumulation of evidence" (Perrett, Oram, & Ashbridge, 1998). More features about the target (target object or target scene) will need to be processed in order to win the competition over the neuronal populations responding to the nonpertinent features. This additional processing will result in a delay to reach decision threshold about the nature of the object or the nature of the scene designed as target.

The model we propose is compatible with a feedforward processing wave within the ventral pathway. The pattern of connectivity in the visual pathway would be shaped by experience-dependent synaptic weight modifications (Masquelier & Thorpe, 2007). When selective populations of neurons are often coactivated, their interconnections will be reinforced, and the activation of one population will lower the threshold of the others. On the other hand, when populations of selective neurons hardly ever fire simultaneously, their interactions will be reduced, and when one of them fires, the threshold of the others will be increased, inducing delayed visual processing (figure 8.5, plate 17). With such a model the processing streams dealing with objects and

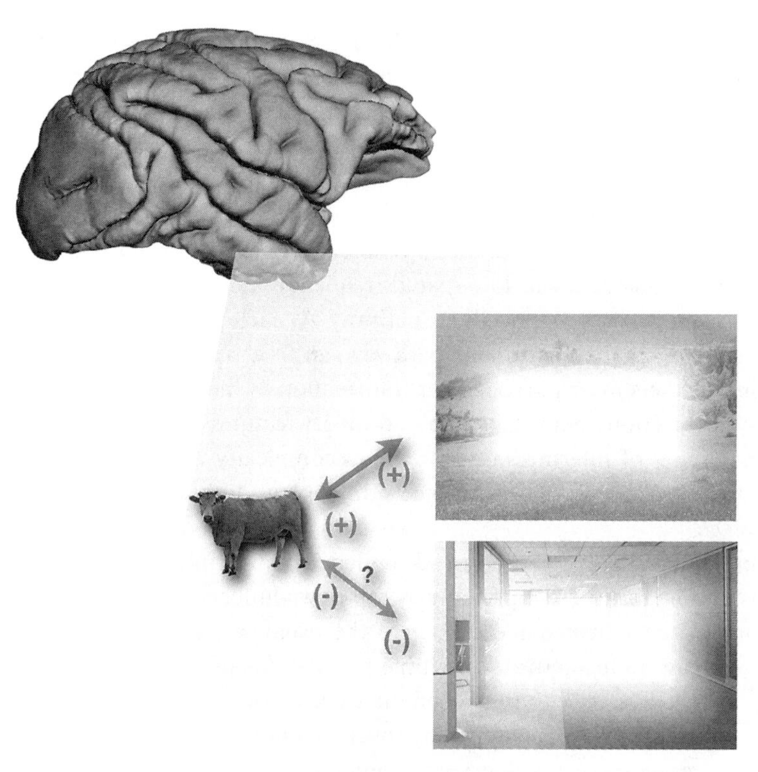

Figure 8.5 (plate 17)
Along the ventral visual pathways the weight of connections could increase or decrease when populations of neurons selective for objects, object parts, or object features are often or seldom coactivated in daily life. From Fabre-Thorpe (2011).

context can interact implicitly very early, and there is no need to make the assumption that scene gist should be extracted first in order to influence object perception (Davenport & Potter, 2004; Joubert et al., 2007).

Fast categorization could be based on the processing of key features of intermediate complexity (Crouzet & Serre, 2011; Delorme, Richard, & Fabre-Thorpe, 2010; Ullman, Vidal-Naquet, & Sali, 2002), and such interactions could thus be already present in early visual cortical areas such as V4, for example (Mirabella et al., 2007).

Exploring an Earlier Temporal Window of Visual Processing

Using the go/no-go rapid visual categorization task, we were never able to show major differences in processing time for scene targets versus object targets. In the case of object categories we found that the accuracy and distribution of reaction times for processing various object-categories at the superordinate level such as animals, vehicles, and human beings were remarkably similar (Rousselet, Macé, & Fabre-Thorpe,

2003; Thorpe et al., 1996; VanRullen & Thorpe, 2001). Even when using human faces as targets in an attempt to shift reactions times toward shorter values (Rousselet et al., 2003), we reported substantially higher accuracy (around 99%) than for other categories, but reaction time distributions were similar to those for other object categories. Reaction times might have reached a limit in categorization tasks requiring manual responses. Indeed, when comparing response latencies to the very first presentation of an image to response latency to the same image after repeated processing, we observed that the repeated training on a given image did not induce any RT decrease except for long latency responses (Fabre-Thorpe, Delorme, Marlot, & Thorpe, 2001). In the case of tasks requiring processing of scene categories, the accuracy scores and distribution of reaction times for scene target at the superordinate level (natural vs. urban) were extremely similar to those observed for object categories (for a review see Joubert et al., 2007). The virtually complete overlap of the RT distributions of the responses produced for three types of targets—animals, natural scenes, and urban scenes—leaves ample time for object/context interactions even in regard to the earliest responses produced.

In recent years we have been using a new task involving saccadic responses that gives access to an even earlier temporal window of visual processing (Crouzet, Kirchner, & Thorpe, 2010; Kirchner & Thorpe, 2006). In this forced-choice saccadic categorization task a fixation cross is displayed, and after a random duration (800–1600 ms), two natural scenes (a target and a distracter) are presented for 400 ms in the left and right hemifields at about 8.5° eccentricity. Subjects are asked to make a saccade as fast as possible to the side of the target. In the original experiment Kirchner and Thorpe (2006) reported that reliable saccades to images containing animals could be initiated as early as 120–130 ms after images onset. If 20 ms is allowed for motor preparation, the latency of these early saccades suggests that the underlying visual processing may need only 100 ms. This result was extended to vehicles and human faces by Crouzet, Kirchner, and Thorpe (2010); moreover, the authors showed a clear ordering of object categories for both accuracy and saccade mean latency. The best performance in accuracy and mean RT was found for face targets (about 94% correct and 150 ms mean RT), and the worst performance for vehicle targets (about 75% correct and 190 ms mean RT); performance on animal targets was intermediate.

Unlike the manual go/no-go visual categorization task, the forced-choice saccadic categorization task was able to rank perceptual category difficulty and provides an opportunity both to investigate an earlier temporal window and to compare the latency of scene versus object categorization. Moreover, it also gives us access to a very early window of processing in order to investigate the earliest latency at which contextual modulation could affect object processing (Crouzet, Joubert, Thorpe, & Fabre-Thorpe, 2012). For the contextual discrimination tasks each trial involved the simultaneous presentation of one natural and one man-made environment (all neutral without foreground objects); thus, with the same stimuli, subjects can be given the task of saccading either toward the natural or toward the man-made environment. To compare object and scene interactions in this forced-choice categorization task, a new stimulus databank was built that contained only unmodified photographs to avoid the effect of stimulus manipulation demonstrated by Joubert et al. (2008). For the object discrimination tasks we used the animal and vehicle object categories as in the study by Crouzet et al. (2010). The animal and vehicle images were selected so that, for each object category, exemplars were extremely varied (mammals, birds, reptiles, fish . . . vs. cars, ships, planes, trains, bikes . . .), with half of the objects presented embedded in a man-made environment and half in a natural environment (figure 8.6, plate 18, 1). Whereas animals are strongly associated with a natural context, the association of vehicles with a given context is much weaker.

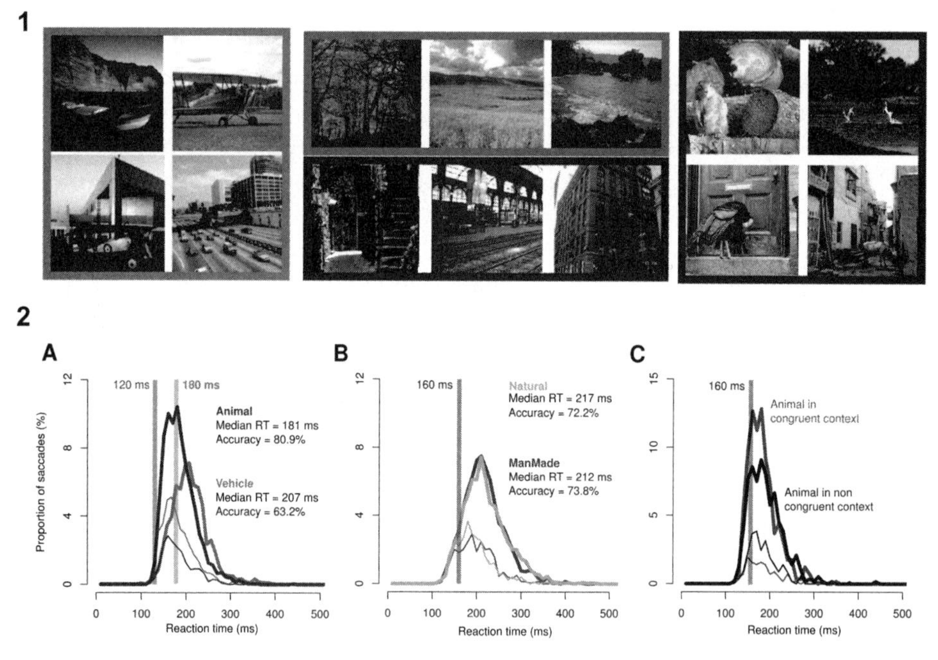

Figure 8.6 (plate 18)
(1) The top row shows examples of the stimuli used in Crouzet et al. (2012). On the left, vehicle objects in natural (top) and man-made (bottom) environments; in the middle, natural scenes (green) and man-made scenes (blue) without foreground objects; on the right, animal in natural congruent (top) and man-made incongruent (bottom) environments. (2) Performance (median RT and percentage correct) and RT distributions of saccadic responses. Adapted from Crouzet et al. (2012). (A) For animals (brown) and vehicles (gray). (B) For man-made (blue) and natural (green) environments. (C) For animal targets when animals are presented in a natural (congruent) background (light trace) or in a man-made (incongruent) environment (dark trace). For A, B, and C, the percentage of correct saccades (thick trace) and incorrect saccades (thin trace) are given as a function of saccade latency (10-ms time bins). The vertical bars indicate the minimal saccadic RT (the first time bin in which correct saccades significantly outnumber incorrect saccades).

A clear ordering appeared again in processing those varied categories. As reported by Crouzet et al. (2010), performance was better for animal targets than for vehicle targets (figure 8.6, plate 18, 2A) for both accuracy and saccade latency (animal, 81% correct, median RT 181 ms; vehicle, 63% correct, median RT 207 ms). It is worth noticing that the very fast saccades toward animals were not completely under top-down control; indeed, the RT distribution shows that, when subjects are instructed to target vehicles, early saccades can be incorrectly made toward animals. The same response bias had been observed with human faces when subjects were instructed to target vehicles (Crouzet et al., 2010). Because animals (and faces) are pertinent biological stimuli—unlike artifactual object categories such as vehicles—they might rely on faster hard-wired neural mechanisms possibly tuned by ancestral priorities (New, Cosmides, & Tooby, 2007).

In the forced-choice saccadic tasks with natural or man-made environments as targets, performance level was similar regardless of the environment category (about 73% correct, median RT 215 ms) and intermediate between animal and vehicle performance (figure 8.6, plate 18, 2A, 2B). This is the first psychophysical experiment that has provided clear evidence for different temporal dynamics in object and gist processing. However, even if scene gist can be accessed using fast processing of global image statistics, it is clear that specific object categories that are biologically pertinent—such as animals (and probably faces when considering the results from Kirchner and Thorpe, 2006)—can be accessed before scene gist.

An analysis of contextual effects failed to reveal any effects on vehicle categorization, which is perhaps not surprising because cars, boats, and planes could be considered to be congruent in both man-made and natural contexts. On the other hand, there was a clear contextual effect with animal targets (figure 8.6, plate 18, 2C), and this effect became statistically significant from 160 ms on. Importantly, with animal targets, saccades with latencies <160 ms were unaffected by contextual cues. This value corresponds also to the minimum latency at which the earliest saccades toward scene targets are observed (figure 8.6, plate 18, 2B). The earliest influence of context on animal detection was thus seen at a latency that matches the earliest saccades performed to global contextual scene targets.

This is an important result because it sheds light on the temporal dynamics of scene processing:

• It first argues against the claim that in fast animal categorization tasks the rapid behavioral responses could rely on global scene statistics. The role of global statistics was also questioned by the fact that computer simulations were at chance for categorizing animals with the manipulated scenes used by Fize et al. (2011).
• It gives a minimal behavioral latency (160 ms) at which scene gist can be accessed and can affect behavior. Note that we have no evidence to conclude whether scene gist is accessed through global scene statistics or via the accumulation of evidence from key features.

• It shows that superordinate categories may not all have the same perceptual saliency and suggests a ranking of object categories in perceptual difficulty from human faces to animals, and then to artifactual targets such as vehicles, scenes gist being of intermediate difficulty between animals and vehicles.

The global results from this series of experiments do not support the strict functional isolation model proposed by Hollingworth and Henderson (Henderson & Hollingworth, 1999; Hollingworth & Henderson, 1998) in that object and context processing clearly interact with each other. However, their model may apply in the initial phase of processing for which we provide clear evidence that at least for some object categories, such as animals, object processing can occur without any effect of context. This context-free processing ends at around 160 ms, the latency at which the earliest saccadic responses to scene gist can be observed. After this latency, object- and context-processing streams can interact in the processing of superordinate object categories. Finally, it appears that the strength of the interactions is significant for object categories such as animals that are strongly associated to certain specific contexts but can be negligible for other object categories such as vehicles.

References

Aminoff, E., Gronau, N., & Bar, M. (2007). The parahippocampal cortex mediates spatial and nonspatial associations. *Cerebral Cortex*, *17*(7), 1493–1503.

Bar, M. (2004). Visual objects in context. *Nature Reviews Neuroscience*, *5*(8), 617–629.

Bar, M., & Aminoff, E. M. (2003). Cortical analysis of visual context. *Neuron*, *38*, 347–358.

Bar, M., Kassam, K. S., Ghuman, A. S., Boshyan, J., Schmidt, A. M., Dale, A. M., et al. (2006). Top-down facilitation of visual recognition. *Proceedings of the National Academy of Sciences of the United States of America*, *103*(2), 449–454.

Bar, M., & Ullman, S. (1996). Spatial context in recognition. *Perception*, *25*(3), 343–352.

Biederman, I. (1972). Perceiving real-world scenes. *Science*, *177*(4043), 77–80.

Biederman, I., Mezzanotte, R. J., & Rabinowitz, J. C. (1982). Scene perception: Detecting and judging objects undergoing relational violations. *Cognitive Psychology*, *14*(2), 143–177.

Boyce, S. J., & Pollatsek, A. (1992). Identification of objects in scenes: The role of scene background in object naming. *Journal of Experimental Psychology. Learning, Memory, and Cognition*, *18*(3), 531–543.

Boyce, S. J., Pollatsek, A., & Rayner, K. (1989). Effect of back-ground information on object identification. *Journal of Experimental Psychology. Human Perception and Performance*, *15*, 556–566.

Brockmole, J. R., Castelhano, M. S., & Henderson, J. M. (2006). Contextual cueing in naturalistic scenes: Global and local contexts. *Journal of Experimental Psychology. Learning, Memory, and Cognition*, *32*(4), 699–706.

Brockmole, J. R., & Henderson, J. M. (2006). Using real-world scenes as contextual cues during search. *Visual Cognition*, *13*, 99–108.

Bruner, J. S. (1957). On perceptual readiness. *Psychological Review*, *64*, 123–152.

Bullier, J. (2001). Integrated model of visual processing. *Brain Research*, *36*, 96–107.

Chun, M. M., & Jiang, Y. (1998). Contextual cueing: Implicit learning and memory of visual context guides spatial attention. *Cognitive Psychology*, *36*, 28–71.

Chun, M. M., & Jiang, Y. H. (2003). Implicit, long-term spatial contextual memory. *Journal of Experimental Psychology. Learning, Memory, and Cognition*, *29*, 224–234.

Crouzet, S. M., Joubert, O. R., Thorpe, S. J., & Fabre-Thorpe, M. (2012). Animal detection precedes access to global image properties. *PLoS ONE*, *7*(12), e51471.

Crouzet, S. M., Kirchner, H., & Thorpe, S. J. (2010). Fast saccades towards face: Face detection in just 100 ms. *Journal of Vision*, *10*(4), 1–17.

Crouzet, S. M., & Serre, T. (2011). What are the visual features underlying rapid object recognition? *Frontiers in Psychology*, *2*, 326.

Davenport, J. L. (2007). Consistency effects between objects in scenes. *Memory & Cognition*, *35*, 393–401.

Davenport, J. L., & Potter, M. C. (2004). Scene consistency in object and background perception. *Psychological Science*, *15*(8), 559–564.

Delorme, A., Richard, G., & Fabre-Thorpe, M. (2000). Ultra-rapid categorisation of natural images does not rely on colour: A study in monkeys and humans. *Vision Research*, *40*, 2187–2200.

Delorme, A., Richard, G., & Fabre-Thorpe, M. (2010). Key visual features for rapid categorization of animals in natural scenes. *Frontiers in Psychology*, *1*(21), 1–13.

Fabre-Thorpe, M. (2011). The characteristics and limits of rapid visual categorization. *Frontiers in Psychology*, *2*(243), 1–12.

Fabre-Thorpe, M., Delorme, A., Marlot, C., & Thorpe, S. J. (2001). A limit to the speed of processing in ultra-rapid visual categorization of novel natural scenes. *Journal of Cognitive Neuroscience*, *13*, 171–180.

Fabre-Thorpe, M., Richard, G., & Thorpe, S. J. (1998). Rapid categorization of natural images by rhesus monkeys. *Neuroreport*, *9*(2), 303–308.

Fiser, J., & Aslin, R. N. (2001). Unsupervised statistical learning of higher-order spatial structures from visual scenes. *Psychological Science*, *12*(6), 499–504.

Fize, D., Cauchoix, M., & Fabre-Thorpe, M. (2011). Humans and monkeys share visual representations. *Proceedings of the National Academy of Sciences of the United States of America*, *108*(18), 7635–7640.

Ganis, G., & Kutas, M. (2003). An electrophysiological study of scene effects on object identification. *Brain Research. Cognitive Brain Research*, *16*(2), 123–144.

Goh, J. O. S., Siong, S. C., Park, D., Gutchess, A., Hebrank, A., & Chee, M. W. L. (2004). Cortical areas involved in object, background and object-background processing revealed with functional magnetic resonance adaptation. *Journal of Neuroscience*, *24*(45), 10223–10228.

Gronau, N., Neta, M., & Bar, M. (2008). Integrated contextual representation for objects' identities and their locations. *Journal of Cognitive Neuroscience*, *20*(3), 371–388.

Henderson, J. M., & Hollingworth, A. (1999). High-level scene perception. *Annual Review of Psychology*, *50*, 243–271.

Hollingworth, A. (2006). Scene and position specificity in visual memory for objects. *Journal of Experimental Psychology. Learning, Memory, and Cognition*, *32*(1), 58–69.

Hollingworth, A., & Henderson, J. M. (1998). Does consistent scene context facilitate object perception? *Journal of Experimental Psychology. General*, *127*(4), 398–415.

Itti, L., & Koch, C. (2001). Computational modelling of visual attention. *Nature Reviews Neuroscience*, *2*(3), 194–203.

Joubert, O. R., Fize, D., Rousselet, G. A., & Fabre-Thorpe, M. (2008). Early interference of context congruence on object processing in rapid visual categorization of natural scenes. *Journal of Vision*, *8*(13), 1–18.

Joubert, O. R., Rousselet, G., Fize, D., & Fabre-Thorpe, M. (2007). Processing scene context: Fast categorization and object interference. *Vision Research*, *47*, 3286–3297.

Kirchner, H., & Thorpe, S. J. (2006). Ultra-rapid object detection with saccadic eye movements: Visual processing speed revisited. *Vision Research*, *46*(11), 1762–1776.

Macé, M. J.-M., Delorme, A., Richard, G., & Fabre-Thorpe, M. (2010). Spotting animals in natural scenes: Efficiency of humans and monkeys at very low contrasts. *Animal Cognition*, *13*, 405–418.

Macé, M. J.-M., Joubert, O. R., Nespoulous, J.-L., & Fabre-Thorpe, M. (2009). Time-course of visual categorizations: You spot the animal faster than the bird. *PLoS ONE*, *4*(6), e5927.

Macé, M. J.-M., Thorpe, S. J., & Fabre-Thorpe, M. (2005). Rapid categorization of achromatic natural scenes: How robust at very low contrasts? *European Journal of Neuroscience*, *21*, 2007–2018.

Mack, M. L., & Palmeri, T. J. (2010). Modeling categorization of scenes containing consistent versus inconsistent objects. *Journal of Vision, 10*(3), 1–11.

Masquelier, T., & Thorpe, S. J. (2007). Unsupervised learning of visual features through spike timing dependent plasticity. *PLoS Computational Biology*, *3*(2), e31.

Mirabella, G., Bertini, G., Samengo, I., Kilavik, B. E., Frilli, D., Della Libera, C., et al. (2007). Neurons in area V4 of the macaque translate attended visual features into behaviorally relevant categories. *Neuron*, *54*, 303–318.

Morton, J. (1969). Interaction of information in word recognition. *Psychological Review*, *76*, 165–178.

Neider, M. B., & Kramer, A. F. (2011). Older adults capitalize on contextual information to guide search. *Experimental Aging Research: An International Journal Devoted to the Scientific Study of the Aging Process*, *37*(5), 539–571.

New, J., Cosmides, L., & Tooby, J. (2007). Category-specific attention for animals reflects ancestral priorities, not expertise. *Proceedings of the National Academy of Sciences of the United States of America*, *104*, 16598–16603.

Norman, D. A. (1968). Toward a theory of memory and attention. *Psychological Review*, *75*, 522–536.

Nunes, A., & Kramer, A. F. (2009). Experience-based mitigation of age-related performance declines: Evidence from air traffic control. *Journal of Experimental Psychology. Applied*, *15*, 12–24.

Oliva, A., & Torralba, A. (2001). Modeling the shape of the scene: A holistic representation of the spatial envelope. *International Journal of Computer Vision*, *42*(3), 145–175.

Oliva, A., & Torralba, A. (2006). Building the gist of a scene: The role of global image features in recognition. *Progress in Brain Research: Visual Perception*, *155*, 23–36.

Oliva, A., & Torralba, A. (2007). The role of context in object recognition. *Trends in Cognitive Sciences*, *11*(12), 520–527.

Palmer, S. E. (1975). The effects of contextual scenes on the identification of objects. *Memory & Cognition*, *3*, 519–526.

Perrett, D. I., Oram, M. W., & Ashbridge, E. (1998). Evidence accumulation in cell populations responsive to faces: An account of generalisation of recognition without mental transformations. *Cognition*, *67*(1–2), 111–145.

Rémy, F., Saint-Aubert, L., Bacon-Macé, N., Vayssière, N., Barbeau, E., & Fabre-Thorpe, M. (2013). Object recognition in congruent and incongruent natural scenes: A life span study. *Vision Research*, *91*, 36–44.

Rousselet, G. A., Joubert, O. R., & Fabre-Thorpe, M. (2005). How long to get to the "gist" of real-world natural scenes? *Visual Cognition*, *12*(6), 852–877.

Rousselet, G. A., Macé, M. J.-M., & Fabre-Thorpe, M. (2003). Is it an animal? Is it a human face? Fast processing in upright and inverted natural scenes. *Journal of Vision*, *3*(6), 440–455.

Smyth, A. C., & Shanks, D. R. (2008). Awareness in contextual cuing with extended and concurrent explicit tests. *Memory & Cognition*, *36*(2), 403–415.

Thorpe, S. J., & Fabre-Thorpe, M. (2001). Seeking categories in the brain. *Science*, *291*, 260–263.

Thorpe, S. J., Fize, D., & Marlot, C. (1996). Speed of processing in the human visual system. *Nature*, *381*(6582), 520–522.

Thorpe, S. J., Gegenfurtner, K., Fabre-Thorpe, M., & Bülthoff, H. H. (2001). Detection of animals in natural images using far peripheral vision. *European Journal of Neuroscience*, *14*, 869–876.

Torralba, A., & Oliva, A. (2003). Statistics of natural image categories. *Network (Bristol, England)*, *14*(3), 391–412.

Ullman, S., Vidal-Naquet, M., & Sali, E. (2002). Visual features of intermediate complexity and their use in classification. *Nature Neuroscience*, *5*, 682–687.

VanRullen, R., & Thorpe, S. J. (2001). Is it a bird? Is it a plane? Ultra-rapid visual categorization of natural and artifactual objects. *Perception*, *30*(6), 655–668.

Walther, D., & Koch, C. (2006). Modeling attention to salient proto-objects. *Neural Networks*, *19*, 1395–1407.

9

Detecting and Remembering Briefly Presented Pictures

Mary C. Potter

During our waking hours we take a new mental snapshot—a fixation—about three times a second. What do we pick up from each glimpse, and for how long do we remember what we saw? What is the form of our memory representation—visual, conceptual, or both—and does it change over time? One method for addressing these questions in the laboratory is to simulate continual shifts of fixation by using rapid serial visual presentation (RSVP) of sequences of unrelated pictures. When viewers are given a target name such as *picnic* or *smiling couple*, they are able to detect a picture in a stream presented for 100 ms per picture, and they do better than chance even at 13 ms/picture. Remarkably, detection is possible even when the name is given only after the sequence has been viewed. These results indicate that understanding may be based initially on feedforward processing, without feedback and without requiring advance information about the target. In contrast to our very rapid comprehension of pictures, we have poor memory for pictures presented for the duration of an average fixation (250 ms). We need 500 ms to view or think about a scene in order to remember it later. Yet long-term memory for pictures viewed for 1 second or more is excellent. The evidence suggests that conceptual information is extracted early and shapes what we remember later.

The paradox of vision is that we make three or four eye fixations each second, all day long, but each glimpse of 250 or 300 ms is too brief to remember later. We need some form of visual short-term memory that spans several fixations to integrate information about the immediate environment, but what we carry over from the preceding fixation lacks detail (e.g., Henderson & Hollingworth, 1999; Irwin, 1992; Irwin & Andrews, 1996). Moreover, studies of change blindness and boundary extension show that we overlook major changes in a scene if the scene is interrupted for as little as 80 ms (e.g., Intraub & Richardson, 1989; Rensink, O'Regan, & Clark, 1997, 2000), suggesting that our immediate memory is incomplete. We do notice changes that affect gist or changes to objects that we are attending or are about to fixate. Thus, the information that we carry over from the previous fixation seems to be meaningful rather than purely visual. However, when unrelated pictures are presented in a continuous sequence at rates in the range of eye fixations (Potter & Levy, 1969), our memory for pictures is poor, implying that one glimpse is not sufficient for later memory.

In contrast, we have good long-term memory for pictures viewed for 1–10 seconds (Nickerson, 1965; Potter & Levy, 1969; Shepard, 1967; Standing, 1973). Pictures viewed for 3 seconds are remembered in detail, whether they represent single objects (Brady, Konkle, Alvarez, & Oliva, 2008) or complex scenes (Konkle, Brady, Alvarez, & Oliva, 2010).

Just How Quickly Do We Understand a Pictured Object or Scene?

Reaction Time

One answer to the question of how quickly a picture is understood is the reaction time (RT) to make a recognition response to a picture. Naming the picture is one such response, but that includes time to retrieve the name after one has already recognized what the object is, and even well-known names take time to retrieve: average RT for naming a familiar object is over 900 ms. A measure of understanding that does not require name retrieval is the time to decide whether the scene or object is a member of a category such as *animal*. This yes-no category detection task turns out to be considerably faster (a mean of about 600 ms) than the time to name a picture (Potter & Faulconer, 1975). These RT measures include the time for the information to pass from the retina to the visual cortex as well as decision and response processes that occur after identification (e.g., Potter, 1983). Research using a go/no-go response gives shorter responses in such category-detection tasks (see the review by Fabre-Thorpe, 2011). Of particular interest is the minimum RT at which performance is above chance, which has been shown to be as short as 150 ms. A still faster response is the initiation of an eye movement to a specified target (e.g., *animal* or *face*) when two pictures are presented simultaneously (e.g., Crouzet, Kirchner, & Thorpe, 2010; Kirchner & Thorpe, 2006): the shortest RT at which performance is above chance can be as little as 100 ms for faces, with a mean time of 140 ms. Another approach is to use measures of brain responses such as event-related potentials (ERPs) that occur before any overt response. In an early go/no-go study in which observers detected animals, the relevant ERP signal was significantly above chance beginning about 150 ms after picture onset (e.g., Thorpe, Fize, & Marlot, 1996).

Masked Stimuli

A different approach to measuring the time to understand a picture is to control the time available for processing the stimulus, measuring the minimum presentation time required to identify it.[1] However, the duration of the physical stimulus is not the same as the effective duration of the stimulus because of visual persistence: a picture presented for only 20 ms followed by a blank screen will persist for 80 ms or more.

A common method to solve that problem is to use a backward pattern mask at a variable delay after the picture (the stimulus onset asynchrony, SOA). Such a mask interrupts processing of the picture, allowing one to determine the minimal viewing time required for identification. For example, in one study (Potter, 1976; see figure 9.1, discussed below), 16 single pictures were each followed by a visual mask with an SOA varying from 50 to 120 ms. In a subsequent yes-no test of recognition memory about half the pictures were remembered at an SOA of 50 ms, rising to 80% at 120 ms.

Questions about Masking

A continuing problem with the logic of the masking procedure, however, is that the neural basis for the effect is not well understood: does the masked stimulus continue to be processed, perhaps unconsciously, after the mask appears, or does processing instantly stop? Macknik and Martinez-Conde (2007) have argued that the mask has an immediate feedforward effect that interrupts processing. But because the extent of masking depends not only on the SOA but also on the stimulus termination asynchrony and the perceptual relation of the mask to the stimulus of interest, the minimal SOA required for identification may not directly measure the time to understand a picture. Moreover, the effect of a following mask also depends on its semantic (conceptual) relation to the target picture. With very short SOAs the visual relation may be the major determinant of the mask's effectiveness, but as the SOA increases, the conceptual relation may be more important, as discussed below.

Context Effects: Perception of Objects and Settings

The role of visual context in perception of objects has long been a topic of interest. A similar question is whether our experience of co-occurrences between objects and settings influences the initial perception of a scene or whether (as suggested by Hollingworth & Henderson, 1998, 1999) objects and settings in a given picture are first understood independently. In one set of studies by Davenport and Potter (2004) pictured objects such as a football player or a priest were superimposed, either congruently or incongruently, on background settings such as a football field or the interior of a cathedral. The pictures were presented for 80 ms with a backward noise mask of the whole picture; the participant was instructed to report the foreground object, the background setting, or both. In each case performance was better in the congruent than the incongruent condition, suggesting that objects and background are processed interactively. When there were two objects in a scene, the likelihood that the two objects would be found together also influenced the report of the objects, an effect that was additive with the effect of congruency with the background (Davenport, 2007). Joubert and colleagues carried out similar studies, finding that objects in congruent contexts were responded to faster than those in incongruous contexts (Joubert, Rousselet, Fize, & Fabre-Thorpe, 2007; Joubert, Fize, Rousselet, &

Fabre-Thorpe, 2008; see also Munneke, Brentari, & Peelen, 2013). These results indicate that objects and settings are processed together.

Rapid Serial Visual Presentation

In studies that use backward masking to limit processing time, each trial typically consists of a single stimulus, such as a picture, followed by a mask. Although the glimpse of the stimulus may be of the same duration as a fixation, in normal vision the eyes make a continuous sequence of fixations, with each fixation presumably masking the previous one. To mimic this effect Potter and Levy (1969) used a method called rapid serial visual presentation (RSVP) (Forster, 1970) to present pictures in a continuous stream at durations in the range of eye fixations, 125–2000 ms/picture. Participants were instructed to attend to and remember all the 16 pictures in a sequence. The pictures were unrelated to each other to enable us to measure memory for information equivalent to that in a single fixation. To test recognition memory following the presentation, the pictures were shown one at a time intermixed with 16 new pictures (distractors). Participants responded yes, maybe, or no. Figure 9.1 shows the proportion of yes responses, corrected for guessing.[2] When the pictures had been shown for the duration of an average fixation, 250 ms, fewer than half the pictures were correctly recognized a minute or two later. With a presentation of 2 seconds,

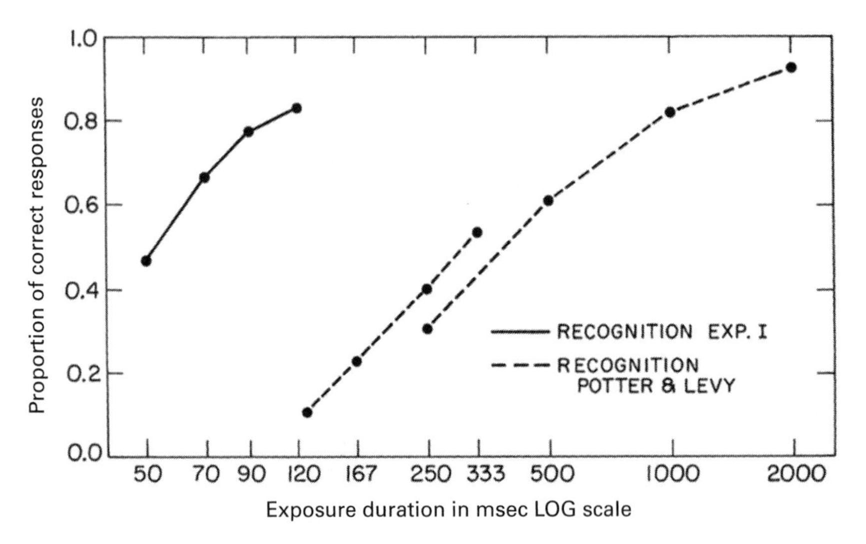

Figure 9.1
Proportion of pictures recognized following single masked presentations (solid curve, Potter, 1976, experiment 3) and proportion recognized after RSVP (dashed curves, two groups with different ranges of presentation durations; Potter & Levy, 1969). Data are corrected for guessing (see note 2). From Potter (1976).

more than 90% of the pictures were remembered, consistent with studies showing that long-term memory for pictures viewed for a few seconds is excellent (e.g., Konkle et al., 2010; Nickerson, 1965; Shepard, 1967; Standing, 1973).

Visual versus Conceptual Masking

Strikingly, however, as shown in the left-hand function in figure 9.1 (Potter, 1976), a single masked picture may be remembered after it is viewed for as little as 50 ms (about 50% were remembered, rising to 80% at 120 ms). It takes four or five times as long, per picture, to process pictures to the same level of accuracy when they are presented in a continuous stream in which all the pictures are to be attended. Pictures in an RSVP sequence are hard to remember not only because of their briefness but also because each picture is immediately followed by another. With a single masked picture, viewers can continue to think about what they saw after the mask appears; that is not possible with a continuous sequence in which all the pictures are potentially relevant. In a study by Intraub (1980) pictures were presented for 110 ms in an RSVP sequence, and only 20% were remembered later, whereas when a blank interstimulus interval (ISI) was added after each picture, the percentage remembered increased steadily as the ISI increased, to 84% with an ISI of 1390 ms. Thus, a viewer can voluntarily continue to process and code into memory a brief picture after it is no longer in view, just as one can continue to think about what one just saw in a brief glimpse. Similarly, another study showed that pictures presented for 173 ms in an RSVP sequence were poorly remembered, but if a blank of 827 ms was added after each picture, memory was almost as good as if the pictures remained in view for the full 1000 ms (Potter, Staub, & O'Connor, 2004).

Voluntary Attention

In a study of the effect of voluntary attention on picture encoding, Intraub (1984) showed a sequence of pictures that alternated between a short duration of 112 ms and a long duration of 1500 ms. When viewers were instructed to attend to all pictures, they remembered about 54% of the short pictures and 73% of the long ones, whereas when instructed to attend only to the brief pictures, they remembered about 63% of the brief pictures and only 54% of the long ones. Altogether, these studies suggest that once the SOA between the picture and the following visual mask is 100 ms or more, memory depends little on the actual duration of presentation but instead on the total time the viewer continues to think about the picture. These results reinforce the distinction between visual and conceptual masking. Visual masking occurs primarily with short SOAs (under 100 ms), whereas conceptual masking (due to attention to a following stimulus) occurs with SOAs up to 500 ms or more (Potter, 1976; see also Intraub, 1980, 1981; Loftus & Ginn, 1984; Loftus, Hanna, & Lester, 1988; Loschky, Hansen, Sethi, & Pydimarri, 2010).

Rapid Memory Loss for Pictures Seen Briefly in RSVP: Serial Position Effects in Memory Testing

People can understand pictures presented briefly but forget most of them a few minutes later. When the recognition test begins immediately, the first one or two pictures tested are likely to be remembered well, but there is rapid loss over the next several seconds of testing (Endress & Potter, 2012; Potter et al., 2004; Potter, Staub, Rado, & O'Connor, 2002); that is, there is a strong serial position effect in the memory test. There is also some loss if there is an unfilled delay of 5 seconds in the start of the memory test, showing that the loss is partly due to the passage of time and partly to interference from testing. Surprisingly, there is no serial position effect in presentation, apart from the known benefit to the final picture, which is unmasked and is not tested. Even with sequences as long as 20 items, there were no primacy or recency effects (Potter et al., 2002). Increasing the memory set size did decrease the extra benefit of early testing somewhat, but not by causing selective forgetting of pictures early in the list.

What Is the Nature of This Short-Lasting Memory for Pictures?

Change Blindness

The time course of forgetting after viewing an RSVP sequence of pictures contrasts with that of *change blindness*, the apparently immediate loss of detailed information about a single picture once it is no longer in view. Change blindness is the inability of viewers to detect a change in one feature of a picture, and the effect has been observed when a blank interval as short as 80 ms intervenes between the initial and changed versions; at longer intervals, the problem is even more acute (see Rensink et al., 1997; 2000; Simons & Levin, 1997). (Imposing a short blank between views is necessary to obscure the transient that would mark the location of the change if there were no interval.) Change blindness can be explained in several ways: the changed details were not perceived in the first place; many specifics of a picture are forgotten immediately; or the next picture updates the similar preceding picture without leaving a record of the changed details. Change blindness is, however, a very different phenomenon than the forgetting observed after an RSVP sequence. Whereas on a change blindness trial there is no question that the picture remains the same in most respects and is thus seen as the same picture, in the RSVP experiments considered here the question is whether a given test picture is one you have ever seen before. Thus, change blindness studies assess the level of detail in immediate memory for a picture, whereas here we are interested in the persistence of a representation sufficient to make the picture as a whole seem familiar.

Other Forms of Brief Visual Memory

Could the short-lasting memory for pictures be *iconic memory* (e.g., Sperling, 1960) or *visual short-term memory* (VSTM) as described by Phillips and his colleagues (Phillips, 1983; Potter & Jiang, 2009)? The answer is, no. Iconic memory is a very brief form of relatively literal perceptual memory (although see Coltheart, 1983, for a somewhat different characterization), but it cannot account for the fleeting picture memory found with an immediate recognition test after an RSVP sequence because iconic memory is eliminated by noise masking, and under photopic conditions it lasts no longer than about 300 ms. VSTM is a form of short-lasting visual memory observed in experiments such as those of Phillips and Christie (1977), who presented viewers briefly with a 4 × 4 matrix in which an average of eight random squares were white and then tested memory by presenting a second matrix that was either identical to the preceding one or had one white cell added or deleted. VSTM, unlike iconic memory, is capacity limited, with an estimated capacity of three or four items. In Phillips and Christie's study the most recent matrix could be maintained for several seconds in VSTM provided that no other such matrices were presented in the interval and the participant continued to attend to the remembered matrix. In contrast, in RSVP studies multiple pictures are presented, and one or more to-be-attended pictures intervene between presentation and testing.

A likely contributor to short-term memory for pictures is *conceptual short-term memory* (CSTM), a short-lasting memory component proposed by Potter (1993, 1999, 2010) that represents conceptual information about current stimuli, such as the meaning of a picture or meanings of words and sentences computed as one reads or listens. The reasons for regarding this brief memory representation as conceptual rather than (say) perceptual include its apparent role in rapid selection between two words on the basis of meaning in relation to context (Potter, Moryadas, Abrams, & Noel, 1993; Potter, Stiefbold, & Moryadas, 1998) and its putative role in sequential visual search tasks like those considered here in which the targets are defined by meaning or category rather than by physical form. During the brief time that information about stimuli is in CSTM, associative links enable extraction of whatever structure is present (such as sentence structure or the gist of a picture) or allow the stimulus to be compared to a target specification in a search task. Any momentarily active information that does not become incorporated into such a structure (such as the irrelevant meaning of an ambiguous word or a nontarget picture) will be quickly forgotten.

Conceptual versus Visual-Perceptual Memory

A critical question is whether the picture representation that persists for several seconds in the studies we have reviewed here is sufficiently abstract to be considered conceptual rather than wholly or partly perceptual. Do viewers remember only the picture's conceptual content or gist, or do they also remember visual features such as color, shape, and layout? Work of Irwin and Andrews (1996), Gordon and Irwin (2000), and Henderson (1997) suggests that the representation of the previous fixation may be at least partially conceptual rather than literal inasmuch as viewers may not notice literal changes that are conceptually consistent with the earlier fixation. Studies of detection to be reviewed below show that the gist of a scene is understood quickly even though the scene may then be forgotten (fairly) rapidly (e.g., Intraub, 1980, 1981; Potter, 1976), which is consistent with the assumption that conceptual information is abstracted rapidly. Intraub (1981) showed, however, that viewers can remember some specific pictorial information, such as the colors and layout, along with the gist.

The relative roles of such specific pictorial information and more abstract conceptual information were explored by Potter et al. (2004). They contrasted a conceptual and a pictorial recognition test of picture memory. In the pictorial test participants made yes-no decisions to five pictures they had just seen (excluding a sixth final picture that was not masked), mixed with five new pictures. In the conceptual test they made yes-no decisions to descriptive verbal titles of the pictures, mixed with titles of unseen pictures. The presentation duration was 173 ms/picture; the 10-item recognition test after each trial took about 8 seconds. The assumption was that test pictures provide both visual and conceptual information, whereas titles provide only conceptual information. If the benefit of immediate testing is that viewers only briefly preserve purely pictorial information, then the title test should reduce the benefit of early testing but should be fairly equivalent to the picture test later in testing. That was just what they found. In a more recent study (Endress & Potter, 2012) the advantage of testing recognition with pictures rather than titles was maintained throughout the test, suggesting that some more detailed information (perceptual or conceptual) beyond that captured by a title does persist over the 8-second test even though memory for both forms of information continues to decline.

In a further test of the conceptual basis of memory, Potter et al. (2004) included in the recognition test occasional *decoy* pictures that matched the title—the gist—of one of the old pictures, replacing that picture in the test. The decoy looked visually different from the old picture it replaced. If viewers rely on a conceptual or gist representation of the presented pictures, they should make more false yeses to decoys than to unrelated new pictures (distractors). Overall, participants recognized 52% of the old pictures, falsely recognized 30% of the decoys, and falsely recognized 15% of the other distractors, showing some susceptibility to conceptual decoys.

Short-Lasting Memory: Summary

Initial memory for a glimpsed picture (seen for the equivalent of a single fixation) is fairly accurate but declines markedly over the first few recognition tests (or across an unfilled delay of 5 seconds). The initial stronger memory may include specifically visual information, whereas after a delay the memory is primarily conceptual. Accurate visual information may be important for maintaining and updating scene representations from one fixation to the next, but conceptual memory seems to be the basis for longer-term organized knowledge. Unlike the rapid forgetting of briefly glimpsed pictures, memory for pictures viewed for a second or more can be highly accurate, at least when viewers are paying attention.

Detecting Pictures to Test Comprehension

Are RSVP Pictures Understood?

The studies of picture memory that I have just reviewed show that pictures presented for durations in the range of typical eye fixations are not well remembered. How do we know whether the forgotten pictures were even understood momentarily? Subjectively, one has the impression that one understands all the pictures when presented up to 10/second, but perhaps that is an illusion. Does it take longer than a single fixation to understand a novel scene? Perhaps viewers fail to remember briefly presented pictures because they did not comprehend them. To discover whether brief pictures are identified but then forgotten, we asked participants to detect target pictures that were shown to them (or named) before the sequence (Potter, 1975, 1976). We used names that captured the conceptual gist of the picture in one to five words but did not give explicit visual information about the picture. Detection was surprisingly good with either kind of cue, even at durations as short as 113 ms/picture (figure 9.2). The results can be compared with the recognition memory results from another group who viewed the same sequences without looking for a target and whose members were tested after each sequence for their recognition memory. That group, also shown in figure 9.2, remembered far fewer pictures than the first group had detected, suggesting that viewers can momentarily understand most of these brief pictures but will then forget many of them as testing begins. A question arises, however, about whether the difference between the two groups in figure 9.2 simply reflects attentional set: having a name presets the visual system to process the scene, which would not be understood otherwise. Intraub (1981) addressed that question by showing that viewers could detect a picture described by a negative category such as "not an animal," although performance was not as good as when given the name. The role of attentional set is considered again below, when we consider the difference between naming a picture before versus immediately after the sequence.

Detecting Two Targets

In another detection study (Potter, Wyble, Pandav, & Olejarczyk, 2010), participants looked for two targets in a category such as "bird" and reported the specific identity

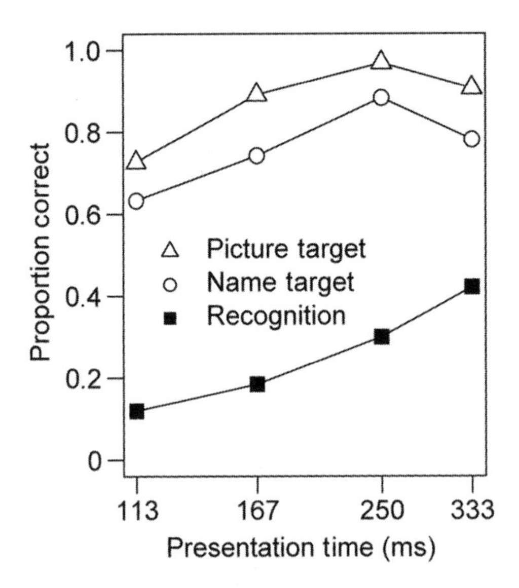

Figure 9.2
Detection of a target picture in an RSVP sequence of 16 pictures, given a picture of the target or a name for the target, as a function of the presentation time per picture. Also shown is later recognition performance in a group that simply viewed the sequence and then was tested for recognition. Results are corrected for guessing (see note 2). From Potter (1976).

of each instance (e.g., *swan* and *eagle*). The RSVP sequence was shown at 107 ms/picture. Figure 9.3 illustrates a trial in which the category was "dinner food." Report of the specific names of both targets (e.g., hamburger, spaghetti) was often successful even when the two targets were presented in immediate succession, although there was an attentional blink (reduced performance) for the second target when the SOA between targets was 213 ms, an effect typically observed in search tasks. Thus, even when given a general name for the target, viewers could detect and retain the specific identities of two targets presented briefly in a sequence.

Detection and Memory when Multiple Pictures Are Presented Simultaneously

Potter and Fox (2009) presented eight successive four-item arrays (figure 9.4) in which each array included none to four pictures, with meaningless texture masks filling the nonpicture locations. The RSVP sequence was presented at 240, 400, or 720 ms per array. When the task was to detect a named target (e.g., *balloons*), detection was relatively successful with up to four simultaneous pictures. Even at 240 ms per array with four simultaneous pictures, 59% of the targets were detected, with 9% false yeses (cf. Rousselet, Fabre-Thorpe, & Thorpe, 2002; Rousselet, Thorpe, & Fabre-Thorpe, 2004a, 2004b). This suggests that detection occurs in parallel with up to four pictures, or detection is extremely fast, or both. When viewers simply tried to remember the

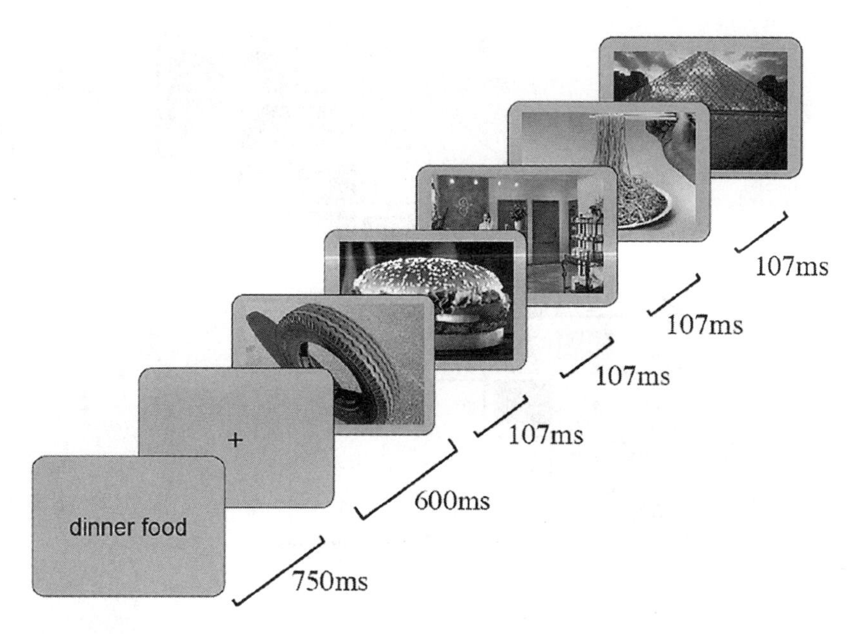

Figure 9.3
An example of an RSVP sequence in a search experiment in which participants reported the specific names of two exemplars of the search category. Here the exemplars are *hamburger* and *spaghetti*. From Potter, Wyble, Pandav, and Olejarczyk (2010).

pictures, later recognition accuracy was much lower overall, particularly when there was more than one picture in the array. We speculate that detection may occur in parallel over the whole array, whether it consists of one picture or up to four. In contrast, memory may require separate attention to each picture.

Detection and Memory with Occlusion, Inversion, and Grayscale Pictures

Meng and Potter (2008) used RSVP to present pictures with or without 30% of the surface randomly occluded by small disks and found that detection (given a name) was well above chance and minimally affected by the disks, even with a duration as brief as 53 ms. When the task was to recognize a picture shown after the sequence, performance was lower than with detection, and the disks significantly interfered. When the pictures were inverted, the disks interfered with detection as well as recognition. Showing the pictures in grayscale did not change performance in the detection condition, and again the occluding disks did not affect performance. When the number of disks was increased to cover 40% of the picture, however, detection did show interference. The results suggest that rapid retrieval of the gist of a picture is based on a global perception of the scene that is robust against local loss of information.

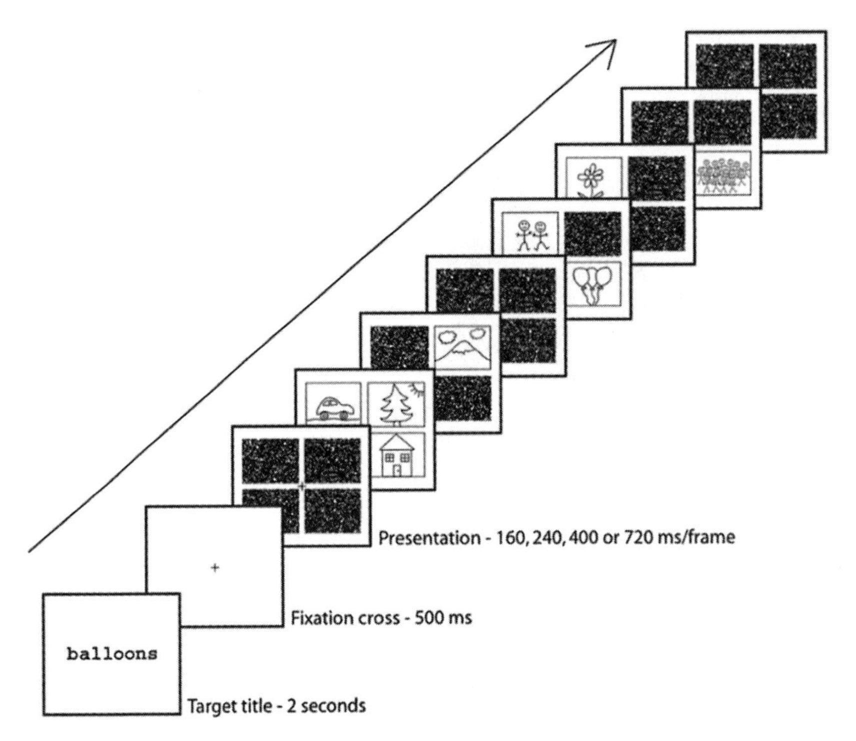

Figure 9.4
Schematic representation of an RSVP sequence in which up to four pictures were presented simultaneously for detection of one named target. From Potter and Fox (2009).

Detecting Pictures at Ultrahigh Rates: Evidence for Feedforward Processing?

Feedforward Processing

In feedforward models of the visual system (Serre, Kreiman, et al., 2007; Serre, Oliva, & Poggio, 2007), units that process the stimulus are hierarchically arranged. Units representing small regions of space (receptive fields) in the retina converge to represent larger and larger receptive fields and more abstract information along a series of pathways from V1 to inferotemporal cortex (IT) and further on to prefrontal cortex (PFC). Visual experience tunes this hierarchical structure, which acts as a filter that permits recognition of a huge range of objects and scenes in a single forward pass of processing. Yet, there is little direct evidence that the feedforward process is able to identify objects and scenes accurately, without feedback. Under normal viewing conditions perception is generally assumed to result from a combination of feedforward and feedback connections (DiLollo, Enns, & Rensink, 2000; Enns & Di Lollo, 2000; Hochstein & Ahissar, 2002; Lamme & Roelfsema, 2000). Feedback from higher to lower levels in the visual system takes time, however. At presentation durations of about 50 ms or less with masking, some have proposed that there would not be time for feedback to arrive before the lower-level activity has been interrupted by the mask, so that perception, if any, would be restricted to the information in the forward pass of neural activity from the retina through the visual system (Hung, Kreiman, Poggio, & DiCarlo, 2005; Liu, Agam, Madsen, & Kreiman, 2009; Perrett, Hietanen, Oram, & Benson, 1992; Thorpe & Fabre-Thorpe, 2001).

Conscious Perception

The ability to identify or remember a stimulus is commonly taken to mean that the viewer was conscious of the stimulus, and here I make the assumption that consciousness is shown by the ability to report on the stimulus by responding to a target picture or by recognizing its title or the picture itself in a memory test. (See, however, evidence for unconscious effects, discussed below.) There is a debate about whether a single forward pass is sufficient for conscious perception. A reentrant process providing feedback may be necessary to achieve understanding and conscious awareness (Dehaene & Naccache, 2001; Hochstein & Ahissar, 2002; Lamme & Roelfsema, 2000). As mentioned above it has been suggested that a threshold duration of about 50 ms must be exceeded if there is a backward mask, or the stimulus will not be consciously perceived. Consciousness of a stimulus may require sufficient time "to establish sustained activity in recurrent cortical loops" (Del Cul, Baillet, & Dehaene, 2007) or to ignite a network required for conscious perception (Deheane, Kergsberg, & Changeux, 1998). These authors thus hypothesize that viewers cannot become conscious of a stimulus on the basis of a single feedforward sweep, without time for any feedback. Detection in RSVP at durations of 50 ms per picture or less should be impossible if there is such a threshold because there is too little time to establish a long-range cortical loop before a picture has been overwritten by subsequent pictures. As reviewed in the next section, however, there is evidence that perception is sometimes possible with very brief masked stimuli, a result that suggests that feedforward processing may be sufficient for conscious perception under some conditions.

Evidence for Processing of Very Brief Stimuli: RSVP Responses by Monkey Neurons and Humans

Recordings of individual neurons in the cortex of the anterior superior temporal sulcus (STSa) of monkeys that viewed a set of pictures of monkey faces and other objects via RSVP at various rates up to 72 per second (14 ms) showed that neurons respond to a preferred picture above chance, even at 14 ms (Keysers, Xiao, Földiák, & Perrett, 2001, 2005). In a detection study with human observers using the same set of pictures but presenting them in seven-picture RSVP sequences, the participants were shown a target picture before each sequence. They detected the target above chance at 14 ms per picture, although detection improved as the duration per picture was increased. In another condition in the same study, recognition of a target picture was tested immediately after the sequence, instead of being shown before the sequence. Participants were still above chance at 14 ms per picture, but performance was not as good as when they saw the target picture in advance. A possible problem with the human study is that the pictures were repeated across trials and hence became familiar, which might have allowed participants to focus on simple features in order to spot the target.

Detection and Immediate Memory for Conceptually Defined Targets

A study by Potter, Wyble, Hagmann, and McCourt (2014) replicated some of the behavioral conditions of Keysers et al. (2001), but crucially, instead of showing the picture target, they gave only a descriptive name for the target (e.g., *smiling couple*), before or immediately after an RSVP sequence of six pictured scenes (figure 9.5 shows the method). Moreover, each picture was presented only once, and none of the pictures was familiar to the participants. Thus, participants had only a conceptual representation of the target they were to detect or recollect. The RSVP sequence was presented at durations between 13 and 80 ms. Even at a presentation duration of 13

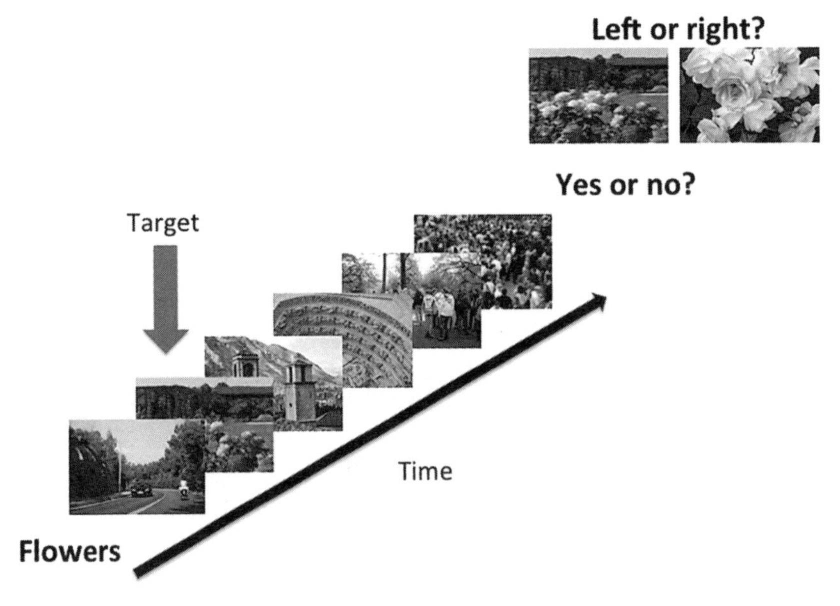

Figure 9.5
Illustration of a six-picture sequence in which the target is named in advance and a yes-no decision is followed by a forced choice between two pictures, both of which match the target name. See Potter, Wyble, Hagmann, and McCourt (2014).

ms, the targets were detected or recognized above chance: that is, the probability of a correct detection on target-present trials was significantly higher than the probability of a false detection response on target-absent trials. In addition, at the end of each trial participants were shown two pictures, both matching the target name, and asked to indicate which one they had seen. They were above chance in selecting the right picture only if they had correctly detected the target; if they missed the target, their forced choice was at chance. Thus, viewers could detect and retain at least briefly information about named targets they had never seen before at an RSVP duration as short as 13 ms. A second experiment replicated those results with sequences of 12 rather than 6 pictures: again, detection and recognition were above chance at all durations, including 13 ms.

These results are consistent with the claim of the feedforward model that pictures can be understood in a single feedforward sweep even when attention has not been directed to a specific category in advance. In the name-after condition the participant had no knowledge of the target at the time he/she viewed the picture sequence, so the pictures had to be processed bottom-up and encoded. Only after the target was named could the participant search recent memory for the target—there was no top-down influence on perception, only on memory search. Performance was somewhat lower when the target name came after the sequence, rather than before, showing that advance information did make detection more likely.

Feedforward Processing and Masked Priming

In masked priming studies a brief presentation of a word becomes invisible when it is followed by a second unmasked word to which the participant must respond (Dehaene et al., 2001; Forster & Davis, 1984). If the prime word is related in some way to the following word, it may increase the accuracy or speed of response to the latter, showing that the prime must have been unconsciously identified. Given that the prime may have been presented for 50 ms or more in typical masked priming experiments (above the threshold for perception with a noise mask), why is the participant not conscious of the prime? In such studies the focus of attention is on the second stimulus, and its longer duration permits it to receive full, recurrent processing that may interfere with retention of the more vulnerable information from the prime that was extracted during the feedforward sweep. When, as in Potter et al. (2014) the masking stimulus has the same duration as the preceding target stimulus and is another picture that is to be attended, a duration of 13 ms is clearly sufficient, on a significant proportion of trials, to drive detection, identification, and (at least briefly) recognition memory for the pictures. It seems likely, however, that the reportable detection observed with RSVP tasks such as those described here has the same neural basis as masked priming.

Discussion

Why are eye fixations so brief? It is clear from the research reviewed here that a typical single fixation of 250 ms is long enough to make it highly likely that the viewer will understand what he or she has looked at, at least momentarily. Yet, normal eye fixations are too brief to guarantee good memory. Why don't we fixate for longer? It appears that the rate at which we move our eyes is just slow enough to allow momentary understanding and to initiate appropriate action if needed (including taking a second look) but still fast enough to keep up with rapid changes in the scene around us, allowing us to dodge a bicycle or catch a ball. If something is important enough to need to be remembered, we can keep looking or keep thinking about it.

How long does it take to understand a pictured scene? To return to a question considered at the beginning of the chapter, what can be concluded about the time required to identify a scene? If the question is the minimum exposure duration (prior to a mask) that is required, 13 ms is sometimes enough when the mask is another scene. But if the question is the time from arrival at the retina to correct categorization, then the most reliable measures available at present are reaction time measures, the most sensitive of which is an eye movement to the appropriate target in a choice situation. For detection of a face (when a picture with a face is presented together with another picture), that time can be as short as 100 ms, with a mean time of 140 ms (Crouzet et al., 2010); detection of a vehicle takes somewhat longer. Momentary comprehension is no guarantee of subsequent memory, however, even seconds later. We comprehend rapidly, but then we forget selectively on the basis of what is relevant to our current goals and needs.

Ultrarapid Presentations and Feedforward Processing

Both the results of Keysers et al. (2001, 2005) with monkey neurons and with humans and those of Potter et al. (2014) with humans show that pictures can be detected and briefly remembered when presented in a short sequence at a rate as high as 75 pictures per second. Even when no target is specified in advance, a name presented immediately after the sequence can prompt memory for the corresponding picture. These results support a feedforward model that can extract a picture's conceptual meaning in a single forward sweep of information with an input as brief as 13 ms without requiring feedback loops from higher to lower levels and back and without requiring a selective attentional set. However, a longer feedforward viewing time of up to 80 ms may be required to grasp the gist of many scenes. When a scene is complex or its components are unfamiliar, we may need more than a single fixation to comprehend it.

Although the results reviewed here indicate that feedforward processing is capable of activating the conceptual identity of a picture even when the picture is briefly presented and is then masked by immediately following pictures, they leave open the possibility that top-down or reentrant loops facilitate processing and may be essential to comprehend details. For example, there is evidence that a rapid but coarse first pass of low-frequency information may provide global category information that is subsequently refined by top-down processing (e.g., Bar et al., 2006). Other work has shown that monkey neurons that are selective for particular faces at a latency of about 90 ms give further information about facial features beginning about 50 ms later (Sugase, Yamane, Ueno, & Kawano, 1999), suggesting reentrant processing (DiLollo, 2012). In any case such reentrant theories rely on feedforward processing to generate tentative interpretations of a picture that are fed back and compared with the representations in earlier levels of processing, suggesting that feedforward processing initiates visual understanding.

But are there other explanations for successful detection when the presentation duration is brief and masked by successive pictures? One possibility is that subsequent pictures do not interrupt processing immediately. As mentioned earlier the neural basis for masking is not well understood. Studies of the monkey visual system using single-cell recordings show that multiple cortical neurons that are selective for different objects can be activated at the same time, suggesting that multiple objects may be "recognized" in parallel at levels as high as the inferior temporal cortex. Something similar in human perception might account for the ability to perceive rapidly presented pictures. In monkeys this initial parallel process is followed within 150 ms by competitive inhibition of all neurons other than those responding to the relevant object in a given receptive field, at least when there is a task that defines the relevant stimulus (e.g., Chelazzi, Duncan, Miller, & Desimone, 1998; see Rousselet et al., 2004a, for a review). The large and overlapping receptive fields found in the inferior temporal cortex may allow for temporary representation in parallel of several successive pictures presented at a high rate, followed by competitive suppression that favors the most salient picture. That could account for the ability on some trials to detect a target by name immediately after the presentation of 6 pictures (although with 12 pictures one would have expected a larger decrement than we observed). If high-level representations of several of the pictures in the sequence were activated, however, it is likely that mutual competition would soon decrease their activation well before the target name was presented. In further experiments with RSVP using very rapid sequences (Potter, Wyble, & Hagmann, unpublished data) a delay of 5 seconds in providing the target name after the sequence did decrease accuracy.

Thus, the feedforward hypothesis remains a strong contender as an explanation of picture identification with very brief presentation durations. In the absence of a specific model for how feedback might assist reportable detection of brief targets, the feedforward hypothesis seems the most plausible account. A lifetime of experience of the world that is built into our visual system appears to allow immediate understanding of most scenes, based on the initial sweep of visual information when the scene is presented.

Acknowledgments

This work was supported by Grant MH47432 from the National Institute of Mental Health. I thank Carl Hagmann for comments and my many collaborators for their participation in the research reported here. Much of this work was also described in Potter (2012).

Notes

1. A measure of the minimal time required for successful processing does not include the time for the retinal signal to arrive at the part of the brain doing the processing, which may take 60–80 ms, or the time to generate an overt response to the stimulus once it is understood (e.g., Potter, 1984).

2. A one-high-threshold formula was used to correct for guessing, $P_{corr} = [P(TY) - P(FY)]/[1 - P(FY)]$, where TY is a correct yes response and FY is a false yes response. This guessing correction is used in all data figures.

References

Bar, M., Kassam, K. S., Ghuman, A. S., Boshyan, J., Schmidt, A. M., Dale, A. M., et al. (2006). Top-down facilitation of visual recognition. *Proceedings of the National Academy of Sciences of the United States of America, 103*(2), 449–454.

Brady, T. F., Konkle, T., Alvarez, G. A., & Oliva, A. (2008). Visual long-term memory has a massive storage capacity for object details. *Proceedings of the National Academy of Sciences of the United States of America, 105*(38), 14325–14329.

Chelazzi, L., Duncan, J., Miller, E. K., & Desimone, R. (1998). Responses of neurons in inferior temporal cortex during memory-guided visual search. *Journal of Neurophysiology, 80*(6), 2918–2940.

Coltheart, M. (1983). Iconic memory. *Philosophical Transactions of the Royal Society of London, 302*(1110), 283–294.

Crouzet, S. M., Kirchner, H., & Thorpe, S. J. (2010). Fast saccades towards face: Face detection in just 100 ms. *Journal of Vision, 10*(4), 1–17.

Davenport, J. L. (2007). Consistency effects between objects in scenes. *Memory & Cognition, 35*, 393–401.

Davenport, J. L., & Potter, M. C. (2004). Scene consistency in object and background perception. *Psychological Science, 15*(8), 559–564.

Dehaene, S., & Naccache, L. (2001). Towards a cognitive neuroscience of consciousness: Basic evidence and a workspace framework. *Cognition, 79*, 1–37.

Dehaene, S., Naccache, L., Cohen, L., LeBihan, D., Mangin, J. F., Poline, J.-B., et al. (2001). Cerebral mechanisms of word masking and unconscious repetition priming. *Nature Neuroscience, 4*, 752–758.

Deheane, S., Kergsberg, M., & Changeux, J. P. (1998). A neuronal model of a global workspace in effortful cognitive tasks. *Proceedings of the National Academy of Sciences of the United States of America, 95*, 14529–14534.

Del Cul, A., Baillet, S., & Dehaene, S. (2007). Brain dynamics underlying the nonlinear threshold for access to consciousness. *PLoS Biology, 5*, 2408–2423.

DiLollo, V. (2012). The feature-binding problem is an ill-posed problem. *Trends in Cognitive Sciences*, *16*(6), 317–321.

DiLollo, V., Enns, J. T., & Rensink, R. A. (2000). Competition for consciousness among visual events: The psychophysics of reentrant visual pathways. *Journal of Experimental Psychology. General*, *129*, 481–507.

Endress, A. D., & Potter, M. C. (2012). Early conceptual and linguistic processes operate in independent channels. *Psychological Science*, *23*, 235–245.

Enns, J. T., & Di Lollo, V. (2000). What's new in visual masking? *Trends in Cognitive Sciences*, *4*, 345–352.

Fabre-Thorpe, M. (2011). The characteristics and limits of rapid visual categorization. *Frontiers in Psychology*, *2*(243), 1–12.

Forster, K. I. (1970). Visual perception of rapidly presented word sequences of varying complexity. *Perception & Psychophysics*, *8*(4), 215–221.

Forster, K. I., & Davis, C. (1984). Repetition priming and frequency attenuation in lexical access. *Journal of Experimental Psychology. Learning, Memory, and Cognition*, *10*, 680–698.

Gordon, R. D., & Irwin, D. E. (2000). The role of physical and conceptual properties in preserving object continuity. *Journal of Experimental Psychology. Learning, Memory, and Cognition*, *26*, 136–150.

Henderson, J. M. (1997). Transsaccadic memory and integration during real-world object perception. *Psychological Science*, *8*(1), 51–55.

Henderson, J. M., & Hollingworth, A. (1999). High-level scene perception. *Annual Review of Psychology*, *50*, 243–271.

Hochstein, S., & Ahissar, M. (2002). View from the top: Hierarchies and reverse hierarchies in the visual system. *Neuron*, *36*(5), 791–804.

Hollingworth, A., & Henderson, J. M. (1998). Does consistent scene context facilitate object perception? *Journal of Experimental Psychology. General*, *127*(4), 398–415.

Hollingworth, A., & Henderson, J. M. (1999). Object identification is isolated from scene semantic constraint: Evidence from object type and token discrimination. *Acta Psychologica*, *102*(2–3), 319–343.

Hung, C. P., Kreiman, G., Poggio, T., & DiCarlo, J. J. (2005). Fast readout of object identity from macaque inferior temporal cortex. *Science*, *310*(5749), 863–866.

Intraub, H. (1980). Presentation rate and the representation of briefly glimpsed pictures in memory. *Journal of Experimental Psychology. Human Learning and Memory*, *6*, 1–12.

Intraub, H. (1981). Rapid conceptual identification of sequentially presented pictures. *Journal of Experimental Psychology. Human Perception and Performance*, *7*, 604–610.

Intraub, H. (1984). Conceptual masking: The effects of subsequent visual events on memory for pictures. *Journal of Experimental Psychology. Learning, Memory, and Cognition*, *10*(1), 115–125.

Intraub, H., & Richardson, M. (1989). Wide-angle memories of close-up scenes. *Journal of Experimental Psychology. Learning, Memory, and Cognition*, *15*(2), 179–187.

Irwin, D. E. (1992). Memory for position and identity across eye movements. *Journal of Experimental Psychology. Learning, Memory, and Cognition*, *18*, 307–317.

Irwin, D. E., & Andrews, R. V. (1996). Integration and accumulation of information across saccadic eye movements. In T. Inui & J. L. McClelland (Eds.), *Attention and Performance XVI: Information integration in perception and communication* (pp. 125–155). Cambridge, MA: MIT Press.

Joubert, O. R., Rousselet, G., Fize, D., & Fabre-Thorpe, M. (2007). Processing scene context: Fast categorization and object interference. *Vision Research*, *47*, 3286–3297.

Joubert, O. R., Fize, D., Rousselet, G., & Fabre-Thorpe, M. (2008). Early interference of context congruence on object processing in rapid visual categorization of natural scenes. *Journal of Vision, 13,* 11.

Keysers, C., Xiao, D. K., Földiák, P., & Perrett, D. I. (2001). The speed of sight. *Journal of Cognitive Neuroscience*, *13*, 90–101.

Keysers, C., Xiao, D.-K., Földiák, P., & Perrett, D. I. (2005). Out of sight but not out of mind: The neurophysiology of iconic memory in the superior temporal sulcus. *Cognitive Neuropsychology*, *22*, 316–332.

Kirchner, H., & Thorpe, S. J. (2006). Ultra-rapid object detection with saccadic eye movements: Visual processing speed revisited. *Vision Research*, *46*(11), 1762–1776.

Konkle, T., Brady, T. F., Alvarez, G. A., & Oliva, A. (2010). Scene memory is more detailed than you think: The role of categories in visual long-term memory. *Psychological Science*, *21*(11), 1551–1556.

Lamme, V. A. F., & Roelfsema, P. R. (2000). The distinct modes of vision offered by feedforward and recurrent processing. *Trends in Neurosciences*, *23*, 571–579.

Liu, H., Agam, Y., Madsen, J. R., & Kreiman, G. (2009). Timing, timing, timing: Fast decoding of object information from intracranial field potentials in human visual cortex. *Neuron*, *62*(2), 281–290.

Loftus, G. R., & Ginn, M. (1984). Perceptual and conceptual masking of pictures. *Journal of Experimental Psychology. Learning, Memory, and Cognition*, *10*, 435–441.

Loftus, G. R., Hanna, A. M., & Lester, L. (1988). Conceptual masking: How one picture captures attention from another picture. *Cognitive Psychology*, *20*, 237–282.

Loschky, L. C., Hansen, B. C., Sethi, A., & Pydimarri, T. N. (2010). The role of higher order image statistics in masking scene gist recognition. *Attention, Perception & Psychophysics*, *72*(2), 427–444.

Macknik, S. L., & Martinez-Conde, S. (2007). The role of feedback in visual masking and visual processing. *Advances in Cognitive Psychology*, *3*, 125–152.

Meng, M., & Potter, M. C. (2008). Detecting and remembering pictures with and without visual noise. *Journal of Vision, 8*(9), 1–10.

Munneke, J., Brentari, V., & Peelen, M. (2013). The influence of scene context on object recognition is independent of attentional focus. *Frontiers in Psychology*, *4*, 552.

Nickerson, R. S. (1965). Short-term memory for complex meaningful visual configurations: A demonstration of capacity. *Canadian Journal of Psychology*, *19*, 155–160.

Perrett, D., Hietanen, J., Oram, M., & Benson, P. (1992). Organization and functions of cells responsive to faces in the temporal cortex. *Philosophical Transactions of the Royal Society of London. Series B, Biological Sciences*, *335*, 23–30.

Phillips, W. A. (1983). Short-term visual memory. *Philosophical Transactions of the Royal Society of London. Series B, Biological Sciences*, *302*, 295–309.

Phillips, W. A., & Christie, D. F. M. (1977). Components of visual memory. *Quarterly Journal of Experimental Psychology*, *29*(1), 117–133.

Potter, M. C. (1975). Meaning in visual scenes. *Science*, *187*, 965–966.

Potter, M. C. (1976). Short-term conceptual memory for pictures. *Journal of Experimental Psychology. Human Learning and Memory*, *2*(5), 509–522.

Potter, M. C. (1983). Representational buffers: The eye-mind hypothesis in picture perception, reading, and visual search. In K. Rayner (Ed.), *Eye movements in reading: Perceptual and language processes* (pp. 423–437). New York: Academic Press.

Potter, M. C. (1984). Rapid serial visual presentation (RSVP): A method for studying language processing. In D. Kieras & M. Just (Eds.), *New methods in reading comprehension research* (pp. 91–118). Hillsdale, NJ: Erlbaum.

Potter, M. C. (1993). Very short-term conceptual memory. *Memory & Cognition*, *21*, 156–161.

Potter, M. C. (1999). Understanding sentences and scenes: The role of conceptual short term memory. In V. Coltheart (Ed.), *Fleeting memories: Cognition of brief visual stimuli* (pp. 13–46). Cambridge, MA: MIT Press.

Potter, M. C. (2010). Conceptual short term memory. *Scholarpedia*, *5*(2), 3334.

Potter, M. C. (2012). Recognition and memory for briefly presented scenes. *Frontiers in Psychology*, *3*, 32.

Potter, M. C., & Faulconer, B. A. (1975). Time to understand pictures and words. *Nature*, *253*, 437–438.

Potter, M. C., & Fox, L. F. (2009). Detecting and remembering simultaneous pictures in a rapid serial visual presentation. *Journal of Experimental Psychology. Human Perception and Performance*, *35*, 28–38.

Potter, M. C., & Jiang, Y. V. (2009). Visual short-term memory. In T. Bayne, A. Cleeremans, & P. Wilken (Eds.), *Oxford companion to consciousness* (pp. 436–438). Oxford: Oxford University Press.

Potter, M. C., & Levy, E. I. (1969). Recognition memory for a rapid sequence of pictures. *Journal of Experimental Psychology*, *81*(1), 10–15.

Potter, M. C., Moryadas, A., Abrams, I., & Noel, A. (1993). Word perception and misperception in context. *Journal of Experimental Psychology. Learning, Memory, and Cognition*, *19*, 3–22.

Potter, M. C., Staub, A., & O'Connor, D. H. (2004). Pictorial and conceptual representation of glimpsed pictures. *Journal of Experimental Psychology. Human Perception and Performance*, *30*, 478–489.

Potter, M. C., Staub, A., Rado, J., & O'Connor, D. H. (2002). Recognition memory for briefly-presented pictures: The time course of rapid forgetting. *Journal of Experimental Psychology. Human Perception and Performance*, *28*(5), 1163–1175.

Potter, M. C., Stiefbold, D., & Moryadas, A. (1998). Word selection in reading sentences: Preceding versus following contexts. *Journal of Experimental Psychology. Learning, Memory, and Cognition*, *24*, 68–100.

Potter, M. C., Wyble, B., & Hagmann, C. E. (unpublished data). A delay of 5 seconds decreases *d'* in memory for rapid sequences.

Potter, M. C., Wyble, B., Hagmann, C. E., & McCourt, E. S. (2014). Detecting meaning in RSVP at 13 ms per picture. *Attention, Perception, & Performance*.

Potter, M. C., Wyble, B., Pandav, R., & Olejarczyk, J. (2010). Picture detection in RSVP: Features or identity? *Journal of Experimental Psychology. Human Perception and Performance*, *36*, 1486–1494.

Rensink, R. A., O'Regan, J. R., & Clark, J. J. (1997). To see or not to see: The need for attention to perceive changes in scenes. *Psychological Science*, *8*, 368–373.

Rensink, R. A., O'Regan, J. K., & Clark, J. J. (2000). On the failure to detect changes in scenes across brief interruptions. *Visual Cognition*, *7*, 127–145.

Rousselet, G., Fabre-Thorpe, M., & Thorpe, S. J. (2002). Parallel processing in high level categorization of natural images. *Nature Neuroscience*, *5*, 629–630.

Rousselet, G., Thorpe, S. J., & Fabre-Thorpe, M. (2004a). How parallel is visual processing in the ventral pathway? *Trends in Cognitive Sciences*, *8*(8), 363–370.

Rousselet, G. A., Thorpe, S. J., & Fabre-Thorpe, M. (2004b). Processing of one, two or four natural scenes in humans: The limits of parallelism. *Vision Research*, *44*(9), 877–894.

Serre, T., Kreiman, G., Kouh, M., Cadieu, C., Knoblich, U., & Poggio, T. (2007). A quantitative theory of immediate visual recognition. *Progress in Brain Research*, *165*, 33–56.

Serre, T., Oliva, A., & Poggio, T. (2007). A feedforward architecture accounts for rapid categorization. *Proceedings of the National Academy of Sciences of the United States of America*, *104*(15), 6424–6429.

Shepard, R. N. (1967). Recognition memory for words, sentences, and pictures. *Journal of Verbal Learning and Verbal Behavior*, *6*, 156–163.

Simons, D. J., & Levin, D. T. (1997). Change blindness. *Trends in Cognitive Sciences*, *1*, 261–267.

Sperling, G. (1960). The information available in brief visual presentations. *Psychological Monographs*, *74*(498), 11.

Standing, L. (1973). Learning 10,000 pictures. *Quarterly Journal of Experimental Psychology*, *25*, 207–222.

Sugase, Y., Yamane, S., Ueno, S., & Kawano, K. (1999). Global and fine information coded by single neurons in the temporal visual cortex. *Nature*, *400*, (869–873).

Thorpe, S. J., & Fabre-Thorpe, M. (2001). Seeking categories in the brain. *Science*, *291*, 260–263.

Thorpe, S., Fize, D., & Marlot, C. (1996). Speed of processing in the human visual system. *Nature*, *381*(6582), 520–522.

Plate 1 (figure 2.4)

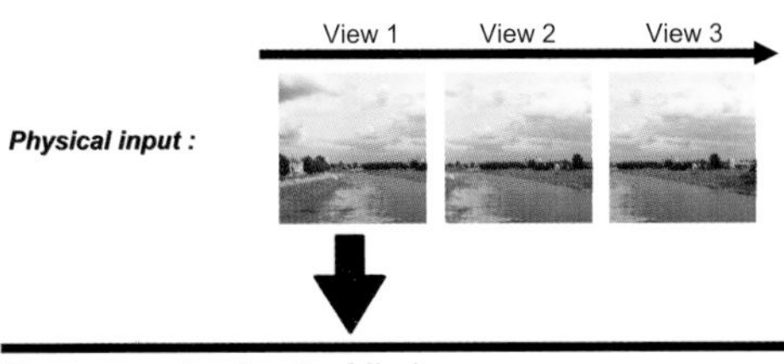

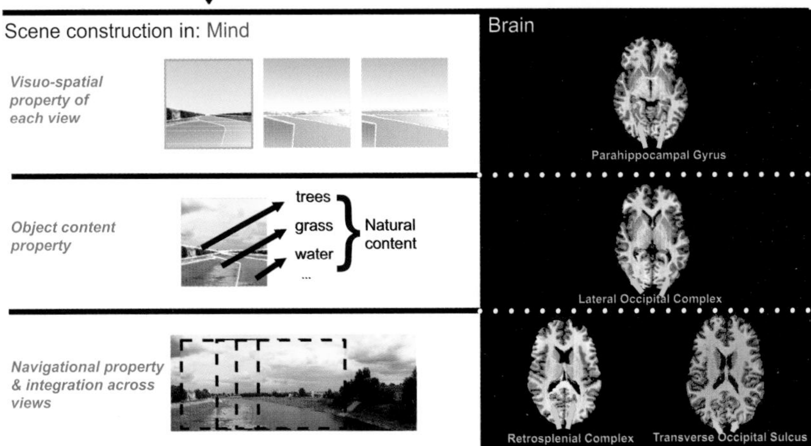

Plate 2 (figure 3.1)

Plate 3 (figure 3.2)

A. Hypothetical patterns of errors

Confusion within the same *Spatial Layout*

Confusion within the same *Content*

Closed Natural
Closed Urban
Open Natural
Open Urban

B. Results

Confusion within the same ***Spatial Layout***

Confusion within the same ***Content***

% of confusion errors

0
10
20
30
40
50
60
70

*

PPA

LOC

Plate 4 (figure 3.3)

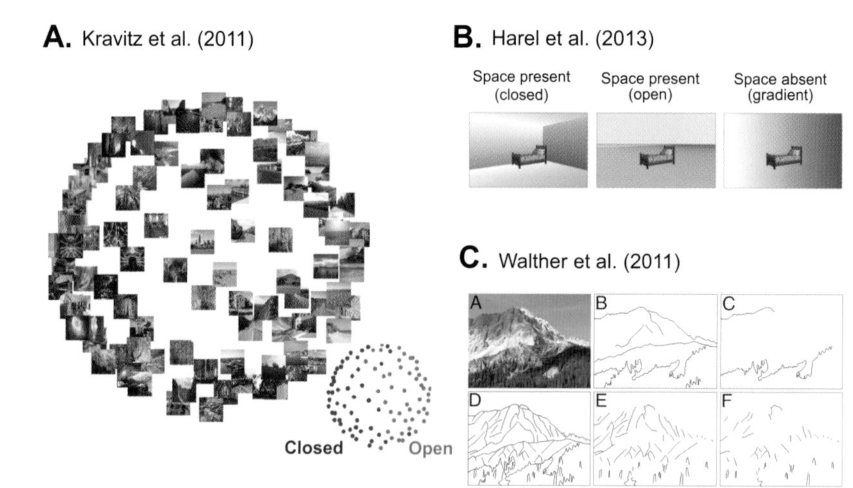

Plate 5 (figure 3.4)

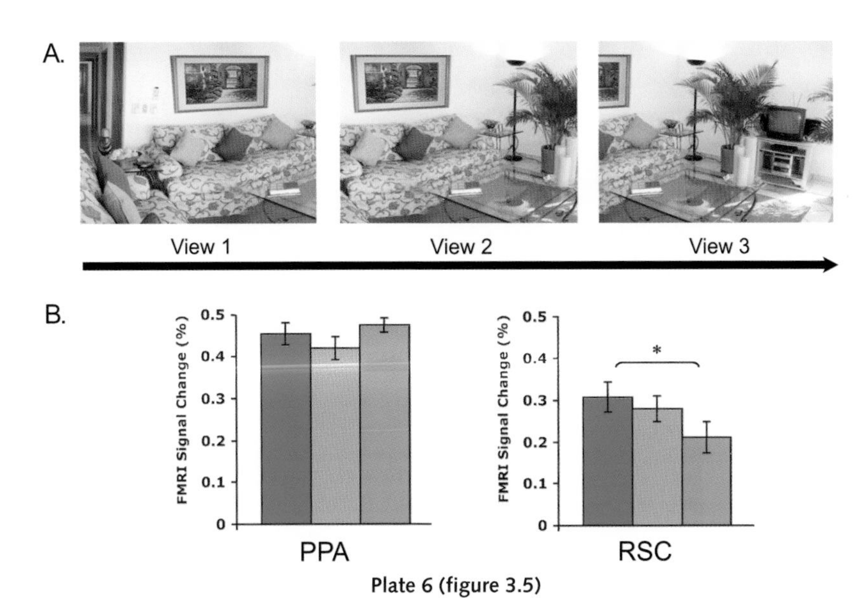

Plate 6 (figure 3.5)

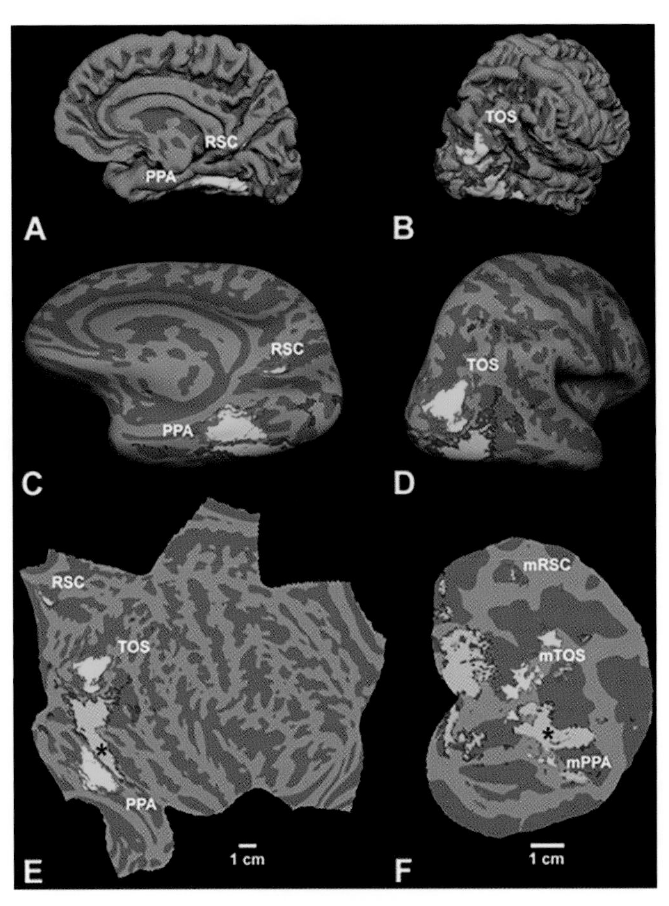

Plate 7 (figure 4.1)

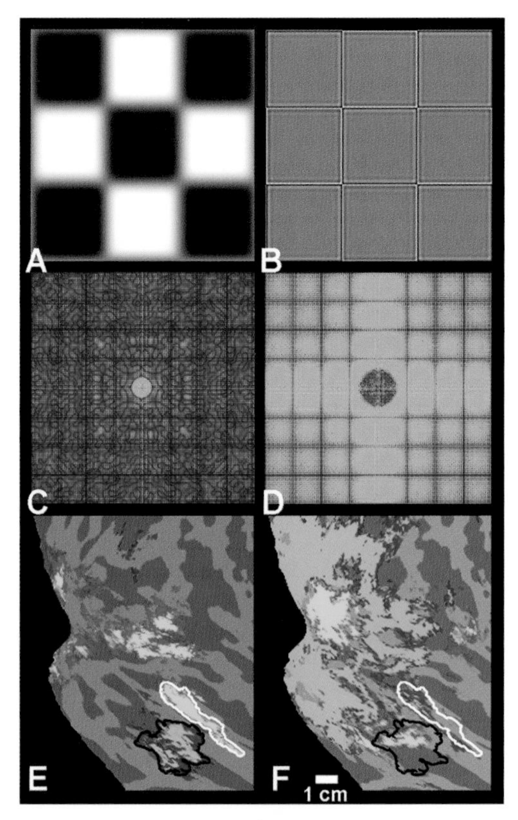

Plate 8 (figure 4.2)

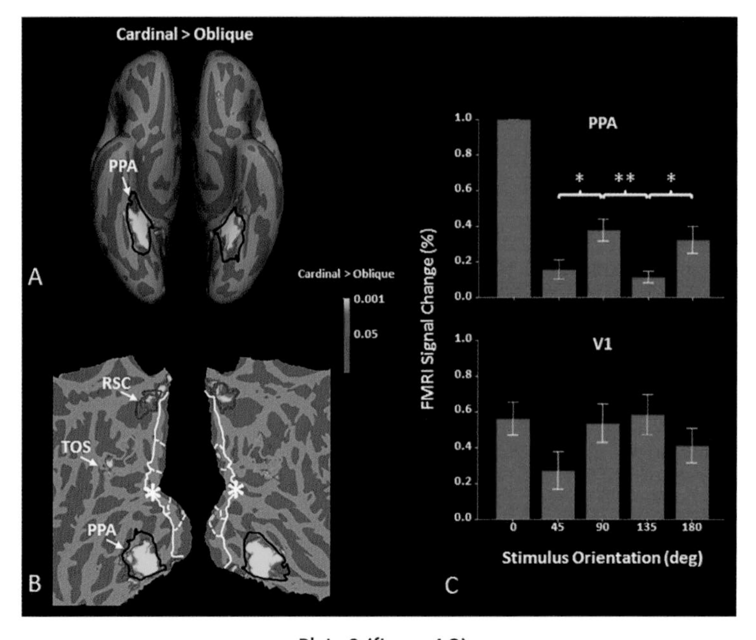

Plate 9 (figure 4.3)

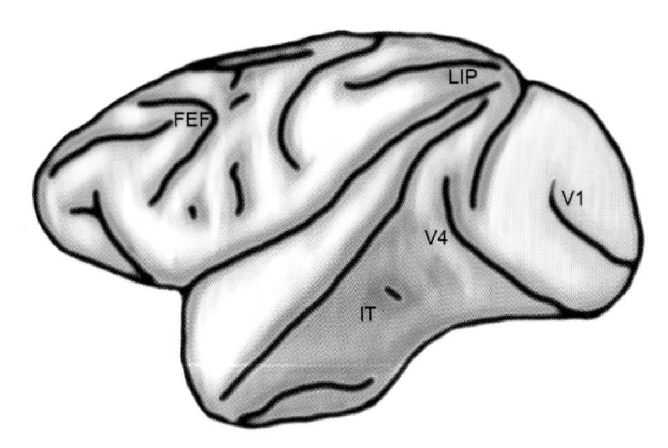

Plate 10 (figure 5.1)

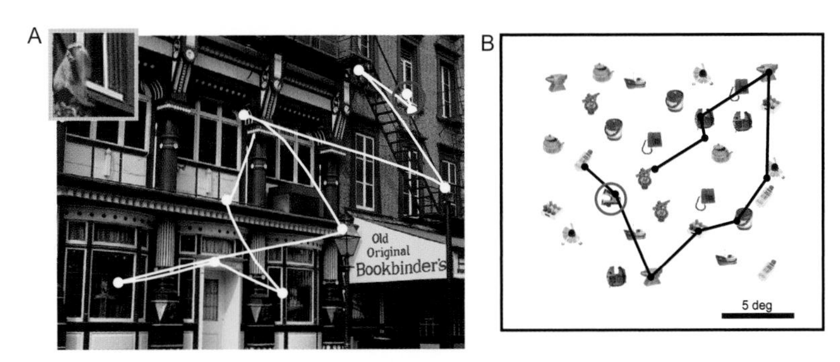

Plate 11 (figure 5.2)

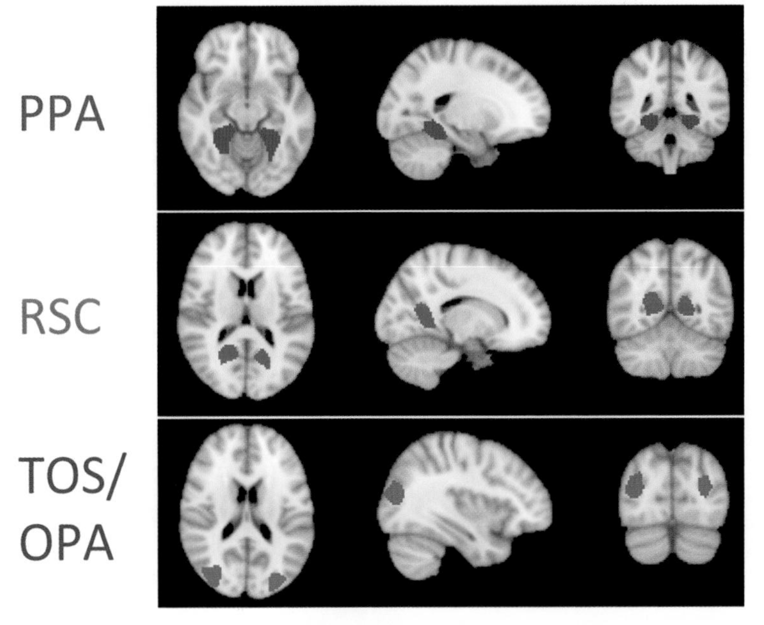

Plate 12 (figure 6.1)

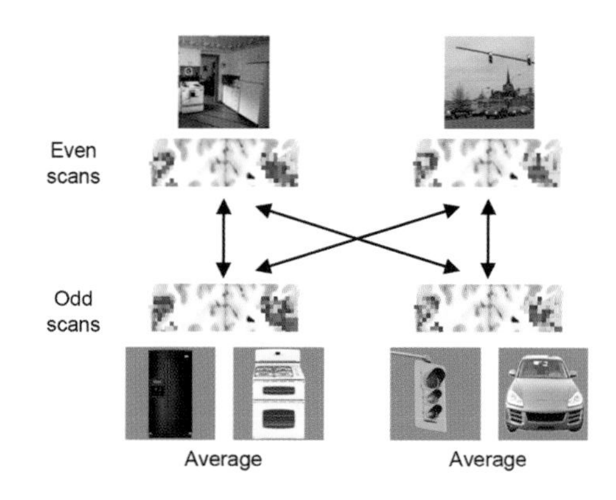

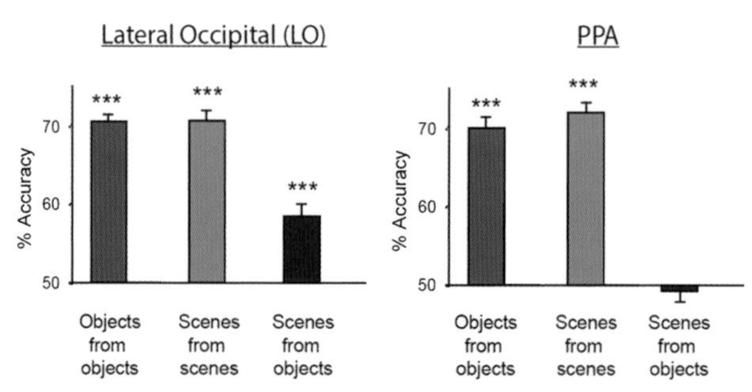

Plate 13 (figure 6.4)

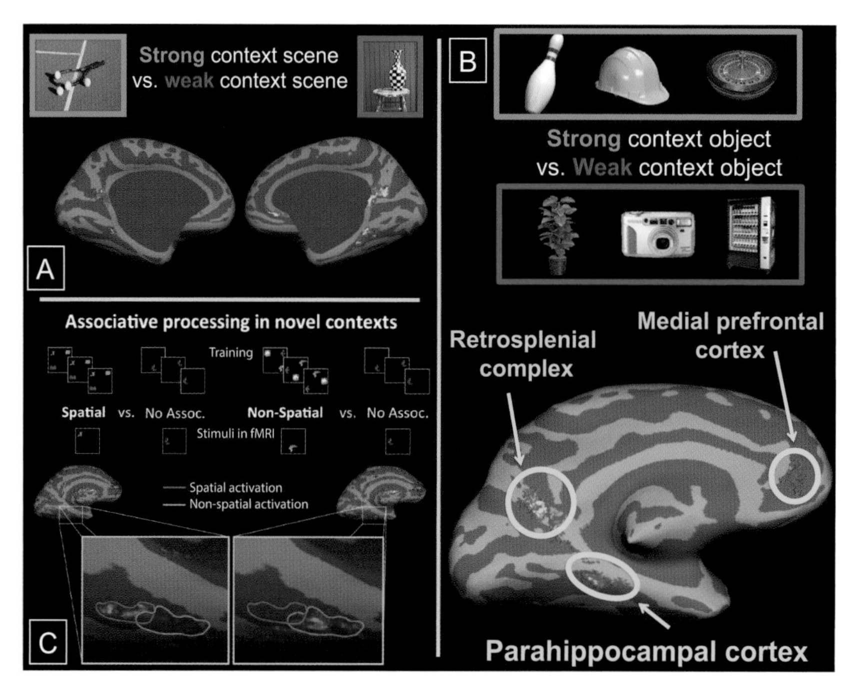

Plate 14 (figure 7.2)

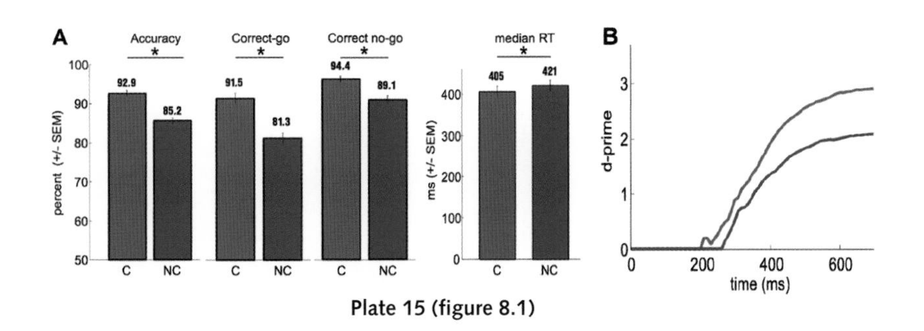

Plate 15 (figure 8.1)

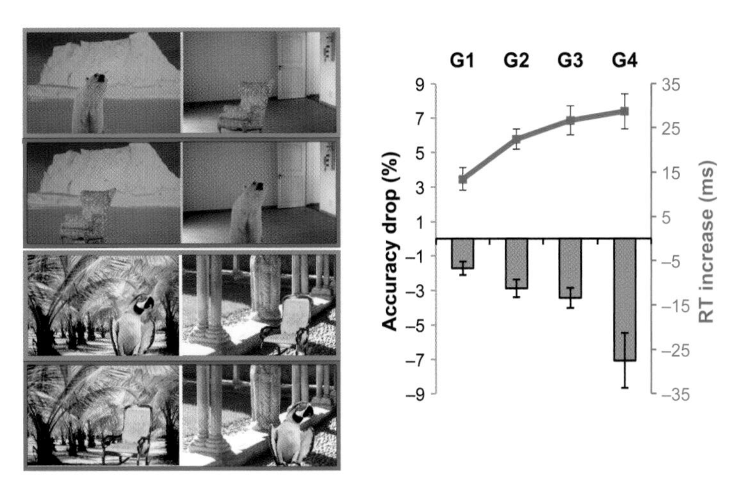

Plate 16 (figure 8.3)

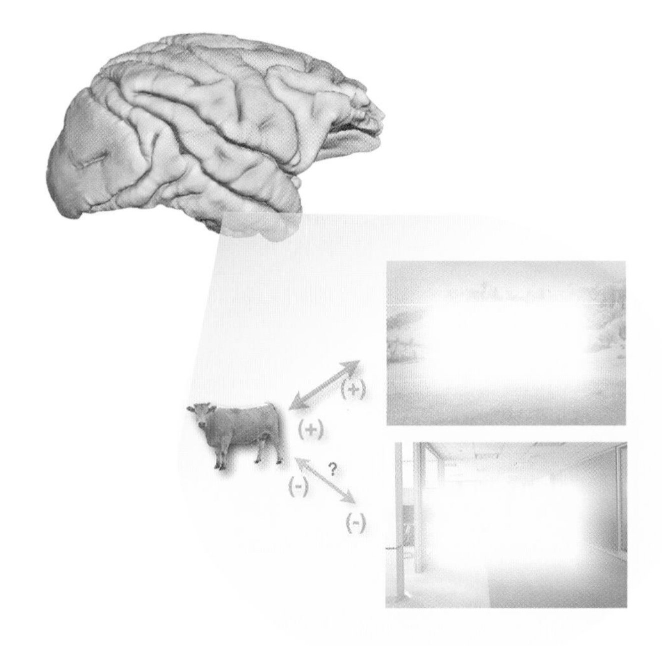

Plate 17 (figure 8.5)

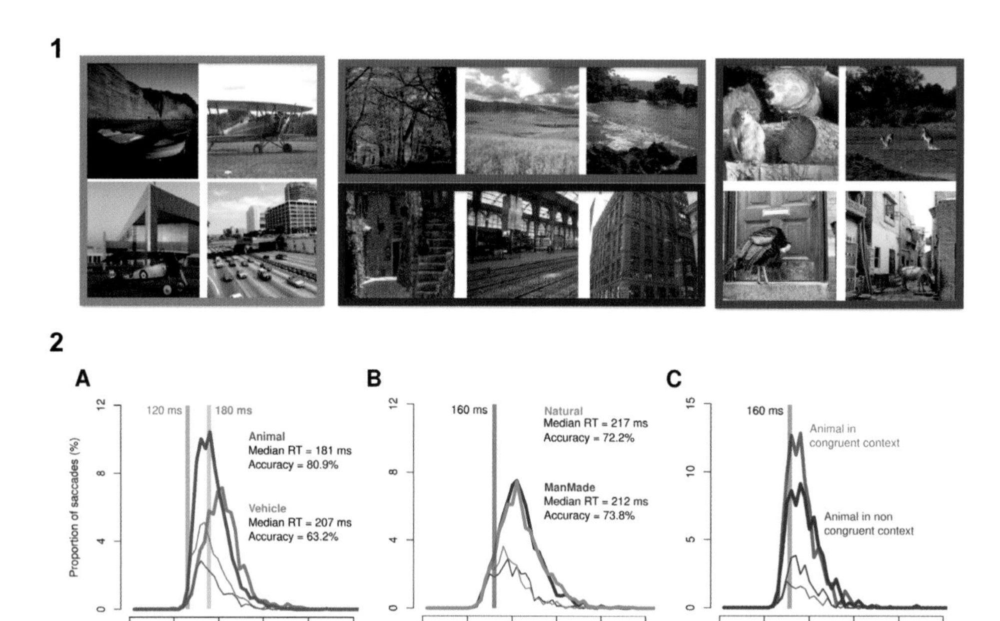

Plate 18 (figure 8.6)

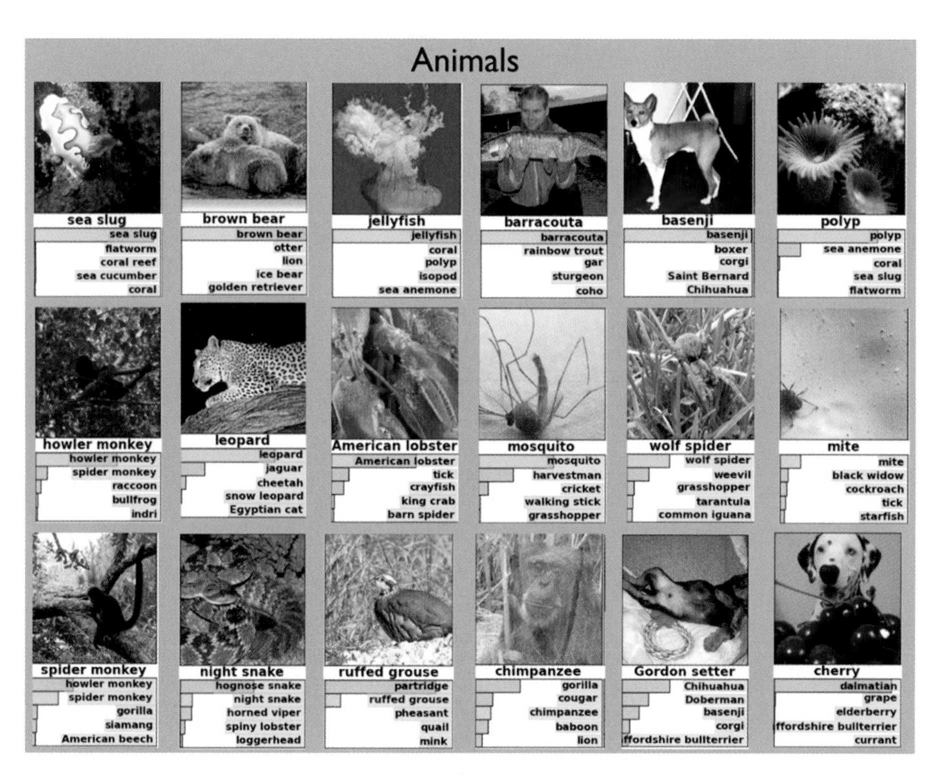

Plate 19 (figure 10.1)

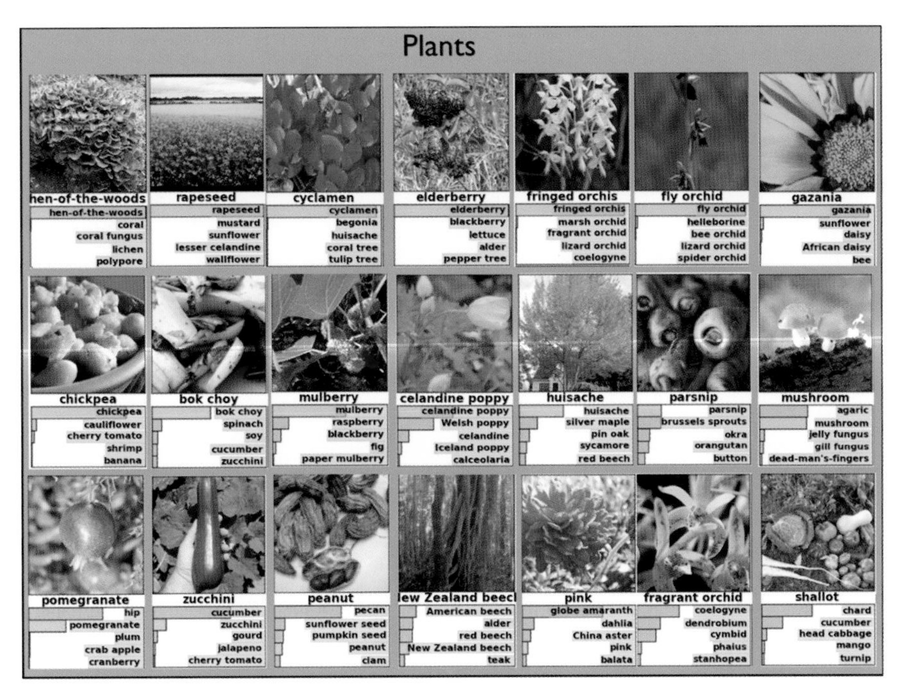

Plate 20 (figure 10.2)

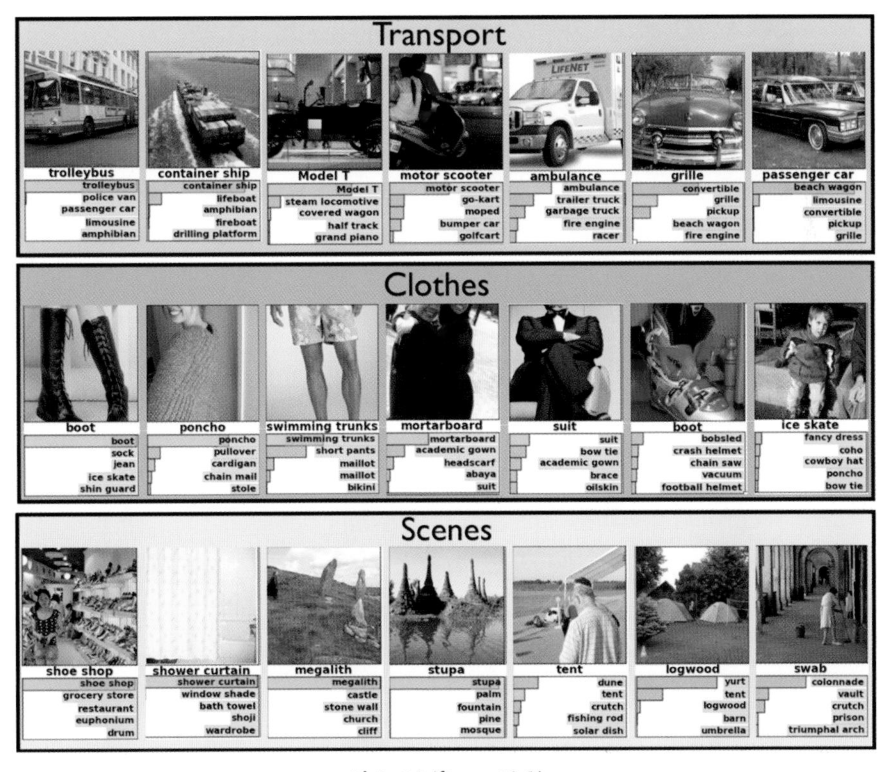

Plate 21 (figure 10.3)

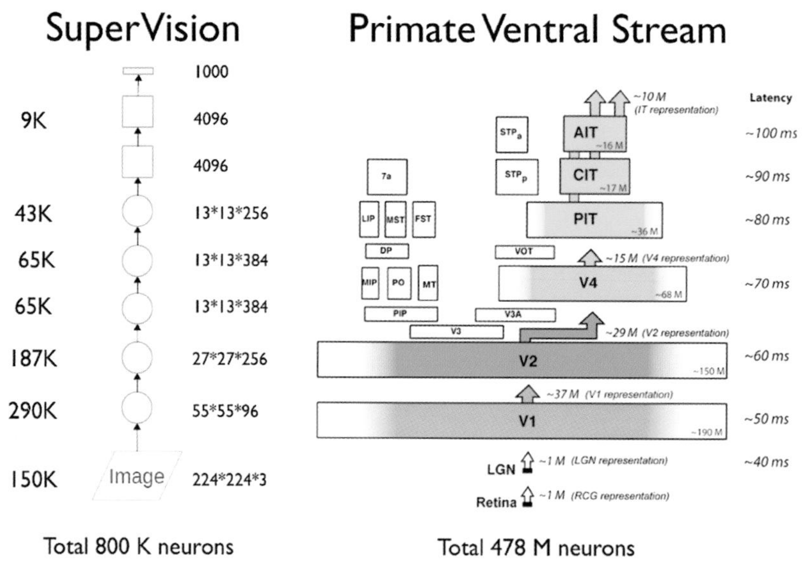

SuperVision

		1000
9K		4096
		4096
43K		13*13*256
65K		13*13*384
65K		13*13*384
187K		27*27*256
290K		55*55*96
150K	Image	224*224*3

Total 800 K neurons

Primate Ventral Stream

Latency

~10 M (IT representation)

STP$_a$ AIT ~16 M ~100 ms

7a STP$_p$ CIT ~17 M ~90 ms

LIP MST FST PIT ~36 M ~80 ms

DP VOT ~15 M (V4 representation)

MIP PO MT V4 ~68 M ~70 ms

PIP V3A

V3 ~29 M (V2 representation)

V2 ~150 M ~60 ms

~37 M (V1 representation)

V1 ~190 M ~50 ms

LGN ~1 M (LGN representation) ~40 ms

Retina ~1 M (RCG representation)

Total 478 M neurons

Plate 22 (figure 10.4)

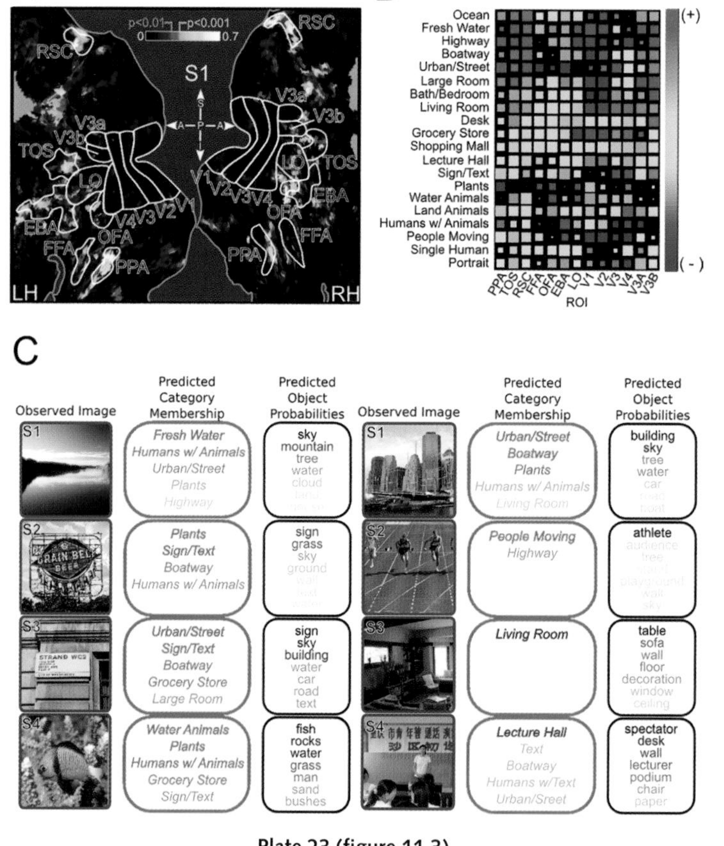

Plate 23 (figure 11.3)

Plate 24 (figure 12.1)

Plate 25 (figure 12.2)

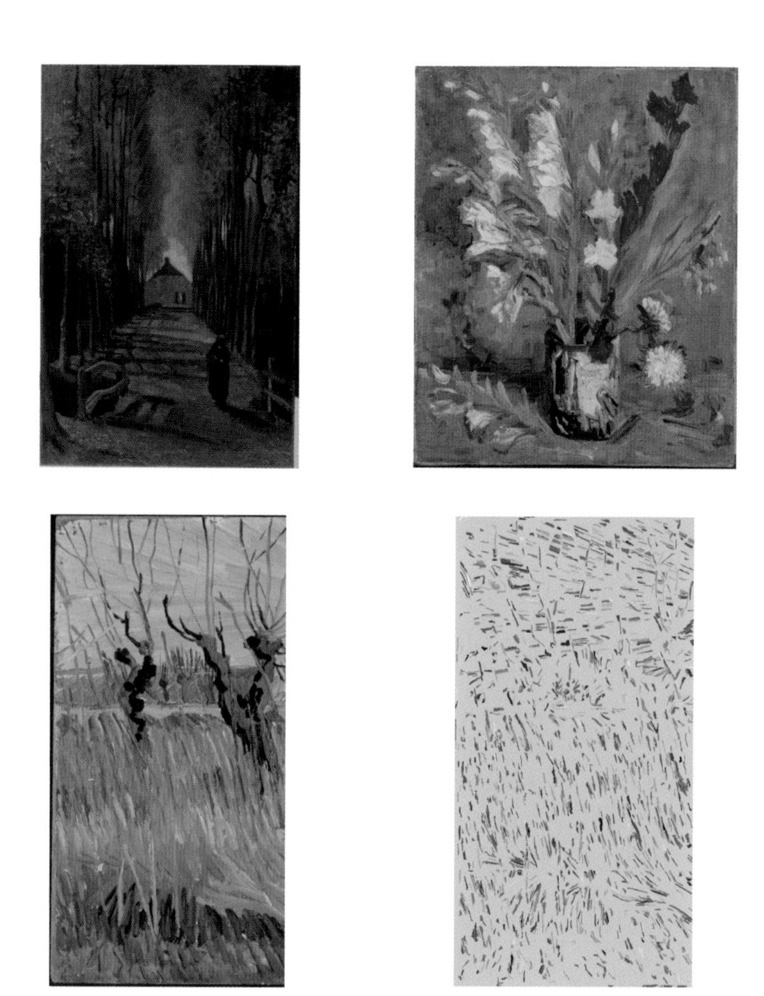

Plate 26 (figure 12.3)

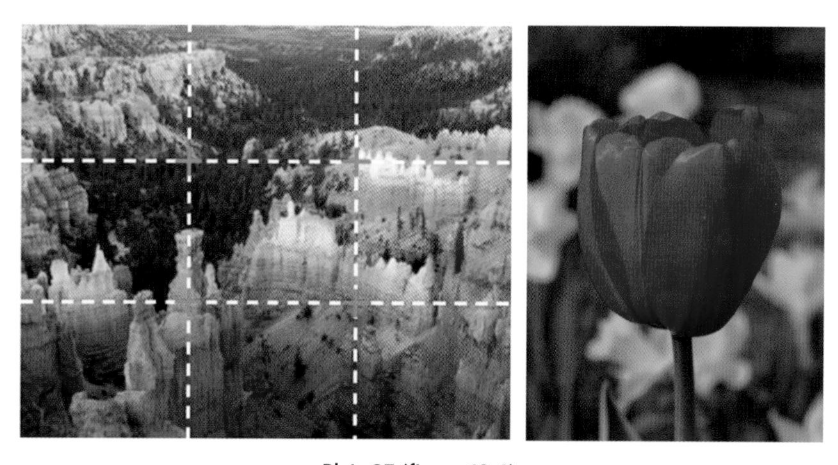

Plate 27 (figure 12.4)

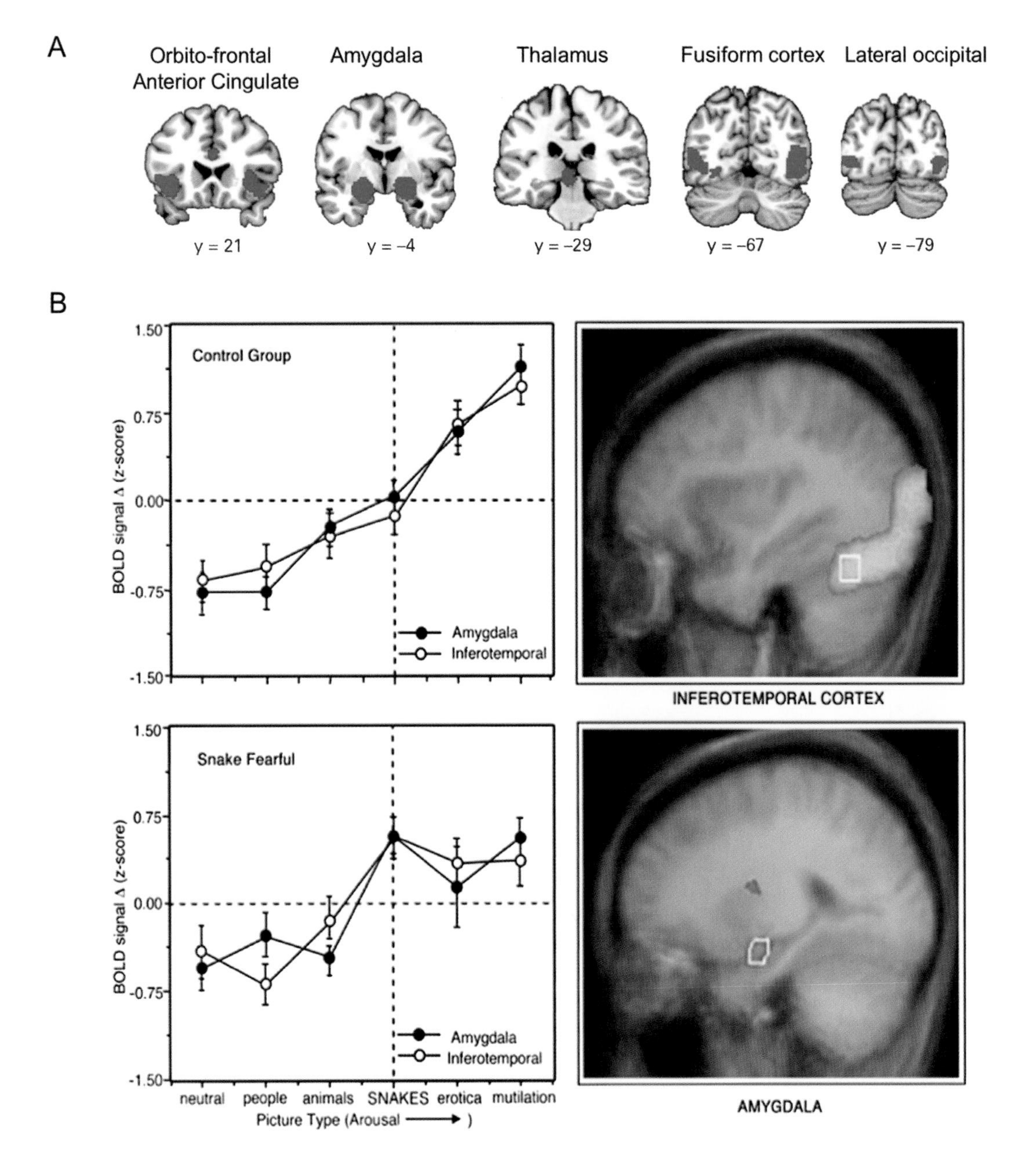

Plate 28 (figure 13.3)

A.

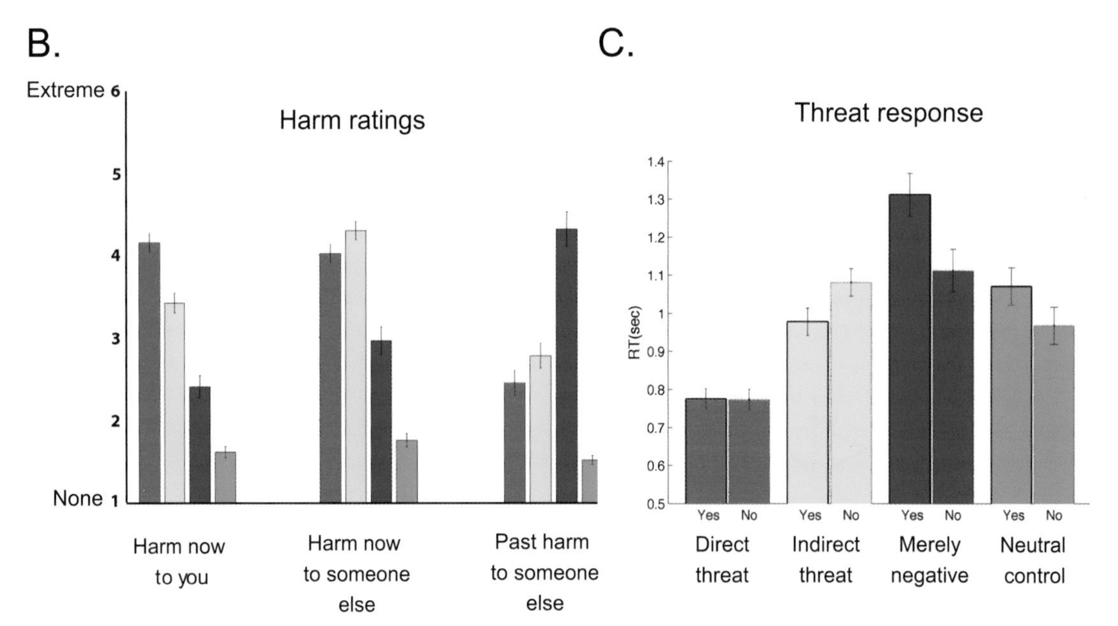

Plate 29 (figure 14.1)

A.

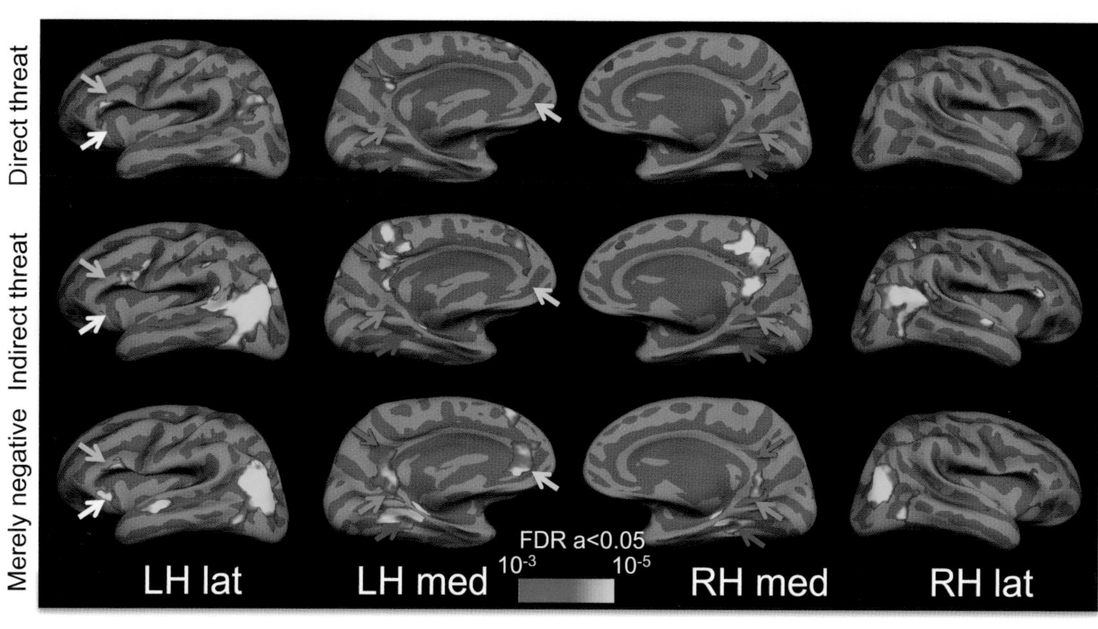

B.

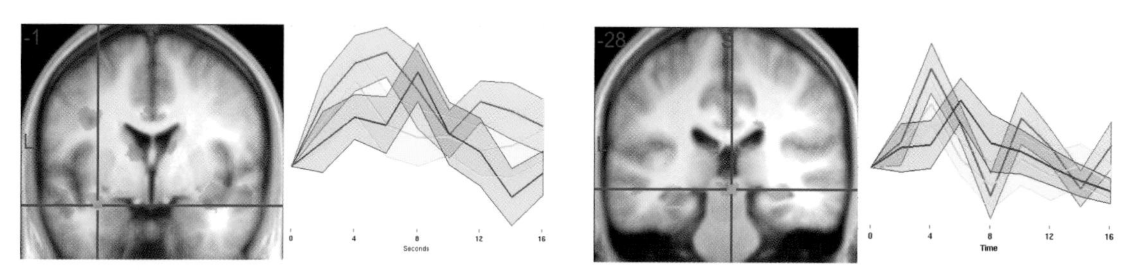

Plate 30 (figure 14.2)

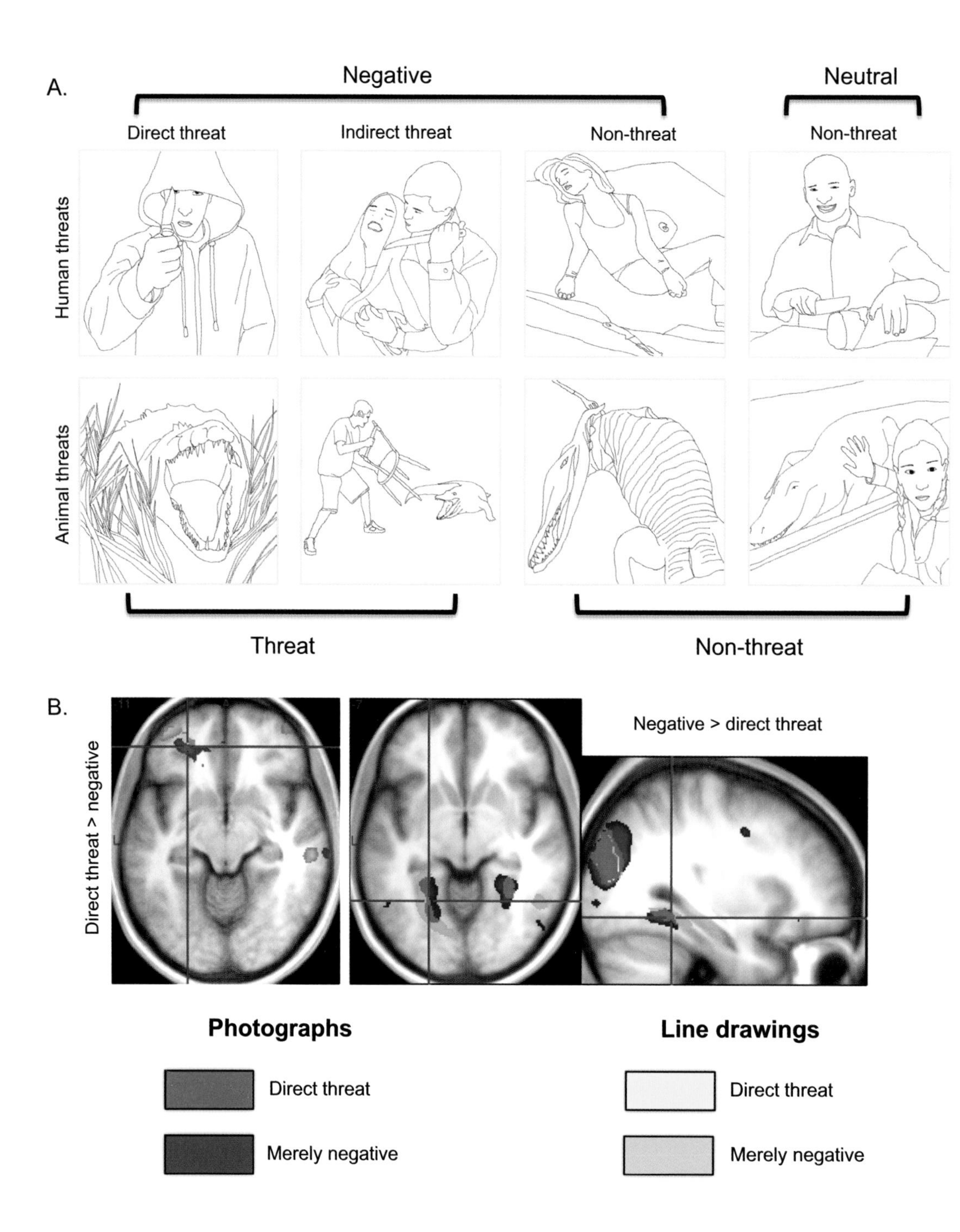

Plate 31 (figure 14.3)

10

Making Sense of Scenes with Spike-Based Processing

Simon Thorpe

Although the ability of the human visual system to process complex natural scenes is very impressive, the state of the art in computer vision is starting to catch up. Interestingly, the best artificial systems use processing architectures built on simple feedforward mechanisms that look remarkably similar to those used in the primate visual system. However, the procedures used for training these artificial systems are very different from the mechanisms used in biological vision. In this chapter I discuss the possibility that spike-based processing and learning mechanisms may allow future models to combine the remarkable efficiency of the latest computer vision systems with the flexible and rapid learning seen in human scene processing.

This chapter discusses how spiking neural networks may provide key insights for understanding how the brain makes sense of the visual world. I start by briefly reviewing some of the experimental findings that demonstrate the remarkable efficiency of scene processing in humans, and I then argue that the speed of processing argues strongly for the use of a largely feedforward processing mode. In a second section, I look at the current state of the art in computer vision, and note the remarkable convergence between the processing architectures employed by the best systems and those used in the primate ventral processing stream. The fact that the state of the art systems use an entirely feedforward architecture provides another argument for the idea that feedforward processing has the power needed to account for a substantial part of image processing. Indeed, the level of performance obtained with the latest systems is so impressive that you might be tempted to say that the problem of rapid scene processing has effectively been solved. However, I argue that the main problem with such models is that the training scheme used is unrealistic in that it requires literally hundreds of millions of training examples using a supervised learning scheme—something that would be out of the question in the case of human vision. In later sections, I sketch a model for learning that could provide a real alternative to the backpropagation learning schemes used in the state-of-the-art computer vision schemes. The proposal uses a simple Spike-based learning scheme based on spike-time-dependent plasticity (STDP) that will naturally generate neurons selective for spatiotemporal spike patterns that occur repeatedly. This means that the visual system will naturally adapt to the statistics of the visual environment by generating selective responses to those objects and configurations that occur most frequently. I will argue that together, feedforward processing architectures coupled with spike-based coding and learning mechanisms can go a long way toward providing an explanation for our remarkable ability to make sense of complex natural scenes.

Performance of the Human Visual System

Everyday experience tells us that the human visual system can understand many complex natural scenes in a fraction of a second. The process is so rapid and effortless that we may be tempted to think that it really is instantaneous. But of course, in reality, it has to take time. In 1996, we published a study that used differential event-related potentials (ERPs) to show that the brain can effectively determine whether a previously unseen natural image contains an animal in around 150 ms (Thorpe, Fize, & Marlot, 1996). This value is already quite short and imposes serious temporal constraints on the underlying hardware. However, more recent work implies that the underlying processing must be even faster. When two images are flashed left and right of fixation, subjects can initiate saccades to the side where an animal is present at latencies that start at just 120–130 ms (Kirchner & Thorpe, 2006). Because this includes the time required to trigger the eye movement, it seems likely that the underlying processing can be done in only 100 ms. The situation is even more remarkable in the case of human faces, which can trigger saccades at latencies of just 100–110 ms (Crouzet, Kirchner, & Thorpe, 2010), implying that the underlying visual processing may take only 80–90 ms.

What sort of mechanisms could allow processing to be done so quickly? For well over two decades, I have been arguing that such fast processing may allow only enough time for a single feedforward pass through the different processing stages involved in scene processing (Thorpe & Imbert, 1989). Although there are large numbers of feedback connections at every level of the visual system (Bullier, Hupe, James, & Girard, 2001; Callaway, 2004), it is clear that every time a computation requires the use of feedback from a later processing stage, this will add significantly to overall computation time. We know, for example, that onset latencies for neuronal responses increase by roughly 10 ms at each level of the cortical processing hierarchy. Thus, onset latencies in primate V1 start at around 40–60 ms, whereas V2 neurons typically need 50–70 ms (see figure 10.3, plate 21, for more details). It follows that if, following a flashed image presentation, processing in V1 depended on the feedback of information computed in V2, this would add roughly 10 ms to overall processing time. If such delays can be avoided by getting as much processing done on the feedforward pass as possible, this would naturally provide a strong selective advantage because any reduction in processing time will naturally increase the speed of behavioral responses.

There are other neurophysiological results that reinforce the view that feedforward processing is at work. First, by analyzing the precise time course of face selectivity in face-selective neurons in primate inferotemporal cortex, Mike Oram and Dave Perrett showed that strong selectivity is already present from the very start of the neural response (Oram & Perrett, 1992). This is not what would be expected if the selectivity depended on the use of feedback loops along the processing hierarchy. A similar point was made by studies demonstrating that it is possible to read out information about object identity using just the very beginning of the neural response in monkey IT (Hung, Kreiman, Poggio, & DiCarlo, 2005) as well as by using intracerebral recordings made in humans (H. Liu, Agam, Madsen, & Kreiman, 2009) and data obtained using magnetoencephalography (MEG; Carlson, Tovar, Alink, & Kriegeskorte, 2013; J. Liu, Harris, & Kanwisher, 2002).

A second neurophysiological argument for feedforward processing comes from a study of neuronal responses to rapid sequential visual presentation (RSVP) sequences at varying rates (Keysers, Xiao, Földiák, & Perrett, 2001). Neurons in the inferotemporal cortex can respond to their preferred stimulus in a selective manner, even when the frame rate was increased to 72 images per second, that is to say, when the duration of each image was only about 14 ms. Given that the response latencies of such neurons were around 100 ms (a value typical for inferotemporal cortex), this implies that as many as seven different images are simultaneously being processed by the visual system in a pipeline. This fits nicely with the number of processing areas involved in the primate ventral stream, which involves a total of seven stages if we include retina, LGN, V1, V2, V4, Posterior inferotemporal cortex, and central inferotemporal cortex en route (see figure 10.3). Again, this result would be very difficult to explain in a model in which feedback from later processing stages was essential.

It is important to note that although feedback from one processing stage to an earlier one would appear to be difficult, given the temporal constraints, it is nevertheless possible that lateral connectivity within a particular cortical stage may operate sufficiently rapidly to influence the initial processing sequence (Panzeri, Rolls, Battaglia, & Lavis, 2001).

Yet another argument for the power of the feedforward pass comes from work in computer vision, where nearly all the state-of-the-art object- and scene-processing systems use architectures that rely almost entirely on a feedforward strategy. This is examined in the next section.

Computer Vision and Biological Vision

Over 30 years ago, David Marr argued that research in computer vision and biological vision was closely related and that it should be possible to build a research community at the interface between the two domains (Marr, 1982). Marr died in 1981, and so we will never know what he thought of the evolution of the field over the last three decades. However, it is likely that he would have been relatively disappointed because in many respects research in computer vision and biological vision has advanced separately. A few researchers attend conferences in both areas, but such people are relatively rare. Most people working in computer vision have little interest in knowing how brains do it—after all, "aircraft don't have to flap their wings to fly." Likewise, researchers working on biological vision often know little about how computer scientists go about trying to recognize objects and process scenes.

Nevertheless, in recent years, there are clear signs that the two areas are starting to converge (Poggio & Ullman, 2013). In particular, the solutions adopted by state-of-the-art vision systems look more and more like the strategies at work in biological vision systems. In 2007, it was reported that a feedforward hierarchical model inspired by the primate visual system could perform a basic animal/nonanimal classification task at a level similar to humans, at least when processing is limited by using strong backward masking (Serre, Oliva, & Poggio, 2007). The specific model used in that work was a development of Riesenhuber and Poggio's HMax model, which had been introduced some years previously (Riesenhuber & Poggio, 1999) but had been developed to take into account a considerable amount of detail concerning the anatomical organization of the primate visual system (Serre, Kreiman, et al., 2007).

Other work in computer vision was less directly concerned with following the blueprints provided by biological vision but nevertheless shared some of the basic design principles. Thus, many successful systems have been based on an essentially feedforward processing sequence (Mel, 1997; Mutch & Lowe, 2008). Many of the most impressive artificial vision systems are based on processing architectures based on "convolutional nets" that involve nothing more than a series of feedforward processing stages one after another, coupled with a final classification stage (Lecun, Kavukcuoglu, & Farabet, 2010). For example, in 2012, the ImageNet Large-scale Visual Recognition Challenge (ILSVRC) was won by a system called SuperVision (Krizhevsky, Sutskever, & Hinton, 2012) that is essentially a feedforward convolutional net very similar to the sort of architectures proposed by researchers such as Fukushima in the 1970s (Fukushima, 1975). The challenge involves training a system with 1.2 million images with roughly 1000 images from each of 1000 different categories. The system then has to correctly label new images that have never been seen before. The SuperVision system achieved an error rate of just 13%, half that of alternative systems. The following figures illustrate the sorts of levels of performance that were obtained with the system. Figure 10.1 (plate 19) shows the labels generated in response to 18 randomly selected animal pictures, figure 10.2 (plate 20) shows the results for 21 randomly selected images of plants, and figure 10.3 (plate 21) shows examples for randomly selected images of means of transport, clothes, and general scenes.

The quality of the labels generated by SuperVision is impressive, as is clear from a cursory inspection of the five labels produced for each image and their relative

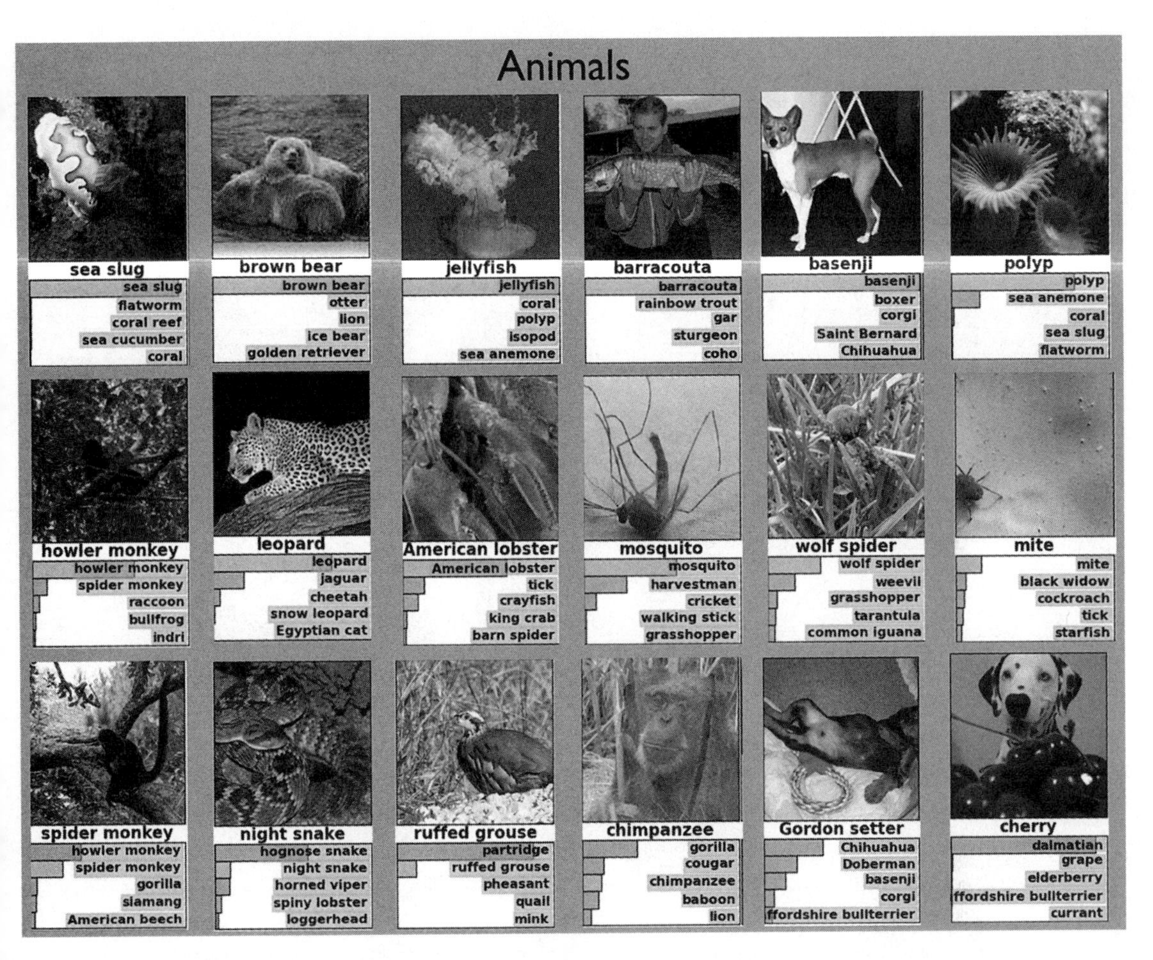

Figure 10.1 (plate 19)
Examples of the sorts of labels generated by "SuperVision," a state-of-the-art computer vision system (Krizhevsky et al., 2012), in response to 18 randomly selected photographs of animal targets from the ILSVRC-2010 test image set. The correct label is written under each image, and the probability assigned to the correct label is also shown with a red bar (if it happens to be in the top five). The images have been ordered with the strongest and most unambiguous matches first. For 12 of the 18 images the most strongly activated label was the "correct" one. For four others the "correct" label came second or third. For the final image SuperVision missed the "correct" label but successfully generated a highly appropriate label—"dalmatian." Images courtesy of Alex Krihzevksy.

Figure 10.2 (plate 20)
Performance of SuperVision on 21 randomly selected photographs of plants. The images have been ordered with respect to the quality of the labels. For 13 images the "correct" label had the highest activation level. For five others the "correct" label had a lower level of activation. Only two images failed to generate the "correct" label. Images courtesy of Alex Krihzevksy.

activation levels. In some cases the labels appear to be even better than the "correct" response—see, for example, the label "Dalmatian" produced in response to the image for which the "correct" response was "cherry" (figure 10.1, plate 19). Furthermore, there can be little doubt that many humans would find some of these distinctions relatively tricky (consider "leopard" vs. "jaguar" vs. "cheetah," for example).

This level of performance was achieved with a system containing the equivalent of roughly 650,000 neurons arranged in five main layers, and around 60 million parameters, which are roughly equivalent to synaptic weight strengths. The fact that such a straightforward architecture can achieve such high levels of performance is very informative. For example, given the debate about the relative importance of feedforward

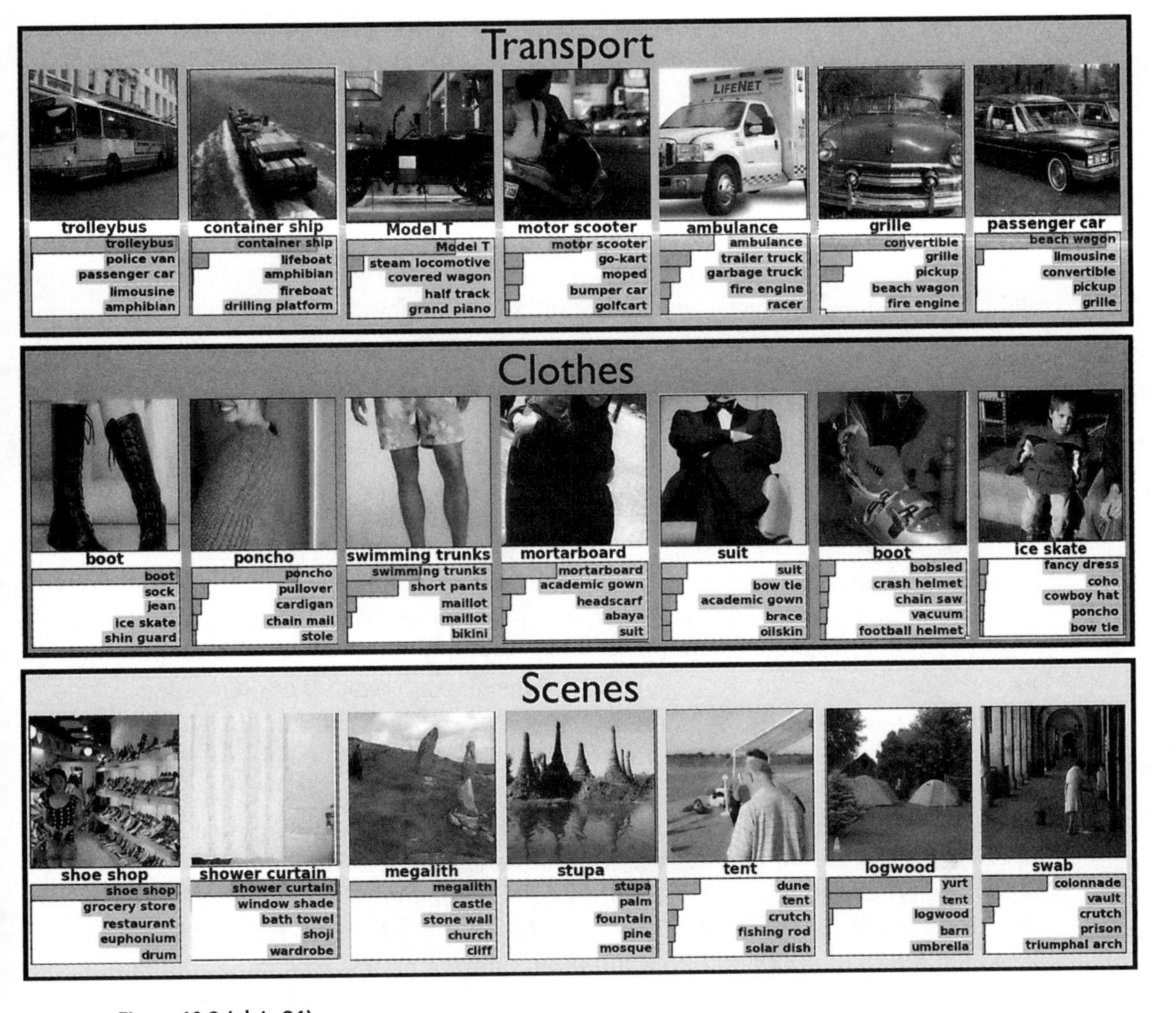

Figure 10.3 (plate 21)
Performance of SuperVision on 21 randomly selected photographs from three general categories—transport, clothes, and scenes. In all three categories the most strongly activated image corresponded to the "correct" label for the majority of the images. Images courtesy of Alex Krihzevksy.

and feedback mechanisms in visual processing, it is striking to note that the processing architecture used in SuperVision is entirely feedforward. Each of the five main processing stages receives information from the previous structure and sends its outputs to the next stage. There are no connections allowing information to be sent from higher-order structures to earlier stages. And there are no horizontal connections within each processing stage that could be used to implement local feedback-like mechanisms. This finding may seem surprising because we know very well that the primate visual system has very extensive feedback connections. Why do these connections exist if complex and challenging tasks such as labeling the objects in complex natural scenes can be achieved using a purely feedforward architecture? One possibility is that the feedback connections may serve other functions. For example, they might be important for scene segmentation, as has been suggested by a number of computational studies (Borenstein & Ullman, 2002; Epshtein, Lifshitz, & Ullman, 2008; Roelfsema, Lamme, Spekreijse, & Bosch, 2002), although even here feedforward mechanisms can be sufficient in some cases (Super, Romeo, & Keil, 2010). Thus, although SuperVision can effectively determine that a given image contains a leopard, determining the borders of the leopard would not be possible simply using the label alone.

Nevertheless, it seems clear that the state of the art really does correspond to a feedforward neural network with an architecture that is essentially very compatible with what we know about the architecture of the human visual system. Indeed, it is particularly interesting to compare the architecture of SuperVision with what is known about the primate visual system. Figure 10.4 (plate 22) compares the two systems, using the numbers for processing units in SuperVision (Krizhevsky et al., 2012) and the numbers of neurons in the different stages of the primate visual system based on a recent review paper (DiCarlo, Zoccolan, & Rust, 2012).

The similarities between the two architectures are striking. Both use a series of hierarchically organized processing stages, and the number of stages is actually very similar. If we ignore the lateral geniculate nucleus, the primate ventral stream has approximately five processing stages on the way to the anterior inferotemporal cortex (AIT), namely V1, V2, V4, posterior inferotemporal cortex (PIT), and central inferotemporal cortex (CIT). Likewise, SuperVision also uses five stages of convolutional networks. The final three stages of SuperVision are somewhat different in that they involve two layers of fully connected nodes that finally give rise to the set of 1000 labels.

Although the number of layers used by each system is similar, the total numbers of neurons used are very different—the "cortical" processing performance by SuperVision involves roughly 650,000 "neurons," whereas the primate ventral stream involves around 478 million neurons, roughly 750 times more. But much of this scaling difference can be simply explained by the fact that the input to SuperVision is a relatively small image that is just 228 pixels across with three different color channels per pixel. Furthermore, those pixels are represented by a first processing stage in which the number of "columns" has been further reduced relative to the original resolution.

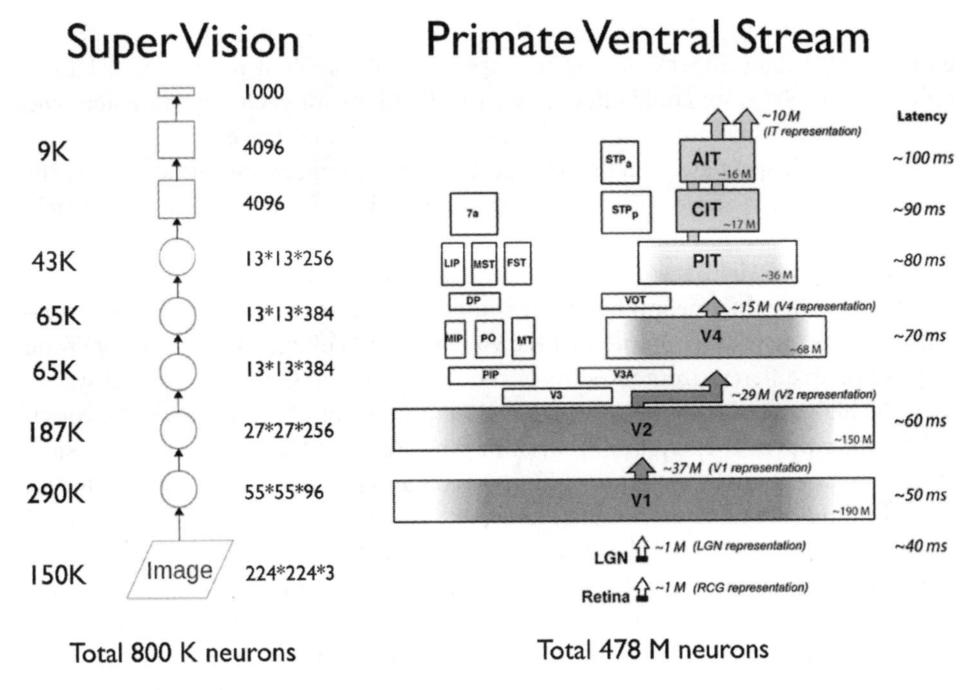

Figure 10.4 (plate 22)
A comparison between the architecture used by SuperVision, a state-of-the-art computer vision system (Krizhevsky et al., 2012), and the ventral stream of the macaque monkey (DiCarlo et al., 2012).

Specifically, there are 55 ×* 55 columns, each with 96 different processing filters. In the primate visual system, the reduction in resolution going from the retina to V1 is much less, meaning that V1 in the primate has a substantially higher resolution than SuperVision's first stage.

We can even give a rough ballpark estimate of the effective size of the retinal input array in the primate. We have roughly 1 million fibers in each optic nerve. In principle this could correspond to an image with 1000 by 1000 pixels if each pixel gave rise to one optic nerve fiber. However, we know that the optic nerve uses a coding scheme with parallel ON- and OFF-center channels for each point in the image (Schiller, 1992), which already cuts down the effective resolution of the image by a factor of two. Furthermore, we also know that the optic nerve also handles color information that presumably produces further reductions in the number of points that can be handled. Add to that the fact that there may be as many as 12 different functionally independent channels in the output of the retina (Werblin & Roska, 2007; Werblin, 2011), and it seems that the effective resolution of the retina may be little more than the 640 × 480 pixels used by a standard VGA webcam. Given these constraints it seems remarkable that we can extract a meaningful impression of the entire visual scene with such a limited number of fibers in each optic nerve. But it does imply that the first stage of cortical processing in the primate visual system may be considerably more detailed than SuperVision's first stage. With 190 million neurons in V1 for 1 million input fibers, we could effectively have 190 filters for each incoming fiber and several hundred filters for each effective point in the retinal image.

Of course, the primate visual system does not use a uniform resolution across the visual field, and a direct comparison with the resolution of a webcam is problematic. There is a great deal of inhomogeneity, with far more resources devoted to processing the foveal region, where there is at least one retinal ganglion cell for every photoreceptor and relatively little neural hardware is used to process higher eccentricities. Nevertheless, it is interesting to note that the 96 filters per "column" used in SuperVision may not be so different from the sorts of processing used in the primate visual system once the higher resolution of processing in the primate has been taken into account.

The other obvious difference between SuperVision and the primate visual system is that the current implementation of SuperVision uses just 1000 output labels, whereas it is likely that humans use a much larger set of labels to describe the content of images. Note, however, that the choice of 1000 output labels was determined by the ImageNet competition. Alex Krizhevsky (personal communication) has successfully extended the system to work with the full 20,000 labels that are available in the original set of training images. Furthermore, other computer vision systems have already been designed that can handle far more output labels. Thus, researchers at Google have recently claimed to be able to run a system capable of generating as many as 100,000 labels on a single machine (Dean et al., 2013). It may be that once the basic processing hierarchy has been established, adding additional labels to the final output stages could be relatively inexpensive because the same underlying processing mechanisms could potentially be reexploited with the new labels.

In this respect it is interesting to consider how many "labels" might be used for scene labeling in the primate visual system. For humans, one way to estimate this would be use the number of nonabstract words in the WordNet lexical database that can be associated with a visual category. This figure is around 75,000 English words (Torralba, Fergus, & Freeman, 2008), but we clearly need to add a substantial number of more personal labels that would be specific for each individual. For example, I can probably recognize and label a few thousand individual people, not including the famous people who would be recognized by many people. I could also recognize a fairly large number of specific geographical locations as well as the exterior and interior of many buildings including the places where I have worked, lived, shopped, eaten, and so forth. Estimating precisely how many such labels we have is clearly not easy, but it is likely that the combined total (when including the 75,000 labels used by Torralba et al.) could be well over 100,000 and potentially several hundred thousand or even a million.

Computer vision may not yet have reached these levels of performance, but the gap is clearly narrowing. If 1000 labels can be successfully handled by a system that effectively has around 650,000 neurons, what would we expect to be able to achieve with the 478 million neurons in the primate ventral stream, a number roughly 750 times larger?

Currently, SuperVision runs on a computer equipped with two powerful graphics processor boards. Training with tens of millions of images takes several days, but this involved using huge numbers of variations for each image in which the image is shifted, zoomed, and rotated to increase the level of invariance. In fact the current implementation of SuperVision is able to process a 24-bit 228 × 228 pixel image and generate the output vector using 1000 labels in about 1 ms per image (Alex Krizhevsky, personal communication). Note that this 1-ms value does not refer to the input–output time. Specifically, it means that you can load up 100 images and process them all in 100 ms. This is because of the way the graphics processors break up the task. It is not possible to process a single image so effectively because the various processing cores are used optimally only when multiple images are processed in a batch.

The latest graphics processor boards used by systems such as SuperVision have phenomenal amounts of computational power. For example, the Nvidia GeForce GTX Titan board achieves 4.5 Teraflops (where a teraflop means 10^{12} floating-point operations per second), has 2668 processing cores, and a memory bandwidth of 288 Gbytes/second. Given that today's fastest supercomputers, such as the Tianhe-2 system, which is capable of nearly 55 Petaflops (5.5×10^{16} floating-point operations per second), are essentially built around tens of thousands of graphic boards linked together, it seems that it may be feasible to imagine building a scaled-up SuperVision-like system that could have performance characteristics that are close to those of the primate ventral stream.

Given this very impressive levels of performance achieved with SuperVision, it is tempting to conclude that the problem of how the human visual system can understand complex natural scenes is close to being solved. After all, we effectively already have artificial systems that can perform the very challenging task of labeling a wide range of objects in complex natural scenes. Furthermore, such systems achieve this using architectures that look remarkably similar to those found in the human visual system. In this respect it is interesting to note that when the SuperVision team designed their processing architecture, they were not using the primate visual system as a blueprint (Alex Krizhevsky, personal communication). The remarkable similarity between the two architectures really does seem to be a beautiful example of convergent evolution. Computer scientists and natural selection seem to have come up with the same basic solution to the same basic problem, namely the need to identify and label objects in natural scenes as fast as possible and with the minimum error rate.

But there are some very major differences between the way in which these sorts of systems work and the way the human visual system operates. The first major difference lies in the fact that in biological systems, the processing units transmit information in the form of all-or-nothing events—spikes—whereas in a system like SuperVision, each of the 650,000 "neurons" transmits a floating-point number that corresponds to its activation level to the "neurons" in the next level. In the next section I discuss how this difference may be functionally very important because there are certain types of computation that can be done much more efficiently using a time-coded spiking architecture than with the continuously variable coding used by convolutional nets.

The second major difference concerns the way in which the system learns. The SuperVision system requires colossal amounts of training: each of the 10 million images is paired with the appropriate label several hundred times, and a modified backpropagation algorithm is used to train representations at the intermediate levels. Few scientists believe that the human brain uses anything directly related to backpropagation. Furthermore, human infants clearly do not have to be given millions of labeled images in order to make sense of the world. For example, a child may learn about dogs and cats over many learning experiences, but it is likely that he or she will need to be told only a few times that "this is a dog" and "this is a cat" for these labels to be extended to a full range of possible viewing angles and breeds of animal. In a later section I discuss how introducing a spike-based learning scheme can radically transform the way the system can learn.

Before moving on to these questions, I would like to stress one important point. The impressive levels of performance achieved by SuperVision do not actually depend specifically on the use of backpropagation and hundreds of millions of training trials with labeled image data. Once the system has been trained, you could deactivate the learning mechanism, and the system will still be able to perform the labeling task with exactly the same level of precision. What are critical are the precise patterns of weights used and the response properties of "neurons" at each level of the system. Quite how those weights were learned is not actually critical. It could be backpropagation. But it could also use some other learning algorithm, perhaps one that has yet to be discovered. Once those weight sets have been learned, you can in principle extract the weight values and use them to program an alternative system that would not have to do the learning. In the most extreme case you could even argue that the weight sets could be learned using natural selection and genetic wiring rules. Indeed, this may even be the case for some biological visual systems, particularly in the case of animals with short life spans, where the opportunities for learning online may be limited. After all, many insects have to be able to correctly label key visual stimuli such as a sexual partner on the very first encounter.

Spike-Based Image Processing

SuperVision uses continuously variable floating-point numbers to represent the activation levels of each of its processing units. Indeed, it is the phenomenal floating-point computational power of the latest generation graphics processors that have made the training feasible in a reasonable amount of time. In fact, the basic architecture used by SuperVision is fairly similar to the sorts of processing architectures that were being proposed in the 1970s and 1980s—it is the availability of systems that achieve 4.5 Teraflops that has made it possible to do the massive amounts of training needed to get such systems to work on challenging real-world problems. In contrast, real neurons do not send floating-point numbers—they send spikes. Is this significant? In this section, I discuss why using spikes could radically transform the way in which computations are performed by the visual system.

For some neurophysiologists, even though neurons send spikes, this fact can be ignored because of the belief that the underlying coding scheme uses firing rates. According to this view all the interesting information can be obtained by measuring the firing rate of each neuron. These firing rates typically vary between 0 and around 200 spikes per second, although for very brief periods, neurons have been described with instantaneous firing rates as high as 700 Hz (Nowak, Azouz, Sanchez-Vives, Gray, & McCormick, 2003). This underlying belief has led to the widespread use of post stimulus time histograms (PSTH) to analyze neuronal response, a technique that averages neuronal responses over large numbers of trials in order to get an estimate of instantaneous firing rate. Another idea commonly seen in the modeling community assumes that once the neuron has integrated its synaptic inputs to generate an activation level, it then proceeds to convert this activation value into a firing pattern using a Poisson process in which the decision about whether to fire is made using a random procedure. Although it is true that the firing of an individual neuron can look roughly Poisson in nature, it seems likely that this apparent variability has more to do with the inability of neurophysiologists to control all the different inputs to the neuron rather than intrinsic noisiness in the spike-initiation process at the level of individual neurons. Indeed, there is evidence that the reliability of the initial part of the response of cortical neurons is actually very high—much higher than would be predicted by a Poisson process (Amarasingham, Chen, Geman, Harrison, & Sheinberg, 2006; Maimon & Assad, 2009). Furthermore, this reliability is much higher for natural movies than for the sorts of bar and grating stimuli conventionally used by neurophysiologists to study visual neurons (Herikstad, Baker, Lachaux, Gray, & Yen, 2011).

Other arguments against the idea that neural firing can be modeled as a time-varying Poisson process come from theoretical studies. Jacques Gautrais used a mathematical analysis to investigate the ability of a Poisson process to transmit information over short periods (Gautrais & Thorpe, 1998). The results were problematic for proponents of Poisson rate coding because the number of neurons required to send even the simplest information in a short period of time was excessive. For example, to be able to determine that the underlying firing rate of a population of identical, redundant neurons was 100 ± 10 Hz in 10 ms with a 90% chance of being correct would require no less than 281 neurons in parallel. Similarly, if you wanted to know which one of two populations of neurons was the more active, and assuming that one population was firing at 100 Hz and the other at 75 Hz, you would need to have 76 neurons in each population to be able to make a choice with a 90% chance of being correct in 10 ms. Such values are clearly incompatible with the ability of the retina to send enough information to the brain in a few tens of milliseconds to allow the scene to be processed with 1 million different channels.

Fortunately, the brain does not have to use Poisson rate coding to send information efficiently. An alternative strategy makes use of the well-known fact that the time taken for a sensory neuron to reach its threshold for firing depends on the strength of the input. As a consequence retinal ganglion cells will fire at a shorter latency when they are stimulated at higher intensities, an effect that was already visible in Lord Adrian's original studies in the 1920s and 1930s but has been largely ignored in most neurophysiological studies until remarkably recently (Gollisch, 2009; Gollisch & Meister, 2008). But, as pointed out as early as 1990, using the relative timing of spikes provides an extremely simple way to perform one of the most vital operations in neural computing, much more simply than would be possible with conventional rate coding (Thorpe, 1990). That operation is the winner-take-all function that compares the activation levels of a set of neurons and recovers the one with the highest value. A related function, namely the MAX operator, is a key element of one of the most popular models of visual processing, the HMAX model (Riesenhuber & Poggio, 1999). In a conventional rate-coded system it is in fact quite difficult to find the most active neuron because typically this would require using inhibitory lateral connections and letting the neurons compete with each other. Because the neuron that initially has the highest firing rate will inhibit its neighbors more effectively, if you leave the system long enough only that neuron will be left firing at the end. However, there is no way of knowing in advance how long this selection process will take to complete because it depends on the relative strengths of activation in the different neurons. In contrast, in a system where the strength of the input is encoded in the latency of firing, finding the most strongly activated neuron can be done by simply picking off the first neuron that fires. To take a particularly extreme example, imagine that you open your eyes and find that you are looking at the thousands of stars in the night sky. To find where the brightest star is would simply involve finding the retinal ganglion cell with the shortest latency response.

Figure 10.5 shows how a relatively simple circuit mechanism based on spike timing can be used to perform some surprisingly sophisticated computations. The left panel (A) shows the basic arrangement in which an output neuron receives connections from 40 different input neurons. However, only 4 of the 40 inputs have an effective synapse (as indicated by the small triangles); the others have weights that are effectively zero. Suppose that we now apply an activation pattern to the input neurons (illustrated by the line on the left) that progressively pushes the input neurons over their threshold.

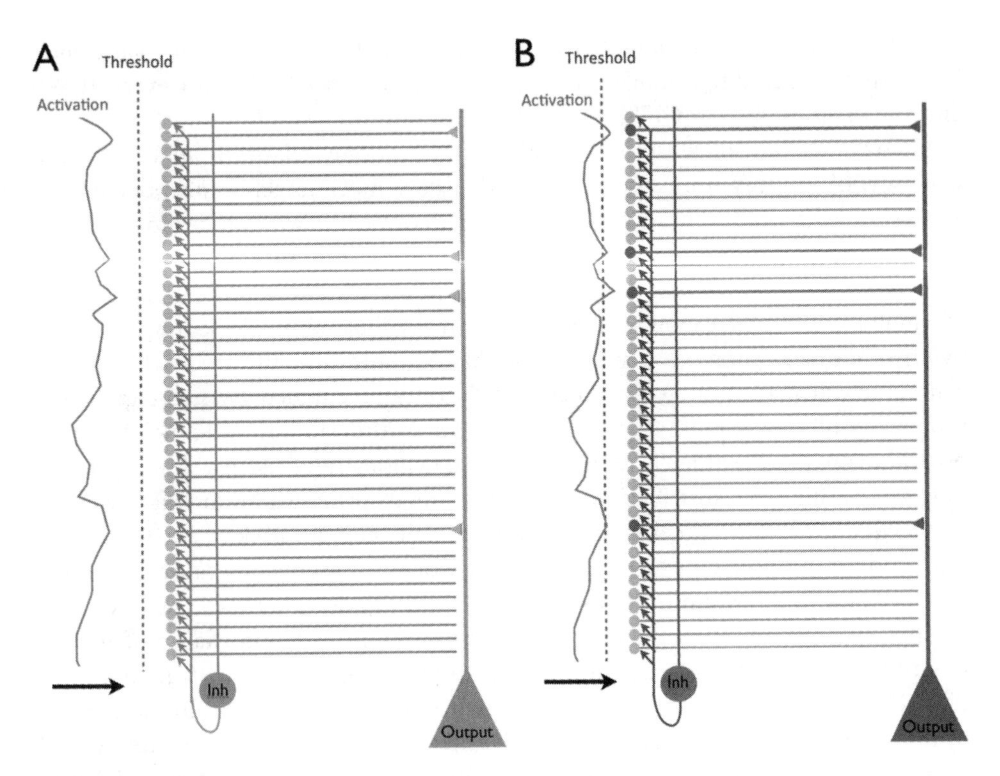

Figure 10.5
A simple neural circuit that allows an output neuron to respond to a specific pattern of activation (see text for details).

The neurons will fire in an order that depends on the local intensity, with the most strongly activated neurons firing first. In addition there is an inhibitory neuron (Inh) that receives strong monosynaptic inputs from each of the input units and provides strong feedback inhibition to all the inputs. If the inhibitory neuron has a threshold of 4, this would mean only the four inputs that fire first will get through because all the other input neurons will be blocked by the feedback inhibition. The right panel of the figure (B) shows the state of the circuit once the four input neurons with the highest activations have fired and where the inhibitory feedback circuit has kicked in to prevent any additional input neurons firing. If we suppose that the output neuron has a threshold of just under 4, it follows that the particular activation pattern will be able to push the neuron over threshold. The remarkable point is that such a circuit will be very selective because it would fire an output spike only if all four of the neuron's "favorite" inputs fired together. But this would occur only if the four highest values in the activation profile happened to be the four with the strong weights. This is effectively the equivalent of putting your hand into a hat containing 36 white balls and 4 black ones, drawing out 4 balls, and discovering that you have only black ones. This would be exceedingly unlikely to occur by chance—specifically once every 91,390 times [= (4/40) × (3/39) × (2/38) × (1/37)].

Although the circuit illustrated in this example is only relatively small scale, the same sort of strategy works with neurons that have much larger numbers of inputs. Suppose that, instead of having 4 strong synapses out of 40, we had the more realistic situation in which a neuron might have 50 strong synapses out of 1000 (i.e., 5% of its inputs). Again, the trick is to use the feedback inhibition circuit to keep the number of simultaneously active inputs under strict control, effectively implementing a *k*-WTA (winner-take-all) operation. If we suppose that no more than 100 of the 1000 inputs can be activated (i.e., 10%), and there are 50 fully potentiated input synapses, the probability of hitting more than a given number of the neuron's "favorite" synapses by chance drops off extremely rapidly, as can be demonstrated using a simple calculation using a binomial distribution. It corresponds to the situation in which you are drawing balls from a hat containing 950 white balls and 50 black ones. If you pick out 100 balls, you would expect to get 5 black ones on average (because 5% of the balls are black). But the probabilities drop off very rapidly. There is a roughly a 1 in 10 chance of getting 10 or more hits, less than a one in a million chance of getting more than 17 hits, and less than a one in a hundred billion chance of getting more than 23 hits. Thus, if the threshold of the neuron were set at 23 (of 50) the chances of it responding to a random input would be effectively zero. The input pattern would have to have an activity profile remarkably similar to the pattern of high weights to have any chance of getting the neuron over threshold.

The use of spike-based coding schemes thus allows neurons to achieve levels of selectivity that would be impossible to obtain in a more conventional neural network architecture in which each neuron has a continuously variable activation level, effectively corresponding to the sorts of floating-point numbers used in systems such as SuperVision. This selectivity may seem remarkable, but it is a natural consequence of the fact that controlling the number of active inputs is simple to do with spiking neurons yet next to impossible to achieve with continuously graded inputs. To see why, consider how difficult it would be to take 1000 neurons coded by floating-point numbers between 0 and 1 and find the 5% with the largest values. Algorithmically, this would normally require scanning though all the values sequentially and maintaining a sorted list of the 50 largest values. In a spiking network using temporal coding, the same operation can be achieved by simply blocking further activation once the first 50 neurons have fired.

Spike-Time-Dependent Plasticity

The simple circuit in figure 10.5 showed how a neuron that has a relatively small number of fully potentiated synapses can be remarkably selective if the proportion of active inputs is kept strictly under control. But how could a neuron end up with just a small number of fully potentiated synapses? In this section I argue that a simple learning mechanism based on STDP will produce precisely this effect.

STDP-based learning was first demonstrated in the late 1990s and has since been the subject of intense experimental and theoretical studies (Bi & Poo, 2001; Feldman, 2012; Markram, Gerstner, & Sjostrom, 2011). But one very interesting feature of this form of learning has remained relatively neglected: this is the fact that STDP will naturally concentrate high synaptic weights on inputs that fire early (Guyonneau, Vanrullen, & Thorpe, 2005). Specifically, if the same pattern of spikes is presented repeatedly, we found that neurons will naturally become selective to that repeating pattern. The reason this occurs can be understood by looking at figure 10.6, which illustrates how a neuron with 12 inputs can rapidly learn to respond to the first 3 inputs of a repeating pattern.

The example illustrated in the figure is clearly very simple, but it is relatively easy to see that the same principle can be extended to much larger problems. Indeed, we

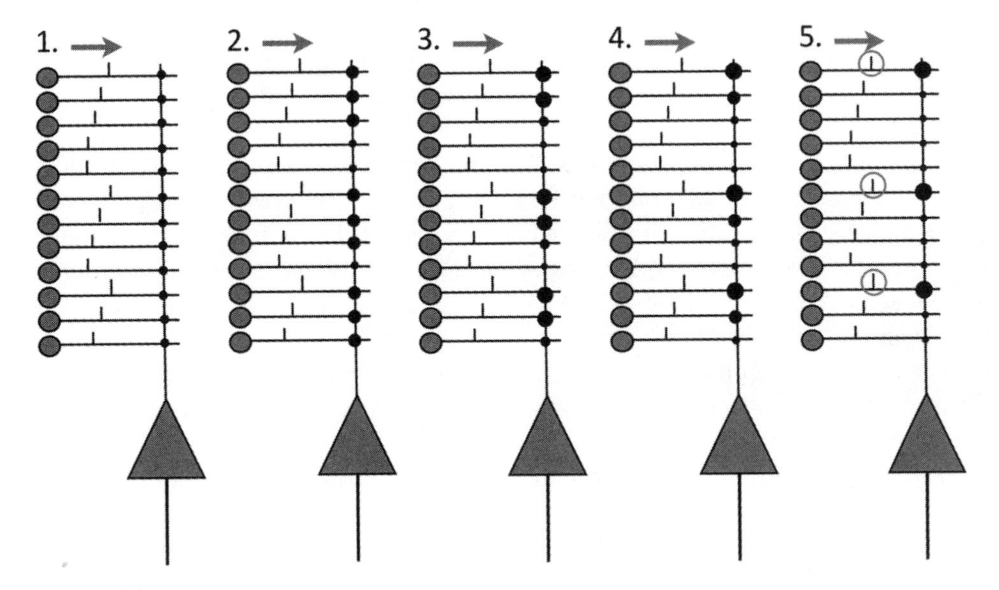

Figure 10.6
A simple illustration of why a simple STDP learning rule will concentrate high weights on the earliest-firing inputs of a repeated pattern. Suppose that a neuron initially has 12 synaptic inputs, each with a weight of 0.25, and that the neuron's threshold is 3.0. When a specific pattern of spikes involving all 12 inputs is repeatedly presented, the neuron will end up with fully potentiated synapses for the 3 inputs that fire first. This happens progressively. On the first presentation (panel 1), the neuron fires only when all the inputs have fired. The STDP rule will potentiate all of them because they all fired before the output neuron. On presentation 2, only the first nine get potentiated, and the last three start dropping out. After presentation 3, only the first six inputs are potentiated, whereas with presentation 4, it is just the first three that are reinforced. After a few more presentations, the first three inputs in the pattern are fully potentiated, and all the others have dropped back to minimal levels.

have used computer simulations to demonstrate that the effect scales up to arbitrarily large numbers of inputs. In one study, we studied the case of a neuron receiving random patterns of activity from 2000 afferents but where a 50-ms segment of activity in a randomly chosen subset of those afferents is copied and then pasted at random intervals throughout the pattern. Remarkably, the receiving neuron can become selective to this repeating pattern after only a few tens of presentations, and within a few minutes the neuron becomes selective to the very beginning of the repeating pattern (Masquelier, Guyonneau, & Thorpe, 2008).

In those original studies we deliberately chose to use totally meaningless patterns of activity. But in other studies we found that a simple hierarchical neural network can learn to create receptive fields to face-like stimuli simply by presenting images from the Caltech face database (Masquelier & Thorpe, 2007). Importantly, this sort of learning occurs without any supervision: at no point is the system told that it has to learn about faces. It learns about faces simply because faces are the most likely stimuli in the inputs. If the same network is trained with photographs of motorcycles, then neurons in the output layer become selective to parts of motorcycles.

In other experiments we simulated a layer of 60 spiking neurons that received inputs from a dynamic vision sensor chip (Bichler, Querlioz, Thorpe, Bourgoin, & Gamrat, 2012; Roclin, Bichler, Gamrat, Thorpe, & Klein, 2013). The dynamic vision sensor developed by Tobi Delbruck's group in Zurich (Lichtsteiner, Posch, & Delbruck, 2008) has an array of 128 × 128 pixels. Each pixel can generate two types of spiking events: one event signals that the local luminance has increased by a fixed percentage (for example, 10%), whereas the other type of event signals that there has been a decrease in luminance. The generation of spikes is completely asynchronous and follows the events in the outside world with microsecond precision. For our simulations we downloaded a set of data obtained with the camera viewing traffic on a six-lane freeway in Pasadena. The data set contains several million spikes generated over a period of roughly 90 seconds. We found that following roughly 3 minutes of presentation (two runs through the complete data set), the 60 neurons in the first layer had developed receptive fields that corresponded to cars at particular positions on the freeway. But we had a second layer of 10 neurons that received connections from the neurons in the first layer. Remarkably, these neurons were found to have learned that there were particular sequences of firing in the first layer and that these sequences correspond to cars going by on each of the six lanes of the freeway. Although simple, the network had effectively learned about the significant events that were occurring in the outside world. Again, it is important to realize that this learning was done in a completely unsupervised way: the system was given no information whatsoever about the environment in which it was placed. It effectively worked out on its own that there were particular events in the environment that were significant simply because they recurred over and over again. This ability could be reasonably thought to correspond to a simple form of intelligence.

The simulations that I have been talking about so far concern artificial neural network systems. So a natural question to ask would be whether or not similar mechanisms exist in the human brain. In a study with my colleagues Trevor Agus and Daniel Pressnitzer in Paris we were able to show that human subjects are capable of forming robust memory traces for meaningless auditory noise stimuli (Agus, Thorpe, & Pressnitzer, 2010). The participants listened to 1-second-long fragments of Gaussian noise and were required to determine whether the first and second halves of the stimulus were identical. It is a difficult task to perform, but we discovered that if the same auditory fragments were repeated during the experiment, the subjects became much more accurate later in the experiment for stimuli that were heard repeatedly. This improvement of performance developed very rapidly, within a few tens of presentations, and was all or none; the subject either was able to learn the stimulus or not. Furthermore, we found that if subjects went away for 2 or 3 weeks and then came back into the lab, their performance on the very first trial with the original stimulus was virtually as good as it had been during the original testing session. It appeared that the subject had developed robust memories for the auditory noise patterns that could last for at least 2 or 3 weeks. One of the striking aspects of this result is that the subjects were able to remember the sensory stimulus even though it was presumably impossible for them to rehearse the pattern in the intervening period.

Normally, with the sorts of material generally used in memory research (such as lists of words), it is usually possible for people to recall the material during the intervening period. But in the particular case of Gaussian auditory noise, this option seems extremely unlikely. It would appear that the memory traces must have been formed during the original experiment. Consolidation during, for example, dream sleep may be involved, but there is effectively no way that the particular pattern of sensory input could be reproduced outside the original learning situation.

Where might this sort of learning be occurring? In a further simulation study, my former student Olivier Bichler looked to see whether a neuron directly connected to the auditory nerve could learn this sort of auditory noise pattern (O. Bichler, unpublished data). Specifically, we modeled the auditory periphery as a set of simple frequency-tuned filters directly connected to a set of leaky integrate-and-fire neurons that correspond approximately to the axons in the auditory nerve. In the simulation there were just 64 channels of frequency-tuned axons connected to a single output neuron. The neuron was initially stimulated with continuous Gaussian noise and was found to fire at approximately 2 spikes per second. Because of the specific learning rule that we used that strengthened synapses that were active just before the neuron fired while depressing all the other ones, the total amount of synaptic input to the neuron remained roughly constant. And, because of this, if the Gaussian noise pattern is continuously renewed, the neuron would continue to fire at roughly 2 spikes a second indefinitely. We then introduced a particular segment of auditory noise that was repeated over and over again, very much like the way in which repeating noise patterns were used in the human psychophysics experiments. Within a few seconds the neuron had started to respond selectively to the repeating noise segment, and within a minute or so the neuron was firing only when the repeating noise pattern was present. Finally, when we switched back to the original continuous noise stimulus, we found that the neuron no longer fired at all, meaning that the cell had become highly selective to the stimuli used for training.

Intriguingly, because the neuron was now extremely unlikely to be activated by the continuous Gaussian noise, it could effectively be considered to be a sort of *grandmother cell* because only stimuli that were very similar to the originally presented stimulus would be capable of driving the cell to threshold. Remember that this pattern of responding was seen for a neuron that was connected directly to just 64 fibers in a simulated auditory nerve. Imagine what would happen with a more complicated input corresponding to the 30,000 fibers in the auditory nerve. And imagine that there was not just one cell listening to the fibers in the auditory nerve but many hundreds or even thousands of cells. If those cells were connected to each other via inhibitory connections, the system would form a sort of competitive learning mechanism in which each cell will effectively learn to respond to any stimulus that has not already been learned by one of the other neurons. Thus, if the auditory system is stimulated with sounds corresponding to English, the neurons would presumably learn to respond to the basic sounds of the English language. In contrast, if the system is stimulated with sounds corresponding to French, the response properties of each neuron would correspond to the acoustic patterns that are most frequently heard in French.

Now consider what would happen if there was not just one layer of neurons but a whole hierarchy of layers. Imagine also that the time constants of neurons in different layers vary with the neurons in the most peripheral layers with those closest to auditory input having very short time constants, whereas those in more central structures might have longer time constants. Under these conditions the neurons in the peripheral structures would learn about very short acoustic features, whereas more centrally located cells would become increasingly selective to longer acoustic events such as phonemes, words, and even word sequences or phrases.

Although this digression into the field of auditory noise learning may seem remote from the general issue of visual scene processing, it is important to stress that the same sorts of principles could easily be applied in the case of visual input.

Relevance to Visual Scene Processing

In the last two sections we have seen how some simple mechanisms based on spike-based processing can provide some remarkably powerful computational features.

First, we saw how a simple mechanism for controlling the percentage of neurons that can fire in an array of input neurons can allow the receiving neuron to show remarkable selectivity. This selectivity would only require the output neuron to select a subset of inputs from the wide range of potential inputs and assign high weights to a relatively small number of them. The precise numbers are not critical, but values of 5% or 10% work well for the numbers of inputs with potentiated weights, and values of 5% to 20% for the percentage of inputs that are allowed to fire are also suitable.

We then saw that a simple STDP learning rule can provide a way for neurons to become selective to patterns of input activation that repeat by concentrating high synaptic weights on the specific set of inputs that fires first during the repeating pattern. In this final section we consider how these mechanisms could be involved in the way in which our visual systems handle complex scenes.

Earlier in the chapter we saw how a state-of-the-art computer vision system could generate thousands of meaningful labels in response to complex natural scenes, and we noted that the basic architecture of the system, namely a pure feedforward network composed of roughly five layers of neurons, is surprisingly similar to the basic design of the primate visual system. We argued that although the performance of such systems implies that a similarly designed system built from real biological neurons should be able to achieve similar feats, the precise learning algorithm and the training procedures used by SuperVision seem unlikely to reflect what happens in our own visual systems. Wiring up the connections in SuperVision requires literally billions of training examples in which a training image is shown together with its labels. Such a learning mechanism seems implausible for humans.

However, there are reasons to believe that the spike-based selectivity and learning mechanisms presented in subsequent sections could provide a biologically plausible way to set up the wiring. According to this view the visual system would naturally develop neurons with selectivity for those input patterns that occur repeatedly, with a natural tendency to allocate more neurons to those stimuli that occur most often. We saw both modeling and experimental studies that imply that this development of selectivity can occur with just a few tens of repetitions.

But there is another difference between SuperVision and the sorts of spike-based learning mechanisms proposed here. SuperVision is a convolutional network, which means that there are "neurons" that perform the same operation for every part of the image. There is thus no specialization for particular parts of the visual field, at least in the first five layers of processing. If there is a "neuron" that is selective for a particular type of visual feature in the top left of the image, there has to be the very same type of unit for every other part of the image—that is indeed the definition of a convolutional network.

In contrast, the spike-based learning scheme would not allow a particular "convolution" to be applied to every point in the image, and indeed, there is no way that a set of weights learned by one location could be copied to all other locations—there can be no "weight sharing" of the type seen in convolutional networks. Neurons with particular patterns of selectivity would occur only if they had actually received the same specific pattern of inputs repeatedly. This fact is clear from the example of the system that was trained using a spiking retina pointed at a six-lane freeway in Pasadena. Because the viewing angle of the camera was fixed, the receptive fields of the neurons and their selectivities matched perfectly the specific scene characteristics that were used during the training period. Indeed, if there is any change to the angle of view or to the zoom factor of the camera, none of the neurons in the system would be able to respond. Each neuron would only be interested in a very specific "car"-like stimulus located in a very specific location—a sort of "grandmother cell" for that specific input. Indeed, it is amusing to imagine trying to work out what the system was interested in if the original training conditions were unknown. Think of a neurophysiologist recording the activity of one of the car-selective neurons and trying to work out the best way to activate the neuron by randomly presenting a wide range of different visual stimuli. Unless the "neurophysiologist" was lucky enough to try a small car-shaped stimulus, moving in the right direction in just the right point of the image, he or she would probably conclude that the neurons were not visually responsive at all.

This simple illustration shows that under normal conditions, neurons in the visual system will learn about the things that reoccur in their particular part of the visual field. If you train the system in a typical natural environment, you might find that the neurons close to the top of the visual field are more likely to respond to blue, because of the higher probability of finding a blue sky in the upper visual field. Likewise dense textures may be more likely to occur toward the lower part of the visual field. Of course, these differences will only be statistical because you may still encounter sky blue objects in the lower part of the visual field and dense textures near the top. However, there is plenty of experimental evidence showing that humans are sensitive to where particular objects occur within the visual field (Chun & Jiang, 1998).

Furthermore, there are both psychophysical data and computational models that show how the configuration of a scene can help identify which particular objects are present (Oliva & Torralba, 2007; Torralba, Oliva, Castelhano, & Henderson, 2006). These sorts of influences are to be expected in a system in which the selectivity of the neural mechanisms is strongly dependent on image statistics and the probability of finding specific combinations of features in particular locations within an image.

Object and Scene Processing

A final comment concerns the question of whether there is really a clear distinction between the processing of objects and scenes. In the literature, object vision and scene vision are often presented as if they are fundamentally different. However, it could be that the underlying mechanisms are essentially identical but simply apply to regions of the visual field that have different sizes. For example, in the real world, we may see a mountain scene that fills up the entire visual field, spanning 180° of horizontal extent. However, we could also take a panoramic image of the same scene and present it on the screen of a smartphone, in which case it might only occupy a horizontal extent of 20° or less. Does the scene become an object because it is being presented to a smaller part of the visual field? Likewise, we can easily recognize an object such as a face presented within an image that takes up 20° of central vision, but we could also blow up the scale so that the face filled virtually the entire visual field. Does it become a scene in that case?

One way to think about the problem would be to consider that the difference between "scenes" and "objects" is somewhat arbitrary and that it may be more useful to talk about those visual patterns that are most likely to occur at different scales. Faces tend to be seen with sizes up to a few tens of degrees at most, simply because it is relatively rare to be very close to a face. Likewise, we normally see entire scenes that include visual features right out to the edge of the visual field, and this sort of information will presumably also contribute to processing of the objects within the visual field.

At the implementation level there may not be any fundamental differences between the neural mechanisms that we use to learn about objects and scenes. Indeed, some of the earliest systems for categorizing scenes simply took the outputs of a 4×4 array of oriented filter banks that covered the entire image (Oliva & Torralba, 2001). This remarkably simple architecture was already capable of making quite subtle distinctions between natural and man-made scenes as well as a number of other judgments including whether or not the scene was navigable, whether it was a long way away, and so forth. In a sense the latest object recognition and labeling systems such as SuperVision are essentially applying the same sort of strategy but with far more detail and with a wider range of intermediate representations.

We are probably approaching a time where it may be possible to implement a complete model of the human visual system that includes the multiple resolutions and organizational features of biological vision. Such a system would have the same sort of hierarchical architecture used by systems such as SuperVision but would develop selectivities at the neuronal level that depended directly on the statistics of the input without simply copying the patterns of weights across the image. Furthermore, it would develop those selectivities without the need for labeling using a mechanism that simply generates selectivities to the most frequently occurring input.

References

Agus, T. R., Thorpe, S. J., & Pressnitzer, D. (2010). Rapid formation of robust auditory memories: Insights from noise. *Neuron*, *66*(4), 610–618.

Amarasingham, A., Chen, T. L., Geman, S., Harrison, M. T., & Sheinberg, D. L. (2006). Spike count reliability and the Poisson hypothesis. *Journal of Neuroscience*, *26*(3), 801–809.

Bi, G., & Poo, M. (2001). Synaptic modification by correlated activity: Hebb's postulate revisited. *Annual Review of Neuroscience*, *24*, 139–166.

Bichler, O., Querlioz, D., Thorpe, S. J., Bourgoin, J. P., & Gamrat, C. (2012). Extraction of temporally correlated features from dynamic vision sensors with spike-timing-dependent plasticity. *Neural Networks*, *32*, 339–348.

Borenstein, E., & Ullman, S. (2002). Class-specific, top-down segmentation. *Computer Vision ECCV*, *2351*, 109–122.

Bullier, J., Hupe, J. M., James, A. C., & Girard, P. (2001). The role of feedback connections in shaping the responses of visual cortical neurons. *Progress in Brain Research*, *134*, 193–204.

Callaway, E. M. (2004). Feedforward, feedback and inhibitory connections in primate visual cortex. *Neural Networks*, *17*(5–6), 625–632.

Carlson, T., Tovar, D. A., Alink, A., & Kriegeskorte, N. (2013). Representational dynamics of object vision: The first 1000 ms. *Journal of Vision*, *13*(10), 1–19.

Chun, M. M., & Jiang, Y. (1998). Contextual cueing: Implicit learning and memory of visual context guides spatial attention. *Cognitive Psychology*, *36*, 28–71.

Crouzet, S. M., Kirchner, H., & Thorpe, S. J. (2010). Fast saccades towards faces: Face detection in just 100 ms. *Journal of Vision, 10*(4), 1–17.

Dean, T., Ruzon, M. A., Segal, M., Shlens, J., Vijayanarasimhan, S., & Yagnik, J. (2013). Fast, accurate detection of 100,000 object classes on a single machine. *IEEE Conference on Computer Vision and Pattern Recognition*, 1814–1821.

DiCarlo, J. J., Zoccolan, D., & Rust, N. C. (2012). How does the brain solve visual object recognition? *Neuron*, *73*(3), 415–434.

Epshtein, B., Lifshitz, I., & Ullman, S. (2008). Image interpretation by a single bottom-up top-down cycle. *Proceedings of the National Academy of Sciences of the United States of America*, *105*(38), 14928–14303.

Feldman, D. E. (2012). The spike-timing dependence of plasticity. *Neuron*, *75*(4), 556–571.

Fukushima, K. (1975). Cognitron: A self-organizing multilayered neural network. *Biological Cybernetics*, *20*(3–4), 121–136.

Gautrais, J., & Thorpe, S. (1998). Rate coding versus temporal order coding: A theoretical approach. *Bio Systems*, *48*(1–3), 57–65.

Gollisch, T. (2009). Throwing a glance at the neural code: Rapid information transmission in the visual system. *HFSP Journal*, *3*(1), 36–46.

Gollisch, T., & Meister, M. (2008). Rapid neural coding in the retina with relative spike latencies. *Science*, *319*(5866), 1108–1111.

Guyonneau, R., Vanrullen, R., & Thorpe, S. J. (2005). Neurons tune to the earliest spikes through STDP. *Neural Computation*, *17*(4), 859–879.

Herikstad, R., Baker, J., Lachaux, J. P., Gray, C. M., & Yen, S. C. (2011). Natural movies evoke spike trains with low spike time variability in cat primary visual cortex. *Journal of Neuroscience*, *31*(44), 15844–15860.

Hung, C. P., Kreiman, G., Poggio, T., & DiCarlo, J. J. (2005). Fast readout of object identity from macaque inferior temporal cortex. *Science*, *310*(5749), 863–866.

Keysers, C., Xiao, D. K., Földiák, P., & Perrett, D. I. (2001). The speed of sight. *Journal of Cognitive Neuroscience*, *13*, 90–101.

Kirchner, H., & Thorpe, S. J. (2006). Ultra-rapid object detection with saccadic eye movements: Visual processing speed revisited. *Vision Research*, *46*(11), 1762–1776.

Krizhevsky, A., Sutskever, I., & Hinton, G. E. (2012). ImageNet classification with deep convolutional neural networks. *Advances in Neural Information Processing Systems*, *25*, 1106–1114.

Lecun, Y., Kavukcuoglu, K., & Farabet, C. (2010). Convolutional networks and applications in vision. *IEEE International Symposium on Circuits and Systems*, 253–256.

Lichtsteiner, P., Posch, C., & Delbruck, T. (2008). Temporal contrast vision sensor. *IEEE Journal of Solid-State Circuits*, *43*(2), 566–576.

Liu, H., Agam, Y., Madsen, J. R., & Kreiman, G. (2009). Timing, timing, timing: Fast decoding of object information from intracranial field potentials in human visual cortex. *Neuron*, *62*(2), 281–290.

Liu, J., Harris, A., & Kanwisher, N. (2002). Stages of processing in face perception: An MEG study. *Nature Neuroscience*, *5*(9), 910–916.

Maimon, G., & Assad, J. A. (2009). Beyond Poisson: Increased spike-time regularity across primate parietal cortex. *Neuron*, *62*(3), 426–440.

Markram, H., Gerstner, W., & Sjostrom, P. J. (2011). A history of spike-timing-dependent plasticity. *Frontiers in Synaptic Neuroscience*, *3*, 1–24.

Marr, D. (1982). *Vision: A computational investigation into the human representation and processing of visual information*. New York: Henry Holt and Co.

Masquelier, T., Guyonneau, R., & Thorpe, S. J. (2008). Spike timing dependent plasticity finds the start of repeating patterns in continuous spike trains. *PLoS ONE*, *3*(1), e1377.

Masquelier, T., & Thorpe, S. J. (2007). Unsupervised learning of visual features through spike timing dependent plasticity. *PLoS Computational Biology*, *3*(2), e31.

Mel, B. W. (1997). SEEMORE: Combining color, shape, and texture histogramming in a neurally inspired approach to visual object recognition. *Neural Computation*, *9*(4), 777–804.

Mutch, J., & Lowe, D. G. (2008). Object class recognition and localization using sparse features with limited receptive fields. *Journal of Computer Vision*, *80*(1), 45–57.

Nowak, L. G., Azouz, R., Sanchez-Vives, M. V., Gray, C. M., & McCormick, D. A. (2003). Electrophysiological classes of cat primary visual cortical neurons in vivo as revealed by quantitative analyses. *Journal of Neurophysiology*, *89*(3), 1541–1566.

Oliva, A., & Torralba, A. (2001). Modeling the shape of the scene: A holistic representation of the spatial envelope. *International Journal of Computer Vision*, *42*(3), 145–175.

Oliva, A., & Torralba, A. (2007). The role of context in object recognition. *Trends in Cognitive Sciences*, *11*(12), 520–527.

Oram, M. W., & Perrett, D. I. (1992). Time course of neural responses discriminating different views of the face and head. *Journal of Neurophysiology*, *68*(1), 70–84.

Panzeri, S., Rolls, E. T., Battaglia, F., & Lavis, R. (2001). Speed of feedforward and recurrent processing in multilayer networks of integrate-and-fire neurons. *Network*, *12*(4), 423–440.

Poggio, T., & Ullman, S. (2013). Vision: Are models of object recognition catching up with the brain? *Annals of the New York Academy of Sciences*, *1305*, 72–82.

Riesenhuber, M., & Poggio, T. (1999). Hierarchical models of object recognition in cortex. *Nature Neuroscience*, *2*(11), 1019–1025.

Roclin, D., Bichler, O., Gamrat, C., Thorpe, S. J., & Klein, J. O. (2013). Design study of efficient digital order based STDP neuron implementations for extracting temporal features. *International Joint Conference on Neural Networks (IJCNN)*, 1–7.

Roelfsema, P. R., Lamme, V. A., Spekreijse, H., & Bosch, H. (2002). Figure-ground segregation in a recurrent network architecture. *Journal of Cognitive Neuroscience*, *14*(4), 525–537.

Schiller, P. H. (1992). The ON and OFF channels of the visual system. *Trends in Neurosciences*, *15*(3), 86–92.

Serre, T., Kreiman, G., Kouh, M., Cadieu, C., Knoblich, U., & Poggio, T. (2007). A quantitative theory of immediate visual recognition. *Progress in Brain Research*, *165*, 33–56.

Serre, T., Oliva, A., & Poggio, T. (2007). A feedforward architecture accounts for rapid categorization. *Proceedings of the National Academy of Sciences of the United States of America*, *104*(15), 6424–6429.

Super, H., Romeo, A., & Keil, M. (2010). Feed-forward segmentation of figure-ground and assignment of border-ownership. *PLoS ONE*, *5*(5), e10705.

Thorpe, S., Fize, D., & Marlot, C. (1996). Speed of processing in the human visual system. *Nature, 381*(6582), 520–522.

Thorpe, S., & Imbert, M. (1989). Biological constraints on connectionist modelling. In R. Pfeifer, Z. Schreter, F. Fogelman-Soulié, & L. Steels (Eds.), *Connectionism in perspective* (pp. 63–92). Amsterdam: Elsevier.

Thorpe, S. J. (1990). Spike arrival times: A highly efficient coding scheme for neural networks. In R. Eckmiller, G. Hartmann, & G. Hauske (Eds.), *Parallel processing in neural systems and computers* (pp. 91–94). Amsterdam: North-Holland/Elsevier.

Torralba, A., Fergus, R., & Freeman, W. T. (2008). 80 million tiny images: A large data set for nonparametric object and scene recognition. *IEEE Transactions on Pattern Analysis and Machine Intelligence, 30*(11), 1958–1970.

Torralba, A., Oliva, A., Castelhano, M., & Henderson, J. M. (2006). Contextual guidance of eye movements in real-world scenes: The role of global features on object search. *Psychological Review, 113*(4), 766–786.

Werblin, F. S. (2011). The retinal hypercircuit: A repeating synaptic interactive motif underlying visual function. *Journal of Physiology, 589*(Pt 15), 3691–3702.

Werblin, F., & Roska, B. (2007). The movies in our eyes. *Scientific American, 296*(4), 72–79.

11

A Statistical Modeling Framework for Investigating Visual Scene Processing in the Human Brain

Dustin E. Stansbury and Jack L. Gallant

An overarching goal of visual neuroscience is to understand how the visual system processes natural scenes. Natural scenes possess statistical structure, and computational models based on this structure have provided numerous insights into the processing mechanisms implemented in the early visual system (Barlow, 1961; Field, 1987; Geisler, Perry, Super, & Gallogly, 2001; Simoncelli & Olshausen, 2001). Despite the success of modeling early visual processing based on natural scene statistics, there are still few studies that take this approach when modeling later stages of visual processing.

In this chapter we present a simple but powerful framework for developing models of visual processing that are based on natural scene statistics. First, we describe a modeling approach that can be used to test a wide range of hypotheses concerning the neural basis of natural scene processing. Next, we discuss how statistical analyses of natural scene properties can generate new hypotheses regarding visual processing. We then show how these two approaches are easily combined into a general framework for investigating natural scene processing throughout the brain. Finally, we use this general framework to study the representation of natural scene categories in later stages of the human visual system.

Linearized Models

A primary goal of visual neuroscience is to develop quantitative models that accurately characterize the computations performed by the visual system during natural operation (Wu, David, & Gallant, 2006). However, many of these computations appear to be quite complex, making modeling efforts difficult. Neurons in early stages of visual processing exhibit nonlinear contrast gain control (Carandini, Heeger, & Movshon, 1997), center-surround interactions (Gilbert & Wiesel, 1990), and temporal summation (Tolhurst, Walker, Thompson, & Dean, 1980). Neurons at later stages of visual processing are even more nonlinear (Carandini et al., 2005; Ito, Tamura, Fujita, & Tanaka, 1995) and are heavily affected by top-down mechanisms such as attention, learning, and memory (Reynolds & Chelazzi, 2004). Consequently, it has been difficult to develop valid models of visual scene processing, particularly at later stages of computation (Marmarelis, 2004; Victor, 2005; Wu et al., 2006). A general framework to facilitate the development and testing of nonlinear models of visual processing is thus in great need.

To address this need, our laboratory has developed a modeling approach adapted from *system identification*. System identification is a set of techniques used in engineering and signal processing to characterize nonlinear input-output systems (Marmarelis, 2004; Wu et al., 2006). Under the system identification approach a (generally nonlinear) parametric function is used to describe the relationship between stimuli and evoked brain responses (David & Gallant, 2005; Theunissen et al., 2001; Wu et al., 2006). The function parameter values are fit by observing N stimulus-response pairs, $\{\mathbf{s}_i, r_i\}$, where $i = 1, 2, \ldots N$. The parameters are adjusted to minimize the difference between the measured brain responses and the responses $\hat{\mathbf{r}}(\mathbf{s})$ predicted by the model. We refer to the stimulus-response function as an *encoding model* because it describes how stimuli are encoded in evoked brain activity. Once the parameters are fit, the encoding model is evaluated in terms of its ability to predict responses evoked by novel stimuli (i.e., stimuli not used in parameter fitting).

The encoding model can take many functional forms (Marmarelis, 2004; Wu et al., 2006). The most common class of encoding models used in sensory and motor neuroscience is the *polynomial encoding model* (PEM). The functional form of the PEM is a linear combination of polynomial functionals applied to the stimulus dimensions. For example, a PEM of order 2 generates predicted responses $\hat{\mathbf{r}}_{\mathrm{PEM}_2}(\mathbf{s})$ to a set of stimuli $\mathbf{s}$ with the following stimulus-response function:

$$\hat{\mathbf{r}}_{\mathrm{PEM}_2}(\mathbf{s}) = \beta_0 + \mathbf{s}^{\mathrm{T}}\boldsymbol{\beta}_1 + \mathbf{s}^{\mathrm{T}}\mathbf{B}_2\mathbf{s} \tag{11.1}$$

The first term in **equation 11.1** is a scalar offset; the second term captures linear relationships between the stimulus dimensions and the responses (the parameter $\boldsymbol{\beta}_1$ is a weight vector); the third term captures pairwise interactions between the stimulus dimensions and the responses (the parameter $\mathbf{B}_2$ is a weight matrix). The PEM is a general function estimator that makes few assumptions about the shape of the stimulus-response function. By adding higher-order terms, a PEM can capture relatively more complicated nonlinear relationships. Many studies have used linear PEMs (consisting of the first two terms in **equation 11.1**) to describe the responses of simple cells in primary visual cortex (e.g., Alonso & Usrey, 2001; Jones & Palmer, 1987; Rust, Schwartz, Movshon, & Simoncelli, 2005; Touryan, Lau, & Dan, 2002). Second-order PEMs (**equation 11.1**) have been used to characterize the properties of complex cells in primary visual cortex (Rust et al., 2005; Touryan, Felsen, & Dan, 2005).

Although PEMs provide good models of low-order stimulus-response functions, they are limited in their usefulness as general nonlinear models of visual processing. One limitation is that PEMs do not scale well and require exponentially more data to fit each higher-order term (Marmarelis, 2004). Thus, it is generally not feasible to fit a PEM beyond second order (but see Oliver, Nishimoto, Naselaris, & Gallant, 2012). Additionally, PEMs generally assume that the stimuli have spherical Gaussian statistics (Chichilnisky, 2001; Paninski, 2003). This makes the PEM unsuitable for modeling how the brain processes natural scenes, which have non-Gaussian statistics (Field, 1987).

To overcome the limitations of PEMs, we have used a linearized modeling approach to capture higher-order nonlinearities (Wu et al., 2006). Linearized modeling breaks the system identification problem into two steps. First, stimuli are transformed (often nonlinearly) into an intermediate feature space $\mathbf{\Phi}(\mathbf{s})$. The feature space is chosen to ensure that the relationship between the features and evoked brain activity is as linear as possible. The feature space can thus be interpreted as an instantiation of an explicit hypothesis regarding the information encoded in evoked brain activity (Naselaris, Kay, Nishimoto, & Gallant, 2011; Wu et al., 2006). In the second step, linear regression is used to fit a set of weights $\mathbf{B}_E$ that best map the intermediate features onto the measured brain activity (figure 11.1A). The resulting *linearized encoding model* (LEM) describes how the intermediate features are explicitly (i.e., linearly) encoded in evoked brain activity (Naselaris et al., 2011; Wu et al., 2006). A fit LEM generates predicted responses $\hat{\mathbf{r}}_{LEM}(\mathbf{s})$ to a set of stimuli $\mathbf{s}$ with the following stimulus-response function:

$$\hat{\mathbf{r}}_{LEM}(\mathbf{s}) = \mathbf{\Phi}(\mathbf{s})^T \mathbf{B}_E \tag{11.2}$$

Although the LEM is less general than the PEM, it offers several important benefits. First, LEM parameters are fit by linear regression, a simple, well-established technique whose theoretical properties are firmly understood. For example, a variety of regularization techniques have been developed for linear regression (Friedman, Hastie, & Tibshirani, 2010). Regularization is important for several reasons: it allows the experimenter to add prior knowledge to the parameter-fitting procedure, it reduces the tendency to overfit the model to noise, and it ensures stable parameter estimates for ill-conditioned problems (i.e., when the number of stimulus-response pairs is fewer than the number of model parameters). Thus, a LEM can generally be fit with fewer data than are required to fit a PEM.

The LEM facilitates model interpretation as well. It is often difficult to interpret or visualize a PEM (Wu et al., 2006), but the linear weights $\mathbf{B}_E$ of a LEM can be viewed directly. One can interpret the weights of an encoding model as estimates of the tuning curve for the set of features under investigation. This tuning curve interpretation can be a significant advantage in functional magnetic resonance imaging (fMRI) studies of higher-order scene processing (Huth, Nishimoto, Vu, & Gallant, 2012; Stansbury, Naselaris, & Gallant, 2013), which have taken little advantage of the tuning curve formalism thus far.

LEMs also do not require stimuli to have spherical Gaussian statistics, so stimuli can be sampled from natural scenes. Using natural scene stimuli is important for a number of reasons. First, the visual system exhibits an array of nonlinear response properties that are absent in the presence of synthetic or noise stimuli (Dan, Atick, & Reid, 1996; David, Vinje, & Gallant, 2004; Wu et al., 2006). Second, natural scenes contain stimulus features that range from low-level pixel structure to high-level semantic concepts. Consequently, natural scenes elicit activity from neurons that represent stimuli at various levels of complexity. If it is possible to record multiple regions of the brain simultaneously (as is the case in fMRI experiments), then natural scene stimuli offer the opportunity to investigate multiple stages of visual processing with a single experiment.

Finally, LEMs can be used to directly compare multiple competing hypotheses on the same data. For example, a classic hypothesis is that neurons (or voxels) in primary visual cortex (area V1) represent spatially localized image structure, such as orientation and spatial frequency (Adelson & Bergen, 1985; Carandini et al., 2005; Jones & Palmer, 1987). This hypothesis can be tested with a LEM by transforming stimuli into a feature space that represents local image structure. One such transformation is a decomposition of visual stimuli using a set of oriented bandpass filters known as Gabor wavelets (Daugman, 1985; figure 11.1). Following stimulus transformation, a set of LEM weights is estimated using linear regression, and model predictions are assessed on a testing set of novel stimuli. The LEMs based on wavelet decomposition provide remarkably accurate predictions of brain activity measured in single neurons (Willmore, Prenger, & Gallant, 2010) and in single voxels (Kay, Naselaris, Prenger, & Gallant, 2008; Nishimoto & Gallant, 2011) in area V1.

A large body of recent work has demonstrated the effectiveness of using LEMs for studying natural scene processing. Neurophysiology studies from our laboratory have used the 2D Fourier power spectrum of images as a feature space to fit LEMs for single neurons in V1 (David et al., 2004) and area V4 (David, Hayden, & Gallant, 2006). The Fourier power spectrum is a spatially global representation of the stimulus. Thus, this class of LEM captures position-invariant properties of these neurons. A neurophysiology study of area MT (Nishimoto et al., 2011) used a feature space consisting of a pyramid of motion-energy filters (i.e., spatiotemporal Gabor wavelets). The motion-energy pyramid describes how spatially localized orientation, spatial frequency, temporal frequency, and direction information represented in area V1 are pooled and encoded in the activity of single MT neurons. This class of LEM accurately predicts responses of individual MT neurons to arbitrary stimuli, including natural movies.

Functional MRI studies from our laboratory have also used feature spaces provided by the Gabor wavelet pyramid (Kay et al., 2008) and the motion-energy pyramid (Nishimoto et al., 2011) to fit LEMs to single voxels in early visual cortex. These LEMs provide a concise account of a wide range of phenomena that have been reported previously in fMRI studies of early visual cortex.

Higher visual areas such as the fusiform face area (FFA) (Kanwisher, McDermott, & Chun, 1997) and the parahippocampal place area (PPA) (Epstein & Kanwisher,

1998) appear to represent more abstract, semantic content of natural scenes. Thus, voxels in these higher visual areas are not well characterized by models based on simple image features. A better approach is to model these areas in terms of semantic features. For example one option is to represent stimuli in terms of the objects (nouns) and actions (verbs) that appear in each scene (Çukur, Nishimoto, Huth, & Gallant, 2013; Huth et al., 2012). These LEMs reveal that semantic information in natural scenes is widely distributed in broad gradients that extend across the higher visual cortex. Furthermore, results from these studies suggest that classically defined regions of interest (ROIs) such as the FFA and PPA are nodal points within these gradients.

Another option is to represent stimuli in terms of abstract scene categories (Naselaris, Prenger, Kay, Oliver, & Gallant, 2009; Naselaris, Stansbury, & Gallant, 2012; Stansbury et al., 2013). Linearized models based on scene category provide an accurate account of semantic tuning in many higher-order visual areas. Results from these studies reveal that the correspondence between selectivity for objects, actions, and scene categories in the human brain likely reflects statistical relationships among objects, actions, and categories that exist in the natural world (Naselaris et al., 2012; Stansbury et al., 2013).

The *linearized decoding model* (LDM) provides a complementary approach to the LEM for studying natural scene processing. As in the LEM approach, the LDM is constructed by mapping stimuli into an intermediate feature space. However, linear regression is performed in the opposite direction, fitting a set of weights $\mathbf{B}_D$ that optimally map brain activity $\mathbf{r}$ onto the intermediate features (Naselaris et al., 2011). Thus, an LDM describes how stimulus features are predicted from brain activity (Naselaris et al., 2009; Nishimoto & Gallant, 2011; Thirion et al., 2006). Given a set of responses, an LDM generates a predicted set of stimulus features $\mathbf{\Phi}(\mathbf{r})$ using the following response-stimulus function:

$$\hat{\mathbf{\Phi}}(\mathbf{s};\mathbf{r}) = \mathbf{r}(\mathbf{s})^T \mathbf{B}_D \tag{11.3}$$

Note that it is also possible to transform a fit LEM into a LDM by using a probabilistic formulation based on Bayes' theorem (Naselaris et al., 2011; Naselaris et al., 2009; Thirion et al., 2006). However, we focus on the formulation in **equation 11.3** without loss of generality. We have shown that LDMs are able to decode the structural content of natural images (Kay et al., 2008), movies (Nishimoto et al., 2011), and the semantic content of natural scenes (Naselaris et al., 2009; Naselaris et al., 2012; Stansbury et al., 2013) with unprecedented accuracy.

Linearized models require a linearizing transformation that maps stimuli into an intermediate feature space (Naselaris et al., 2011). Thus, a primary goal when using linearized models is to develop new feature spaces (and the associated linearizing transformations) that capture novel hypotheses about brain function. To date most linearized models have used feature spaces developed by hand. Some of these feature

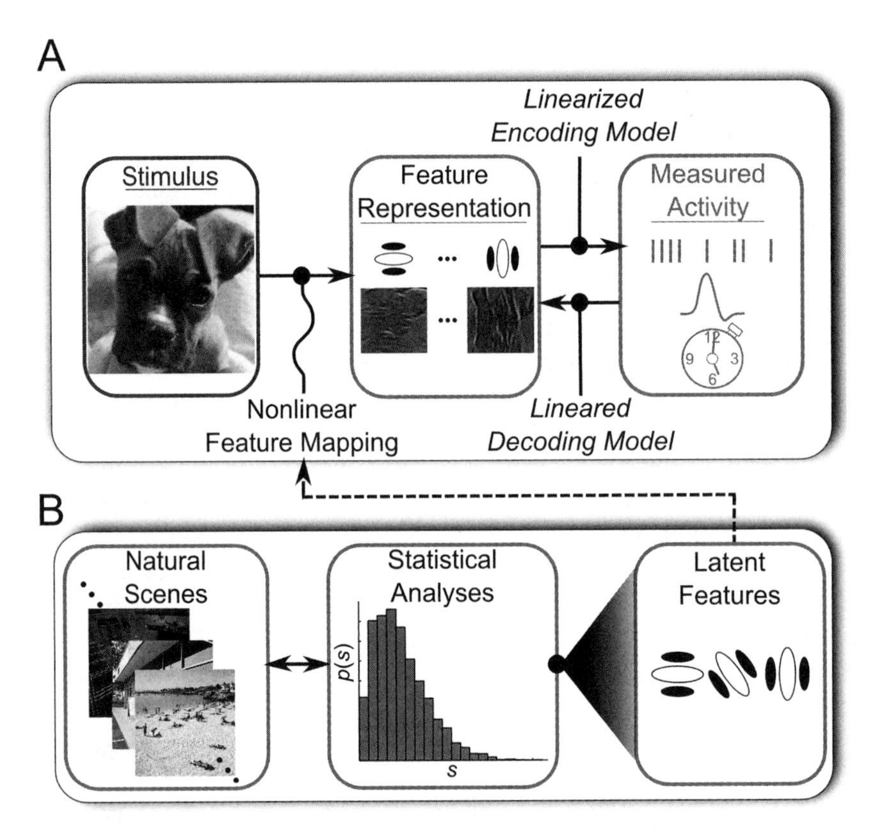

Figure 11.1
A framework combining statistical analysis of natural scenes and linearized modeling. (A) A linearized encoding model maps a stimulus (left box) into an intermediate feature representation that captures specific properties of the stimulus (center box). The feature-mapping transformation is usually nonlinear. In the example shown here the stimulus image is transformed by projecting each spatial location onto a set of filters that capture local orientation and spatial frequency. This transformation linearizes the relationship between the feature representation and evoked brain activity (right box). Brain activity can be measured using many different methods, such as neurophysiology (top), functional MRI (middle), or reaction times (bottom). Whereas the linearized encoding model maps stimulus features onto brain activity, the linearized decoding model works in the opposite direction, mapping activity onto features. (B) Natural scenes (left box) have a specific statistical structure that can be analyzed using quantitative techniques (center box). These statistical analyses can identify latent features in natural scenes that are theoretically important for neural representation. For example, oriented bandpass filters emerge from analysis of the pixel intensity distributions of small image patches extracted from natural scenes (right box). When these filters are used as a basis for the feature transformation in the linearized modeling framework, the resulting models accurately account for brain activity measured in early stages of the primate visual system.

spaces reflect documented filtering computations associated with the neural population under investigation (Kay et al., 2008; Nishimoto et al., 2011). Others capture intuitive linguistic or cognitive labels (Huth et al., 2012; Naselaris et al., 2009). However, it would be preferable to develop new feature spaces in an objective, principled manner. Here we propose that statistical analysis of natural scenes provides a principled means to generate novel feature spaces and thus new hypotheses regarding visual scene processing.

Statistical Analysis of Natural Scenes

It is widely accepted that the brain is specialized to represent the statistical structure of natural scenes (Barlow, 1961; Field, 1987; Simoncelli & Olshausen, 2001). This claim is supported by numerous predictions provided by models based on *statistical analysis of natural scenes* (SANS). One simple SANS approach is to compile a series of empirical measurements of some specific property of natural scenes. The descriptive statistics of the resulting distribution of measurements (e.g., mean, variance, mode, shape) often provide accurate predictions of sensory and perceptual phenomena. For example, the distribution of local contrasts in natural images predicts the shape of the contrast response function observed in the retina (Laughlin, 1981), the lateral geniculate nucleus (LGN) (Tadmor & Tolhurst, 2000), and area V1 (Brady & Field, 2000; Clatworthy, Chirimuuta, Lauritzen, & Tolhurst, 2003). Disparity tuning in area MT is predicted by the distribution of binocular disparities in natural scenes (Liu, Bovik, & Cormack, 2008; DeAngelis & Uka, 2003). The distribution of colors in natural scenes predicts human performance in segmentation tasks (Fine, MacLeod, & Boynton, 2003), and the distribution of trichromatic photoreceptors in the primate retina (L-M cones) appears to facilitate object segmentation in natural scenes (Párraga, Troscianko, & Tolhurst, 2002). Furthermore, human performance in contour grouping (Geisler et al., 2001) and contour completion tasks (Geisler & Perry, 2009) is predicted by the distribution of contours in natural scenes.

A second SANS approach is to develop explicit statistical models whose parameters are optimized by analyzing natural scenes under the constraints of some plausible neural coding criterion. Such models are capable of identifying emergent features in natural scenes that are theoretically important for neural computation. For example, primary visual cortex contains a highly overcomplete representation of the visual input, consisting of hundreds of V1 neurons for each retinal ganglion cell (Olshausen & Field, 1997). It has also been observed that V1 neurons are sparsely activated in the presence of natural scenes (Daugman, 1985; Field, 1987). Olshausen and Field (1996) developed a statistical model of local regions in natural scenes based on these two observations. Specifically, the model incorporates an overcomplete representation in which the number of units that code for the visual input is larger than the number of inputs. Additionally, the model includes a sparseness constraint on the number of coding units that are active at one time. The resulting "sparse coding" model identifies latent features with properties that are strikingly similar to the response profiles of "simple" cells observed in V1 (figure 11.1B). Similar models have also identified features that accurately account for neuronal response properties in early stages of visual processing, including the retina (Atick & Redlich, 1992; Srinivasan, Laughlin, & Dubs, 1982) and the LGN (Dong & Atick, 1995).

Although the early stages of visual processing appear to be optimized to process the low-level statistical structure of natural scenes, there is less evidence that this optimization persists at later stages of processing. However, the visual system is organized hierarchically and later visual areas receive substantial input from early areas (Felleman & Van Essen, 1991). Therefore, it seems reasonable to assume that later areas are also affected by low-level scene statistics (Hebb, 1949).

Additionally, natural scenes have statistical structure at relatively higher levels of abstraction. For example, natural scenes consist of collections of objects that interact in regular ways. Cars are usually found on streets, and fish are usually found underwater. Thus, when one is crossing the street it is more likely that one will encounter a car than a shark. Of course, these relationships are not absolute, but probabilistic. Cars can be found in showrooms, in museums, or in a garage. Fish sometimes jump out of the water or are found on a dinner plate. It is likely that the brain represents these higher-order statistical relationships and that it exploits this information during natural vision. Thus, statistical analyses of natural scenes may also provide a means for developing models of processing in later stages of the visual system.

A General Framework Combining Linearized Modeling and Statistical Analyses of Natural Scenes

Thus far we have discussed two complementary approaches for studying visual scene processing: linearized modeling and statistical analyses of natural scenes. These two approaches can be combined in order to study natural scene processing across the entire visual system. To do this, SANS is used to model the distribution of some specific natural scene property. The resulting model is then used to implement the transformation required for linearized modeling.

For example, if the SANS method involves compiling a distribution of empirical measurements, the resulting (normalized) histogram can be used as a lookup table that implements the required linearizing transformation (the corresponding feature space thus represents the likelihood of a stimulus, conditioned on the distribution of the natural scene property modeled). For SANS methods that use an explicit statistical model, the emergent features identified by the model can be used as a basis for the linearizing transformation. Consider, for example, the sparse coding model discussed earlier. The features identified by the sparse coding model are strikingly similar to V1 neuron spatial response profiles. These features are local, oriented bandpass filters akin to Gabor wavelets (figure 11.1B). (However, whereas Gabor wavelets are handcrafted mathematical objects, the features identified by SANS are entirely data driven.) Projecting stimuli onto the identified features thus provides a linearizing transformation that is comparable to that provided by the Gabor wavelet decomposition. Consequently, an LEM based on these emergent features provides very accurate predictions of neural activity in V1 (Vu et al., 2011).

Although it is apparent that this framework is effective for modeling processing in V1, it is unclear if this approach will be effective for modeling processing in later visual areas. Below we summarize the results of one study demonstrating how this framework can be used to model the representation of natural scene categories in human anterior visual cortex.

Scene Category Representation in Anterior Visual Cortex

During natural vision, humans categorize the scenes they encounter: "an office," "the beach," and so on. We hypothesized that the visual system exploits the co-occurrence statistics of objects in natural scenes in order to represent scene categories. To test this hypothesis we first used SANS to analyze the co-occurrences of objects in natural scenes in order to recover a set of latent scene categories. We then used LEMs to determine if and where these latent categories are represented in the human visual system (figure 11.2).

We first labeled the objects in a large collection of natural scenes. We then analyzed the frequency counts of the object labels using a statistical model called *latent Dirichlet allocation* (LDA) (Blei, Ng, & Jordan, 2003; figure 11.2A). LDA was originally developed to identify latent topics in large corpora of text from word frequency data. When applied to the frequency counts of labeled objects in natural scenes, LDA identifies a set of latent scene categories. Each latent category is defined as a probability distribution over all the objects in an available vocabulary. Objects that are more likely to occur in a given category are assigned high probability, and unlikely objects are assigned low probability (examples of the latent categories are shown in the right of figure 11.2B).

Next, we used a set of natural scene photographs as stimuli in an fMRI experiment conducted on four human subjects. We then used the trained LDA model to calculate the probability that each stimulus scene belonged to each of the latent categories, conditioned on the objects occurring in the scene (figure 11.2B, center box). This procedure produced a set of category probability features for each stimulus scene. Finally, we fit LEMs based on the categorical features to the blood-oxygen-level-dependent (BOLD) activity measured within each voxel.

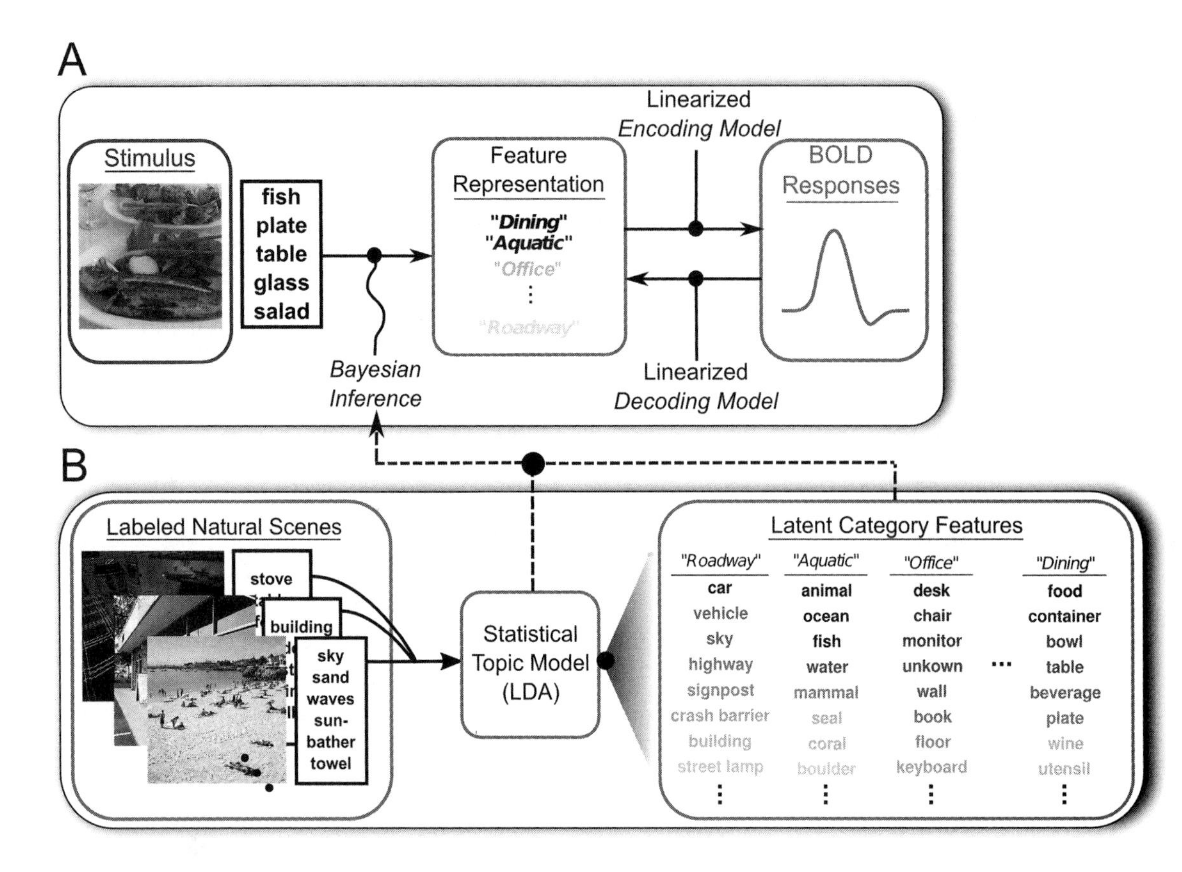

Figure 11.2
Modeling the representation of natural scene categories in the human brain. (A) The framework consists of using SANS to identify the latent categorical structure of natural scenes and then using these categories as the feature space for LEMs fit to recorded fMRI brain activity. First, the objects in a series of scenes are labeled by hand (left boxes). These are submitted to LDA (panel B) in order to calculate a set of probabilistic features (center box), which describe the probability that each scene belongs to each member in a set of latent scene categories. The LEMs predict BOLD responses evoked by each scene from the category features associated with the scene. The LDMs predict scene category membership from evoked BOLD responses. (B) The latent categories are identified by latent Dirichlet allocation (LDA, center box), a statistical model developed originally to identify latent topics in large text corpora. The latent scene categories (right box) are represented as a set of probabilities assigned to the objects in an available vocabulary. Here each list of object labels represents a latent category; label saturation indicates an object's probability for the category. LDA also provides a Bayesian inference algorithm that probabilistically assigns a new scene to each of the latent scene categories, conditioned on the objects in the scene. Here, this inference procedure is used as the nonlinear feature transformation for the linearized models in panel A (dashed black line).

The accuracy with which each model predicted responses to novel scenes (using a separate data set reserved for this purpose) is summarized in figure 11.3 (plate 23). Evoked BOLD activity across much of anterior visual cortex is accurately predicted by the latent scene categories (figure 11.3A, plate 23). Thus, voxels in these regions of the visual system represent information about the latent categories. Voxels located in several visual areas identified using standard functional localizers (Spiridon, Fischl, & Kanwisher, 2006) are accurately characterized by the model. However, the model does a poor job at predicting activity in early visual areas. These areas are known to be selective for low-level image structure. Thus, the failure to predict activity in early visual area voxels confirms that the model does not simply capture low-level structure that is correlated with scene categories.

As mentioned earlier, one advantage of using LEMs is that the model weights can be interpreted as tuning curves for the intermediate features. Typical tuning curves for the latent scene categories are shown in figure 11.3B (plate 23; here they are averaged across voxels within specific functional areas). These tuning curves are generally consistent with tuning reported in previous studies (Spiridon et al., 2006). However, tuning also appears to be more complicated than suggested by previous functional localizer studies. This finding is consistent with recent results from our group (Çukur et al., 2013; Huth et al., 2012; Naselaris et al., 2012).

We also constructed an LDM to predict the category probabilities of novel scenes from BOLD responses evoked by those scenes. As shown in figure 11.3C (plate 23), the LDM accurately predicts the category probabilities from brain activity evoked by viewing a diversity of novel natural scenes. Remarkably, the LDM can also recover the likely objects in each scene, even though the intermediate feature space does not contain any explicit information about objects. During the identification of the latent scene categories, the LDA model establishes the statistical relationship between objects and the latent scene categories. Therefore, the decoded scene categories can be mapped directly into object probabilities, as shown in figure 11.3C (plate 23). Taken together, these results suggest that the human brain represents scene categories that capture the co-occurrence statistics of objects in the natural world.

Conclusions

In this chapter we present a general framework for developing, testing, and comparing nonlinear models of visual processing. The linearized modeling component of this framework offers several benefits over traditional nonlinear modeling methods: experiments are simple to design and can even include complex, naturalistic stimuli; fitting model parameters is straightforward and requires relatively little data; and the fit models are simple to interpret. Incorporating SANS into the framework provides an objective method for developing hypothetical feature spaces that complement linearized modeling.

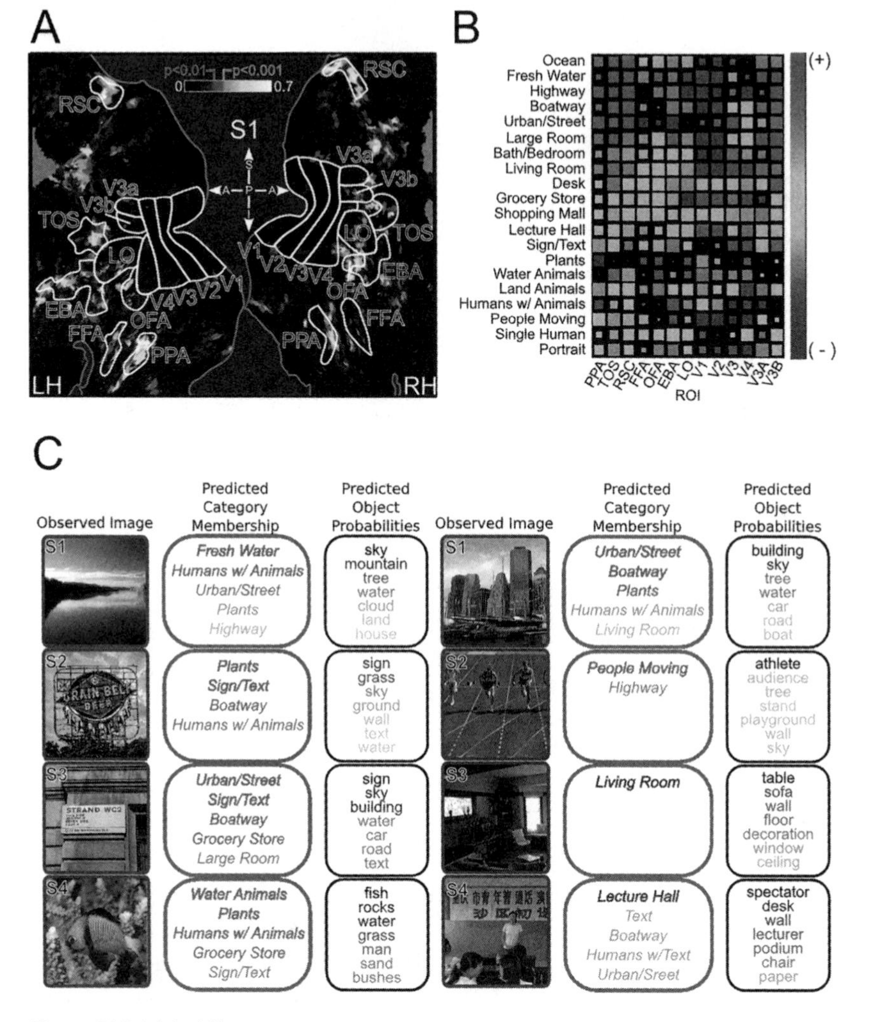

Figure 11.3 (plate 23)
Representation of natural scene categories in the human brain. (A) Prediction accuracy of voxelwise LEMs based on categorical features for one representative subject (S1). Both left (LH) and right (RH) cortical surfaces are shown; flat gray indicates areas outside of the scan window. Bright locations indicate voxels that are predicted accurately by the LEM. The color bar highlights prediction accuracy at two levels of statistical significance: $p < 0.01$ (0.21) and $p < 0.001$ (0.28). Standard ROIs identified using functional localizers are outlined in white. The bright regions overlap with several ROIs in the anterior visual cortex that have been implicated in scene representation. However, the models do not predict activity in retinotopic visual areas (V1, V2, V3, V4, V3a, and V3b). ROI abbreviations: retinotopic visual areas 1–4 (V1–V4); parahippocampal place area (PPA); fusiform face area (FFA); parahippocampal place area (PPA); extrastriate body area (EBA); occipital face area (OFA); retrosplenial cortex (RSC); transverse occipital sulcus (TOS). Center key: A, anterior; P, posterior; S, superior; I, inferior. (B) LEM weights for voxels within distinct functional ROIs. Weights have been averaged within ROIs and across the four subjects. Each row shows the average LEM weights for the scene category feature listed along the left margin. Each column shows average weights for an individual ROI. The color and saturation of each square reflect the magnitude and sign of the weight for the corresponding category. Large positive weights are given by saturated red squares; large negative weights by saturated blue squares. The size of each square is proportional to the confidence. Standard errors are scaled according to the data within each ROI (columns). (C) Examples of scene category membership and object probabilities predicted by LDMs from the LEMs fit to the same four subjects as shown in panels A and B. Rows 1–4 show data for subjects 1–4, respectively. Columns 1 and 4 (blue boxes) show the novel scenes used as stimuli. Columns 2 and 5 (red boxes) show category membership probabilities predicted by the decoding model for the corresponding scenes. The saturation of each category represents the predicted probability that the observed scene belongs to the corresponding category. Columns 3 and 6 (black boxes) show the objects with the highest estimated probability of occurring in each scene. The saturation of each label represents the estimated probability that the corresponding object occurs in the scene.

Developing predictive models of brain systems requires quantitative measurements of both stimulus and response. There are many different ways to measure brain activity (e.g., fMRI BOLD responses, spike rates, reaction times) (figure 11.1A). Likewise, there are many ways to measure low-level stimulus statistics (e.g., pixel intensity, contrast, etc.). However, measuring high-level scene statistics is more difficult. The range of objects, events, and contexts that might occur in a scene is enormous, and it is unclear how these properties and their relationships should be quantified. For these reasons there have been few serious attempts to develop predictive models of high-level scene processing.

Two successful models of high-level scene processing were proposed previously. The *gist* model (Oliva & Torralba, 2001) accounts for scene categorization, and the *saliency* model (Itti, Koch, & Niebur, 1998) accounts for bottom-up attention mechanisms. These two models provide good predictions of human behavioral data, including performance on scene classification tasks (Greene & Oliva, 2009) and the direction of eye movements when one is viewing natural scenes (Itti, 2005). However, both models are based on low-level scene properties, so it is unclear whether they really capture high-level processing mechanisms or merely correlations between low-level image structure and behavior. The framework proposed in this chapter provides a straightforward way to test this important issue. For example, the gist and saliency image descriptors associated with each model could be used as feature spaces for LEMs. The predictions of these competing LEMs could then be compared directly.

Our proposed framework provides a data-driven approach for developing, testing, and comparing models of visual scene processing. However, several technical issues must be addressed when using these techniques. For example, the choice of SANS method will depend on the domain of the scene properties in question (Field, 1987). The LDA model discussed above requires discrete (i.e., multinomial-distributed) data and is thus an appropriate choice for analyzing categorical features. However, LDA would be a poor choice for analyzing continuous (i.e., Gaussian-distributed) quantities such as pixel luminance. Thus, domain knowledge of the scene properties under investigation is important for applying SANS methods.

Other factors must be considered when using linearized models. Linear regression often requires additional preprocessing, such as centering and rescaling the intermediate features and responses. The likelihood function of the linear regression must also be chosen appropriately (Friedman et al., 2010). For example, in the case study described earlier, a multinomial likelihood was chosen to enforce categorical predictions when fitting the LDM. If we had desired predictions of continuous-valued BOLD responses rather than categorical predictions, then a Gaussian likelihood function would have been more appropriate. To ensure optimal results, exploratory data analyses are generally required prior to using this framework.

Our framework leverages two exciting new approaches used in computational neuroscience: SANS and LEM. As methods for measuring neural function, performing statistical learning, and solving parameter optimization problems improve, so will the power of this framework. Thus, this approach is likely to become more common in future studies of sensory and cognitive processing in the human brain.

References

Adelson, E. H., & Bergen, J. R. (1985). Spatiotemporal energy models for the perception of motion. *Journal of the Optical Society of America. A, Optics, Image Science, and Vision*, *2*(2), 284–299.

Alonso, J. M., & Usrey, W. M. (2001). Rules of conectivity between geniculate cells and simple cells in cat primary visual cortex. *Journal of Neuroscience*, *21*(11), 4002–4015.

Atick, J., & Redlich, N. (1992). What does the retina know about natural scenes? *Neural Computation*, *4*(2), 196–210.

Barlow, H. (1961). Possible principles underlying the transformation of sensory messages. In W. A. Rosenblith (Ed.), *Sensory communication* (pp. 217–234). Cambridge, MA: MIT Press.

Blei, D. M., Ng, A. Y., & Jordan, M. I. (2003). Latent Dirichlet allocation. *Journal of Machine Learning Research*, *3*, 993–1022.

Brady, N., & Field, D. J. (2000). Local contrast in natural images: Normalisation and coding efficiency. *Perception*, *29*(9), 1041–1055.

Carandini, M., Demb, J. B., Mante, V., Tolhurst, D. J., Dan, Y., Olshausen, B. A., et al. (2005). Do we know what the early visual system does? *Journal of Neuroscience*, *25*(46), 10577–10597.

Carandini, M., Heeger, D. J., & Movshon, J. A. (1997). Linearity and normalization in simple cells of the macaque primary visual cortex. *Journal of Neuroscience*, *17*(21), 8621–8644.

Chichilnisky, E. J. (2001). A simple white noise analysis of neuronal light responses. *Network*, *12*(2), 199–213.

Clatworthy, P. L., Chirimuuta, M., Lauritzen, J. S., & Tolhurst, D. J. (2003). Coding of the contrasts in natural images by populations of neurons in primary visual cortex (V1). *Vision Research*, *43*(18), 1983–2001.

Çukur, T., Nishimoto, S., Huth, A. G., & Gallant, J. L. (2013). Attention during natural vision warps semantic representation across the human brain. *Nature Neuroscience*, *16*(6), 763–770.

Dan, Y., Atick, J. J., & Reid, R. C. (1996). Efficient coding of natural scenes in the lateral geniculate nucleus: Experimental test of a computational theory. *Journal of Neuroscience*, *16*(10), 3351–3362.

Daugman, J. G. (1985). Uncertainty relation for resolution in space, spatial frequency, and orientation optimized by two-dimensional visual cortical filters. *Journal of the Optical Society of America. A, Optics, Image Science, and Vision*, *2*(7), 1160–1169.

David, S. V., & Gallant, J. L. (2005). Predicting neuronal responses during natural vision. *Network*, *16*(2–3), 239–260.

David, S. V., Hayden, B. Y., & Gallant, J. L. (2006). Spectral receptive field properties explain shape selectivity in area V4. *Journal of Neurophysiology*, *96*(6), 3492–3505.

David, S. V., Vinje, W. E., & Gallant, J. L. (2004). Natural stimulus statistics alter the receptive field structure of V1 neurons. *Journal of Neuroscience*, *24*(31), 6991–7006.

DeAngelis, G. C., & Uka, T. (2003). Coding of horizontal disparity and velocity by MT neurons in the alert macaque. *Journal of Neurophysiology*, *89*(2), 1094–1111.

Dong, D. W., & Atick, J. J. (1995). Temporal decorrelation: A theory of lagged and nonlagged responses in the lateral geniculate nucleus. *Network (Bristol, England)*, *6*(2), 159–178.

Epstein, R. A., & Kanwisher, N. (1998). A cortical representation of the local visual environment. *Nature*, *392*(6676), 598–601.

Felleman, D. J., & Van Essen, D. C. (1991). Distributed hierarchical processing in the primate cerebral cortex. *Cerebral Cortex*, *1*(1), 1–47.

Field, D. J. (1987). Relations between the statistics of natural images and the response properties of cortical cells. *Journal of the Optical Society of America. A, Optics, Image Science, and Vision*, *4*(12), 2379–2394.

Fine, I., MacLeod, D. I. A., & Boynton, G. M. (2003). Surface segmentation based on the luminance and color statistics of natural scenes. *Journal of the Optical Society of America. A, Optics, Image Science, and Vision*, *20*(7), 1283–1291.

Friedman, J., Hastie, T., & Tibshirani, R. (2010). Regularization paths for generalized linear models via coordinate descent. *Journal of Statistical Software*, *33*(1), 1–22.

Geisler, W. S., & Perry, J. S. (2009). Contour statistics in natural images: Grouping across occlusions. *Visual Neuroscience*, *26*(1), 109–121.

Geisler, W. S., Perry, J. S., Super, B. J., & Gallogly, D. P. (2001). Edge co-occurrence in natural images predicts contour grouping performance. *Vision Research*, *41*(6), 711–724.

Gilbert, C. D., & Wiesel, T. N. (1990). The influence of contextual stimuli on the orientation selectivity of cells in primary visual cortex of the cat. *Vision Research*, *30*(11), 1689–1701.

Greene, M. R., & Oliva, A. (2009). The briefest of glances: The time course of natural scene understanding. *Psychological Science*, *20*(4), 464–472.

Hebb, D. (1949). *The organization of behavior: A neuropsychological theory*. New York: Wiley.

Huth, A. G., Nishimoto, S., Vu, A. T., & Gallant, J. L. (2012). A continuous semantic space describes the representation of thousands of object and action categories across the human brain. *Neuron*, *76*(6), 1210–1224.

Ito, M., Tamura, H., Fujita, I., & Tanaka, K. (1995). Size and position invariance of neuronal responses in monkey inferotemporal cortex. *Journal of Neurophysiology*, *73*(1), 218–226.

Itti, L. (2005). Quantifying the contribution of low-level saliency to human eye movements in dynamic scenes. *Visual Cognition*, *12*(6), 1093–1123.

Itti, L., Koch, C., & Niebur, E. (1998). A model of saliency-based visual attention for rapid scene analysis. *IEEE Transactions on Pattern Analysis and Machine Intelligence*, *20*(11), 1254–1259.

Jones, J. P., & Palmer, L. A. (1987). The two-dimensional spatial structure of simple receptive fields in cat striate cortex. *Journal of Neurophysiology*, *58*(6), 1187–1211.

Kanwisher, N., McDermott, J., & Chun, M. M. (1997). The fusiform face area: A module in human extrastriate cortex specialized for face perception. *Journal of Neuroscience*, *17*(11), 4302–4311.

Kay, K. N., Naselaris, T., Prenger, R. J., & Gallant, J. L. (2008). Identifying natural images from human brain activity. *Nature*, *452*(7185), 352–355.

Laughlin, S. (1981). A simple coding procedure enhances a neuron's information capacity. *Zeitschrift fur Naturforschung. Section C. Biosciences*, *36*(9–10), 910–912.

Liu, Y., Bovik, A. C., & Cormack, L. K. (2008). Disparity statistics in natural scenes. *Journal of Vision*, *8*, 19, 1–14.

Marmarelis, P. V. Z. (2004). *Nonlinear dynamic modeling of physiological systems*. San Francisco, CA: Wiley-IEEE Press.

Naselaris, T., Kay, K. N., Nishimoto, S., & Gallant, J. L. (2011). Encoding and decoding in fMRI. *NeuroImage*, *56*(2), 400–410.

Naselaris, T., Prenger, R. J., Kay, K. N., Oliver, M. D., & Gallant, J. L. (2009). Bayesian reconstruction of natural images from human brain activity. *Neuron*, *63*, 902–915.

Naselaris, T., Stansbury, D. E., & Gallant, J. L. (2012). Cortical representation of animate and inanimate objects in complex natural scenes. *Journal of Physiology, Paris*, *106*(5–6), 239–249.

Nishimoto, S., & Gallant, J. L. (2011). A three-dimensional spatiotemporal receptive field model explains responses of area MT neurons to naturalistic movies. *Journal of Neuroscience*, *31*(41), 14551–14564.

Nishimoto, S., Vu, A. T., Naselaris, T., Benjamini, Y., Yu, B., & Gallant, J. L. (2011). Reconstructing visual experiences from brain activity evoked by natural movies. *Current Biology*, *21*(19), 1641–1646.

Oliva, A., & Torralba, A. (2001). Modeling the shape of the scene: A holistic representation of the spatial envelope. *International Journal of Computer Vision*, *42*(3), 145–175.

Oliver, M. D., Nishimoto, S., Naselaris, T., & Gallant, J. L. (2012). *Using natural movies to map foveal V1 spatial-temporal-chromatic receptive fields in awake animals*. Paper presented at the Society for Neuroscience, New Orleans, LA.

Olshausen, B. A., & Field, D. J. (1996). Emergence of simple-cell receptive field properties by learning a sparse code for natural images. *Nature*, *381*(6583), 607–609.

Olshausen, B. A., & Field, D. J. (1997). Sparse coding with an overcomplete basis set: A strategy employed by V1? *Vision Research*, *37*(23), 3311–3325.

Paninski, L. (2003). Convergence properties of some spike-triggered analysis techniques. *Network (Bristol, England)*, *14*(3), 437–464.

Párraga, C. A., Troscianko, T., & Tolhurst, D. J. (2002). Spatiochromatic properties of natural images and human vision. *Current Biology*, *12*(6), 483–487.

Reynolds, J. H., & Chelazzi, L. (2004). Attentional modulation of visual processing. *Annual Review of Neuroscience*, *27*, 611–647.

Rust, N. C., Schwartz, O., Movshon, J. A., & Simoncelli, E. P. (2005). Spatiotemporal elements of macaque V1 receptive fields. *46, 6*(945–956).

Simoncelli, E. P., & Olshausen, B. A. (2001). Natural image statistics and neural representation. *Annual Review of Neuroscience*, *24*, 1193–1216.

Spiridon, M., Fischl, B., & Kanwisher, N. (2006). Location and spatial profile of category-specific regions in human extrastriate cortex. *Human Brain Mapping*, *27*(1), 77–89.

Srinivasan, M. V., Laughlin, S. B., & Dubs, A. (1982). Predictive coding: A fresh view of inhibition in the retina. *Proceedings of the Royal Society of London, B, 216*(1205), 427–459.

Stansbury, D. E., Naselaris, T., & Gallant, J. L. (2013). Natural scene statistics account for the representation of scene categories in human visual cortex. *Neuron*, *79*(5), 1025–1034.

Tadmor, Y., & Tolhurst, D. J. (2000). Calculating the contrasts that retinal ganglion cells and LGN neurones encounter in natural scenes. *Vision Research*, *40*(22), 3145–3157.

Theunissen, F. E., David, S. V., Singh, N. C., Hsu, A., Vinje, W. E., & Gallant, J. L. (2001). Estimating spatio-temporal receptive fields of auditory and visual neurons from their responses to natural stimuli. *Network*, *12*(3), 289–316.

Thirion, B., Duchesnay, E., Hubbard, E., Dubois, J., Poline, J.-B., Lebihan, D., et al. (2006). Inverse retinotopy: Inferring the visual content of images from brain activation patterns. *NeuroImage*, *33*(4), 1104–1116.

Tolhurst, D. J., Walker, N. S., Thompson, I. D., & Dean, A. F. (1980). Non-linearities of temporal summation in neurones in area 17 of the cat. *Experimental Brain Research*, *38*(4), 431–435.

Touryan, J., Felsen, G., & Dan, Y. (2005). Spatial structure of complex cell receptive fields measured with natural images. *Neuron*, *45*(5), 781–791.

Touryan, J., Lau, B., & Dan, Y. (2002). Isolation of relevant visual features from random stimuli for cortical complex cells. *Journal of Neuroscience*, *22*(24), 10811–10818.

Victor, J. D. (2005). Analyzing receptive fields, classification images and functional images: Challenges with opportunities for synergy. *Nature Neuroscience*, *8*(12), 1651–1656.

Vu, V. Q., Ravikumar, P., Naselaris, T., Kay, K. N., Gallant, J. L., & Yu, B. (2011). Encoding and decoding V1 fMRI responses to natural images with sparse nonparametric models. *Annals of Applied Statistics*, *5*, 1159–1182.

Willmore, D. B., Prenger, R. J., & Gallant, J. L. (2010). Neural representation of natural images in visual area V2. *Journal of Neuroscience*, *30*(6), 2102–2114.

Wu, M. C.-K., David, S. V., & Gallant, J. L. (2006). Complete functional characterization of sensory neurons by system identification. *Annual Review of Neuroscience*, *29*, 477–505.

12

On Aesthetics and Emotions in Scene Images: A Computational Perspective

Dhiraj Joshi, Ritendra Datta, Elena Fedorovskaya, Xin Lu, Quang-Tuan Luong, James Z. Wang, Jia Li, and Jiebo Luo

In this chapter we discuss the problem of computational inference of aesthetics and emotions from images. We draw inspiration from diverse disciplines such as philosophy, photography, art, and psychology to define and understand the key concepts of aesthetics and emotions. We introduce the primary computational problems that the research community has been striving to solve and the computational framework required for solving them. We also describe data sets available for performing assessment and outline several real-world applications for which research in this domain can be employed. This chapter discusses the contributions of a significant number of research articles that have attempted to solve problems in aesthetics and emotion inference in the last several years. We conclude the chapter with directions for future research.

The image-processing community together with vision and computer scientists has, for a long time, attempted to solve image quality assessment (Daly, 1993; Ke, Tang, & Jing, 2006; Sheikh, Bovik, & Cormack, 2005; Watson, 1998) and image semantics inference (Datta, Joshi, Li, & Wang, 2008). More recently researchers have drawn ideas from the aforementioned to address yet more challenging problems such as associating pictures with aesthetics and emotions that they arouse in humans, with low-level image composition (Datta, Joshi, Li, & Wang, 2006; Datta, Li, & Wang, 2007; Valenti, Jaimes, & Sebe, 2010; Valenti, Sebe, & Gevers, 2007). Figure 12.1 (plate 24) shows an example of state-of-the-art automatic aesthetics assessment. Because emotions and aesthetics also bear high-level semantics, it is not a surprise that research in these areas is heavily intertwined. Besides, researchers in aesthetic quality inference also need to understand and consider human subjectivity and the context in which the emotion or aesthetics is perceived. As a result, ties between computational image analysis and psychology, study of beauty (Lang, Greenwald, Bradley, & Hamm, 1993; Perrett, May, & Yoshikawa, 1994), and aesthetics in visual art, including photography, are also natural and essential.

Despite the challenges, various research attempts have been made and are increasingly being made to address basic understanding and solve various subproblems under the umbrella of aesthetics, mood, and emotion inference in pictures. The potential

Figure 12.1 (plate 24)
Pictures with high, medium, and low aesthetics scores from ACQUINE, an online automatic photo aesthetics engine.

beneficiaries of this research include general consumers, media management vendors, photographers, and people who work with art. Good shots or photo opportunities may be recommended to consumers; media personnel can be assisted with good images for illustration, and interior and healthcare designers can be helped with more appropriate visual design items. Picture editors and photographers can make use of automated aesthetics feedback when selecting photos for photo clubs, competitions, portfolio reviews, or workshops. Similarly, from a publication perspective, a museum curator may be interested in assessing if an artwork is enjoyable by a majority of the people. Techniques that study similarities and differences among artists and artwork at the aesthetic level could be of value to art historians.

We strongly believe that computational models of aesthetics and emotions may be able to assist in such expert decision making and perhaps, with time and feedback, learn to adapt to expert opinion better. Figure 12.2 (plate 25) shows user-rated emotions under the framework of web image search that can potentially be used for learning emotional models. Computational aesthetics does not intend to obviate the need for expert opinion. On the other hand, automated methods would strive toward

Figure 12.2 (plate 25)
Pictures and emotions rated by users from ALIPR.com, a research site for machine-assisted image tagging.

becoming useful suggestion systems for experts, systems that can be personalized (to one or few experts) and improved with feedback over time (as also expressed in Stork, 2009).

In this chapter we have attempted to introduce components that are essential for the broader research community to get involved and excited about this field of study. We next discuss aesthetics with respect to philosophy, photography, art, and psychology. The following section introduces a wide spectrum of research problems that have been attempted in computational aesthetics and emotions. The computational framework in the form of feature extraction, representation, and modeling is the topic of the next section. Data sets and other resources available for aesthetics and emotions research are then reviewed, and the last section takes a futuristic stance and discusses potential research directions and applications.

Background

The word "aesthetics" originates from the Greek word *aisthētikos,* sensitive, derived from *aisthanesthai,* "to perceive, to feel." The *American Heritage Dictionary of the English Language* provides the following currently used definitions of aesthetics:

1. The branch of philosophy that deals with the nature and expression of beauty, as in the fine arts. In Kantian philosophy, the branch of metaphysics concerned with the laws of perception.
2. The study of the psychological responses to beauty and artistic experiences.
3. A conception of what is artistically valid or beautiful.
4. An artistically beautiful or pleasing appearance.

Philosophical studies have resulted in formation of two views on beauty and aesthetics: the first view considers aesthetic values to exist objectively and be universal, and the second position treats beauty as a subjective phenomenon, depending on the attitude of the observer.

A Perspective on Photographs

Although aesthetics can be colloquially interpreted as a seemingly simple matter determining what is beautiful, few can meaningfully articulate the definition of aesthetics or how to achieve a high level of aesthetic quality in photographs. For several years *Photo.net* has been a place for photographers to rate the photos of peers. Here a photo is rated along two dimensions, aesthetics and originality, each with a score between 1 and 7. Sample reasons for a high rating include "looks good, attracts/holds attention, interesting composition, great use of color, (if photo journalism) drama, humor, and impact, and (if sports) peak moment, struggle of athlete."

Ideas of aesthetics emerged in photography around the late nineteenth century with a movement called Pictorialism. Because photography was a relatively new art at that time, the Pictorialist photographers drew inspiration from paintings and etchings to the extent of emulating them directly. Photographers used techniques such as soft focus, special filters, lens coatings, special darkroom processing, and printing to achieve desired artistic effects in their pictures. By around 1915 the widespread cultural movement of Modernism had begun to influence photographic circles. In Modernism, ideas such as formal purity, medium specificity, and originality of art became paramount. Postmodernism rejected ideas of objective truth in art. Sharp classifications into high art and low art became defunct.

In spite of these differing factors certain patterns stand out with respect to photographic aesthetics. This is especially true in specific domains of photography. For example, in nature photography it can be demonstrated that the appreciation of striking scenery is universal. Nature photographers often share common techniques or rules of thumb in their choices of colors, tonality, lighting, focus, content, vantage point, and composition. One such accepted rule is that the purer the primary colors, red (sunset, flowers), green (trees, grass), and blue (sky), the more striking the scenery is to viewers. In terms of composition there are again common and not so common theories or rules. The *rule of thirds* is the most widely known and states that the most important part of the image is not the exact center of the image but rather at the one third and two third lines (both horizontal and vertical) and their four intersections. A less common rule in nature photography is to use diagonal lines (such as a railway, a line of trees, a river, or a trail) or converging lines for the main objects of interest to draw the attention of the human eye. Another composition rule is to frame the shot so that there are interesting objects in both the close-up foreground and the far-away background. However, great photographers often have the talents to know when to break these rules to be more creative. Ansel Adams said, "There are no rules for good photographs, there are only good photographs."

A Perspective on Paintings

Painters in general have a much greater freedom to play with the palette, the canvas, and the brush to capture the world and its various seasons, cultures, and moods. Photographs at large represent true physical constructs of nature (although film photographers sometimes aesthetically enhanced their photos by dodging and burning). Artists, on the other hand, have always used nature as a base or as a "teacher" to create works that reflected their feelings, emotions, and beliefs.

History abounds with many influential art movements that dominated the world art scene for certain periods of time and then faded away, making room for newer ideas. It would not be incorrect to say that most art movements (and sometimes individual artists) defined characteristic painting styles that became the primary determinants of art aesthetics of the time. One of the key movements of Western art, Impressionism, started in the late nineteenth century with Claude Monet's masterpiece *Impression, Sunrise, 1872*. Impressionist artists focused on ordinary subject matter, painted outdoors, used boldly visible brushstrokes, and employed colors to emphasize light and its effect on their subjects. A derivative movement, Pointillism, was pioneered by Georges Seurat, who mastered the art of using colored dots as building blocks for paintings. Early twentieth-century Post-impressionist artists digressed from the past and introduced a personal touch to their world depictions, giving expressive effects to their paintings. Van Gogh is especially known for his bold and forceful use of colors in order to express his artistic ideas (figure 12.3, plate 26). Van Gogh also developed a bold style of brush strokes, an analysis of which can perhaps offer newer perspectives into understanding his work and that of his contemporaries. Figure 12.3, plate 26, shows an example of automatic brushstroke extraction research presented by Johnson et al. (2008).

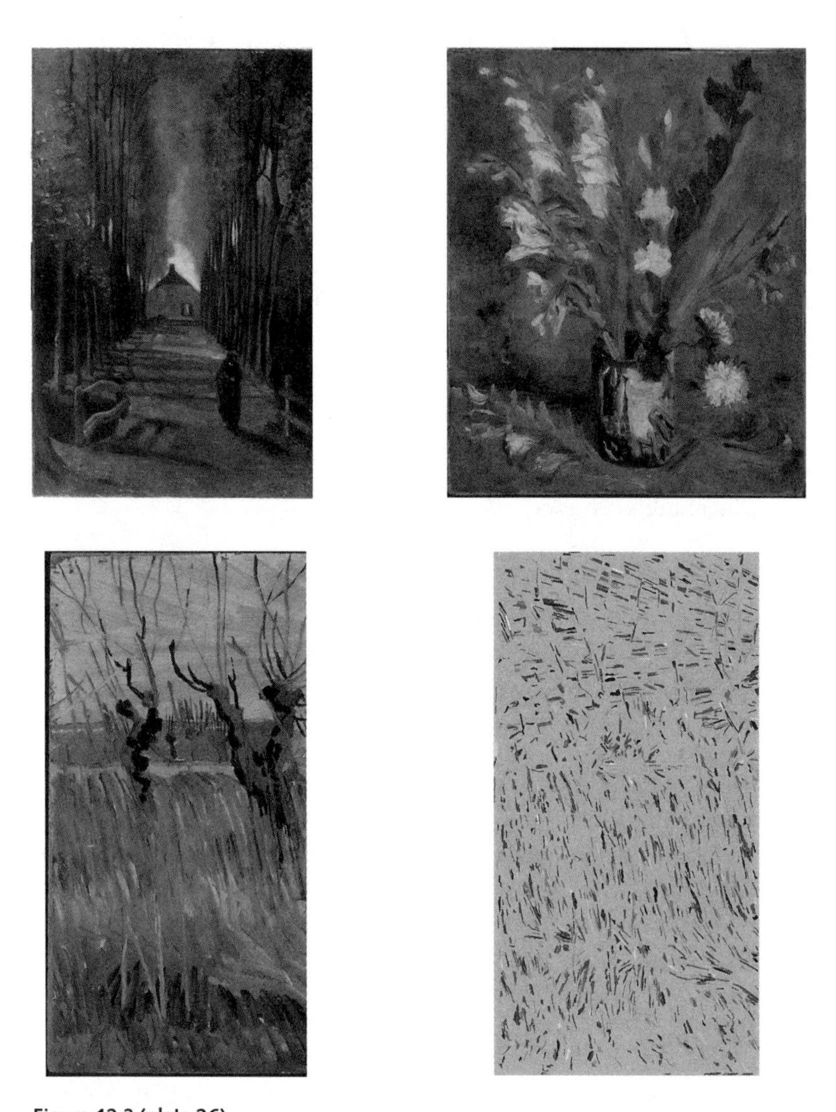

Figure 12.3 (plate 26)
Paintings by Vincent Van Gogh (1853–1890): (Top left) *Avenue of Poplars in Autumn*, 1884 Nuenen. (Top right) *Still Life: Vase with Gladioli and Chinese Asters*, 1886 Paris. (Bottom left) *Willows at Sunset*, 1888 Arles. (Bottom right) Automatically extracted brushstrokes for *Willows at Sunset*. Notice the widely different nature and use of colors in the paintings. Top images courtesy of Van Gogh Museum Amsterdam (Vincent van Gogh Foundation). Bottom images courtesy of Collection Kröller-Müller Museum, Otterlo, The Netherlands (left) and James Z. Wang Research Group at Penn State (right).

With the rise of Expressionism, the blending of reality and artists' emotions became vogue. Expressionist artists freely distorted reality into a personal emotional expression. Abstract expressionism, a post–World War II phenomenon, put the United States in the center stage of art for the first time in history. Intense personal expression combined with spontaneity and hints of subconscious and surreal emotion gave a strikingly new meaning to art, and possibilities of creation became virtually unbounded. Although there has recently been some work on analyzing aesthetics in paintings (C. C. Li & Chen, 2009; Taylor, 2004; Taylor, Micolich, & Jones, 1999), such work is usually limited to a small-scale specific experimental setup. One such work (Taylor, 2004) scientifically examines the works of Mondrian and Pollock, two stalwarts of modern art with drastically distinct styles (the former attempted to achieve spiritual harmony in art via geometrics and the latter became known for his mixing of sand and broken glass with paint as well as for his unconventional paint drip technique).

Aesthetics, Emotions, and Psychology

There are several main areas and directions of experimental research related to psychology that focus on art and aesthetics: *experimental aesthetics* (psychology of aesthetics), *psychology of art*, and *neuroaesthetics*. These fields are interdisciplinary and draw on knowledge in other related disciplines and branches of psychology.

Experimental aesthetics is one of the oldest branches of experimental psychology, which officially begins with the publishing of Fechner's "Zur experimentalen Ästhetik" in 1871, and *Vorschule der Aesthetik* in 1876 (Fechner, 1871, 1876). Fechner suggested three methods for use in experimental aesthetics, (1) including the method of choice, where subjects are asked to compare objects with respect to their "pleasingness"; (2) the method of production, where subjects are required to produce an object that conforms to their tastes by drawing or other actions; and (3) the method of use, which analyzes works of art and other objects on the assumption that their common characteristics are those that are most approved in society.

Developments in other areas of psychology of the early decades of the twentieth century contributed to the psychology of aesthetics. Gestalt psychology produced influential ideas such as the concept of goodness of patterns and configurations emphasizing regularity, symmetry, simplicity, and closure (Koffka, 1935). In the 1970s Berlyne revolutionized the field of experimental aesthetics by bringing to the forefront of investigation psychophysiological factors and mechanisms underlying aesthetic behavior. In his seminal book *Aesthetics and Psychobiology* (Berlyne, 1971), Berlyne formulated several theoretically and experimentally substantiated ideas that helped shape modern experimental research in aesthetics into the science of aesthetics (Palmer, 2009).

Berlyne's ideas and research directions together with the advances in understanding of neural mechanisms of perception, cognition, and emotion in psychology (Solso,

2003), psychophysiology, and neuroscience and facilitated by modern imaging techniques led to the emergence of neuroaesthetics in the 1990s (Kawabata & Zeki, 2004; Kirk, Skov, Hulme, Christensen, & Zeki, 2009; Ramachandran & Hirstein, 1999; Zeki, 1999). Recent studies associated with the processing fluency theory by Reber et al. (2004) suggest that aesthetic experience is a function of the perceiver's processing dynamics: the more fluently the perceiver can process an image, the more positive is her or his aesthetic response.

Key Problems in Aesthetics and Emotions Inference

Many different problems have been studied under the umbrella of aesthetics and emotions evoked from pictures and paintings. Although different problem formulations are focused on achieving different high-level goals, the underlying process is always aimed at modeling an appeal, aesthetics, or emotional response that a picture, a collection of pictures, or a piece of art evokes in people. We divide this discussion into two sections. The first section is devoted to mathematically formulating the core aesthetics and emotions prediction problems. In the second section we discuss some problems that are directly or indirectly derived from the core aesthetics or emotions prediction problems in their scope or application.

Core Problems

Aesthetics Prediction

We assume that an image I has associated with it a true aesthetics measure, $q(I)$, which is the asymptotic average if the entire population rated it. The average over the size n sample of ratings, given by $\hat{q}(I)=\frac{1}{n}\sum_{i=1}^{n} r_i(I)$, is an estimator for the population parameter $q(I)$ where $r_i(I)$ is the ith rating given to image I. Intuitively, a larger n gives a better estimate. A formulation for aesthetics score prediction is therefore to infer the value of $\hat{q}(I)$ by analyzing the content of image I, which is a direct emulation of humans in the photo-rating process. This lends itself naturally to a regression setting whereby some abstractions of visual features act as predictor variables and the estimator for $\hat{q}(I)$ is the dependent variable. An attempt at regression-based score prediction has been reported by Datta et al. (2006), where the quality of score prediction is assessed in the form of rate or distribution of error.

It has been observed by both Datta et al. (2006) and Ke et al. (2006) that score prediction is a highly challenging problem, mainly due to noise in user ratings. To make the problem more solvable, the regression problem is changed to one of classification by thresholding the average scores to create high- versus low-quality image classes (Datta et al., 2006) or professional versus snapshot image classes (Ke et al., 2006). An easier problem, but one of practical significance, is that of selecting a few representative high-quality or highly aesthetic photographs from a large collection. In this case it is important to ensure that most of the selected images are of high quality even though many of those not selected may be of high quality as well. An attempt at this problem (Datta et al., 2007) has proven to be more successful than the general classification problem. The classification problem solutions can be evaluated by standard accuracy measures (Datta et al., 2006; Ke et al., 2006). Conversely, the selection of high-quality photos needs only to maximize the precision in high quality within the top few photos, with recall being less critical.

An aesthetics score can potentially capture finer gradations of aesthetics values, and hence a score predictor would be more valuable than an aesthetics class predictor. However, score prediction requires training examples from all spectrums of scores in the desired range, and hence the learning problem is much more complex than the class prediction (which can typically be translated into a multiclass classification problem well known in machine learning). Opportunities lie in learning and predicting "distributions of aesthetics values" instead of singular aesthetics classes or scores. Scores or values, being ordinal rather than categorical in nature, can be mapped to the real number space. Learning distribution of aesthetics on a per-image basis can throw useful light on human perception and help algorithmically to segment people into "perception categories." Such research can also help characterize various gradations of "artist aesthetics" and "consumer aesthetics" and study how they influence one another, perhaps over time. An effort in this direction has been made by Wu, Hu, and Gao (2011).

Emotion Prediction

If we group emotions that natural images arouse into categories such as "pleasing," "boring," and "irritating," then emotion prediction can be conceived as a multiclass classification problem (Yanulevskaya et al., 2008). Consider that there are K emotion categories, and people select one or more of these categories for each image. If an image I receives votes in the proportion, $\Pi_1(I),\ldots,\Pi_K(I)$, then two possible questions arise:

Most dominant emotion: We wish to predict, for an image I, the most voted emotion category $k(I)$ as given by $k(I) = armax_i \Pi_i(I)$. The problem is meaningful only when there is clear dominance of $k(I)$ over others.
Emotion distribution: We wish to predict the distribution of votes (or an approximation) that an image receives from users, that is, $\Pi_1(I),\ldots,\Pi_K(I)$, which is well suited when images are fuzzily associated with multiple emotions.

The "most dominant emotion" problem is assessed as a standard multiclass classification problem. For "emotion distribution," assessment requires a measure of similarity between discrete distributions, for which Kullback-Leibler (KL) divergence is a possible choice.

Whereas the most dominant emotion prediction translates the problem into a multiclass classification problem that has successfully been attempted in machine learning, emotion distribution would be more realistic from a human standpoint. Human beings rarely associate definitive emotions with pictures. In fact, it is believed that great works of art evoke a "mix of emotions," leaving little space for emotional purity, clarity, or consistency. However, learning a distribution of emotions from pictures requires a large and reliable emotion ground truth data set. At the same time, emotional categories are not completely independent (e.g., there may be correlations between "boring" and "irritating"). One of the key open issues in this problem is settling on a set of plausible emotions that are experienced by human beings. Opportunities also lie in attempting to explore the relationships (both causal and semantic) between human emotions and leveraging them for prediction.

Associated Problems

Image Appeal, Interestingness, and Personal Value

Often, the appeal that a picture makes on a person or a group of people may depend on factors not easily describable by low-level features or even image content as a whole. Such factors could be sociocultural, demographic, purely personal (e.g., "a grandfather's last picture"), or influenced by important events, vogues, fads, or popular culture (e.g., "a celebrity wedding picture"). In the age of ever-evolving social networks, "appeal" can also be thought of as being continually reinforced within a social media framework. Facebook allows users to "like" pictures, and it is not unusual to find "liking" patterns governed by one's friends and network (e.g., a person is likely to "like" a picture in Facebook if many of her friends have done so). Flickr's interestingness attribute is another example of a community-driven measure of appeal based on user-judged content and community reinforcement.

A user study to determine factors that would prevent people from including a picture in their albums was reported by Savakis, Etz, and Loui (2000). Factors such as "not an interesting subject," "a duplicate picture," "occlusion," or "unpleasant expression" were found to dominate the list. Attributing multidimensional image value indices (IVI) to pictures based on their technical and aesthetic qualities and social relevance has been proposed by Louis, Wood, Scalise, and Birkelund (2008). While technical and aesthetic IVIs are driven by learned models based on low-level image information, an intuitive social IVI methodology can be adherence to social rules learned jointly from users' personal collections and social structure. An example could be to give higher weights to immediate family members than to cousins, friends, and neighbors in judging a picture's worth (Louis et al., 2008).

Although a personal or situational appeal or value would be of greater interest to a nonspecialist user, generic models for appeal may be even more short-lived than those for aesthetics. In order to make an impact the problems within this category must be carefully tailored toward learning personal or situational preferences. From an algorithmic perspective, total dependence on visual characteristics for modeling and predicting consumer appeal is a poor choice, and it is desirable to employ image metadata such as tags, geographical information, time, and date. Inferring relationships between people based on the faces and their relative geometric arrangements in photos could also be a very useful exercise (Gallagher & Chen, 2009).

Aesthetics and Emotions in Artwork Characterization

Artistic use of paint and brush can evoke a myriad of emotions among people. These are tools that artists employ to convey their ideas and feelings visually, semantically, or symbolically. Thus, they form an important part of the study of aesthetics and emotions as a whole. Painting styles and brushstrokes are best understood and explained by art connoisseurs. However, research in the last decade has shown that models built using low-level visual features can be useful aids to characterize genres and painting styles or for retrieval from large digitized art galleries (Berezhnoy, Postma, & van den Herik, 2005, 2007; Falco, 2007; Kroner & Lattner, 1998; Kushki, Androutsos, Plataniotis, & Venetsanopoulos, 2004; Rockmore, Lyu, & Farid, 2006). In an effort to encourage computational efforts to analyze artwork, the Van Gogh and Kröller-Müller museums in the Netherlands have made 101 high-resolution grayscale scans of paintings available to several research groups (Johnson et al., 2008).

Brushstrokes provide reliable modeling information for certain types of paintings that do not have colors. In J. Li and Wang (2004), mixtures of stochastic models have been used to model an artist's signature brushstrokes and painting styles. The research provides a useful methodology for art historians who study connections among artists or periods in the history of art. Another important formulation of this characterization problem has been discussed by Bressan, Cifarelli, and Perronnin (2008). The work constructs an artists' graph wherein the edges between two nodes are representative of some measure of collective similarities between paintings of the two artists (and in turn influence of artists on one another). A valuable problem to the commercial art community is to model and predict a common man's perception and appreciation of art as opposed to that of art connoisseurs (C. C. Li & Chen, 2009).

An interesting application of facial expression recognition technology has been shown to be the decoding of the expression of portraits such as the Mona Lisa to get an insight into the artist's mind (Lupsa, 2006). Understanding the emotions that paintings arouse in humans is yet another aspect of this research. A method that categorizes emotions in art based on ground truth from psychological studies has been described by Yanulevskaya et al. (2008), wherein training was performed using a well-known image data set in psychology with the approach demonstrated on art masterpieces.

Problems discussed within this category range from learning nuances of brush-strokes to analyzing emotions that artworks arouse in humans and even emotions depicted in the artworks themselves. This is a challenging area, and the research is expected to be helpful to curators of art as well as to commercial art vendors. However, contribution here would in most scenarios benefit from direct inputs of art experts or artists themselves. Because most of the paintings that are available in museums today were done before the twentieth century, obtaining first-hand inputs from artists is impossible. However, such research aims to build healthy collaborations between the art and computer science research communities, some of which are already evident today (Johnson et al., 2008).

Aesthetics, Emotions, and Attractiveness

Another manifestation of emotional response is attraction among human beings, especially to members of the opposite sex. The psychology of attraction may be multidimensional, but an important aspect of attraction is the perception of a human face as beautiful. Understanding beauty has been an important discipline in experimental psychology (Valentine, 1962). Traditionally, beauty was synonymous with perfection and hence, symmetric or perfectly formed faces were considered attractive. In later years psychologists conducted studies that indicated subtle asymmetry in faces is perceived as beautiful (Scheib, Gangestad, & Thornhill, 1999; Swaddle & Cuthill, 1995; Zaidel & Cohen, 2005). Therefore, it seems that computer vision research on asymmetry in faces (such as that of Liu, Schmidt, Cohn, & Mitra, 2003) can be integrated with psychological theories to computationally understand the dynamics of attractiveness. Another perspective is the theory that facial expression can affect the degree of attractiveness of a face (Doherty et al., 2003). The cited work uses advanced MRI techniques to study the neural response of the human brain to a smile. The current availability of Web resources has been leveraged to formulate judging facial attractiveness as a machine learning problem (Davis & Lazebnik, 2008).

Research in this area is tied to work in face and facial expression recognition. There are controversial aspects of this research in that it tries to prototype attraction or beauty by visual features. Although the problem is approached here purely from a research perspective, the overtones of the research may not be well accepted by the community at large. Beauty and attraction are personal things, and many people would dislike having these attributes rated on a scale. It should also be noted that beauty contests also assess the complete personality of participants and do not judge merely by visual aspects.

Aesthetics, Emotions, and Image Retrieval

Although image retrieval largely involves generic semantics modeling, certain interesting offshoots that involve feedback, personalization, and emotions in image retrieval have also been studied (Wang & He, 2008). Human factors such as those mentioned above largely provide a way to rerank images or search among equals for matches closer to the heart of a user. Bianchi-Berthouze (2003) describe an image-filtering system that uses the Kansei methodology to associate low-level image features with human feelings and impressions. Another work (Fang, Geman, & Boujemaa, 2005) attempts to model the target image within the mind of a user using relevance feedback to learn a distribution over the image database. In a recent work the attractiveness of images was used to enhance the performance of a Web image search engine in terms of online ranking, interactive reranking, and offline index selection (Geng, Yang, Xu, Hua, & Li, 2011). Along similar lines, Redi and Merialdo (2012) integrated semantic, aesthetic, and affective features to achieve a significant improvement in the task of scene recognition on various diverse and large-scale datasets.

Of late there is emphasis on human-centered multimedia information processing, which also touches aspects of retrieval. However, such research is not easily evaluable or verifiable because, again, the level of subjectivity is very high. One potential research direction is to assess the tradeoff between personalization of results and speed of retrieval.

Computational Framework

From a computational perspective, we need to consider steps that are necessary to obtain a prediction (some function of the aesthetics or emotional response) from an input image. We divide this discussion into two distinct sections, feature representation and modeling and learning, and elucidate how researchers have approached each of these computational aspects with respect to the current field. However, before we move forward, it is important that we understand and appreciate certain inherent *gaps* when any image-understanding problem is addressed in a computational way. Smeulders et al. introduced the term *semantic gap* in their pioneering survey of image retrieval to summarize the technical limitations of image understanding (Smeulders, Worring, Santini, Gupta, & Jain, 2000). In an analogous fashion the technical challenge in automatic inference of aesthetics is defined by Datta, Li, and Wang (2008) as the *aesthetics gap*, as follows: the aesthetics gap is the lack of coincidence between the information that one can extract from low-level visual data (i.e., pixels in digital images) and the aesthetics response or interpretation of emotions that the visual data may arouse in a particular user in a given situation.

Features and Representation

In the last decade and a half there have been significant contributions to the field of feature extraction and image representation for semantics and image understanding (Datta et al., 2007). Aesthetics and emotional values of images have bearings on their semantics, and so it is not surprising that feature extraction methods are borrowed or inspired from the existing literature. There are psychological studies that show that aesthetic response to a picture may depend on several dimensions such as composition, colorfulness, spatial organization, emphasis, motion, depth, or presence of humans (Axelsson, 2007; Freeman, 2007; Peters, 2007). Conceiving meaningful visual properties that may have a correlation with perceived aesthetics or an emotion is itself a challenging problem. In literature we notice a spectrum from very generic color, texture, and shape features to specifically designed feature descriptors to model the aesthetic or emotional value of a picture or artwork. We do not intend to provide an exhaustive list of feature descriptors here but rather attempt to discuss significant feature usage patterns.

Photographers generally follow certain principles that can distinguish professional shots from amateur ones. A few such principles are the rule of thirds, use of complementary colors, and close-up shots with high dynamic ranges. The rule of thirds is a popular one in photography. It specifies that the main element or the center of interest in a photograph should lie at one of the four intersections (figure 12.4, plate 27). Datta et al. (2006) defined the degree of adherence to this rule as the average hue, saturation, and intensities within the inner third region of a photograph. They also noted that pictures with simplistic composition and a well-focused center of interest are more pleasing than pictures with many different objects. Professional photographers often reduce the *depth of field* (DOF) to shoot single objects by using larger aperture settings, macro lenses, or telephoto lenses. DOF is the range of distance from a camera that is acceptably sharp in a photograph (figure 12.4, plate 27). Datta et al. (2006) used wavelets to detect a picture with a low DOF. However, a low DOF has a positive aesthetic

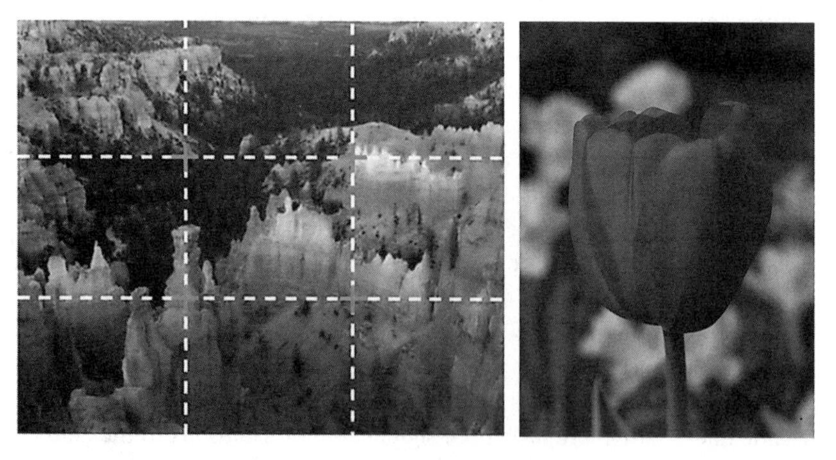

Figure 12.4 (plate 27)
(Left) The rule of thirds in photography. (Right) A low-depth-of-field picture.

appeal only in an appropriate context and may not always be desirable (e.g., in photography, landscapes with narrow DOF are not considered pleasing; instead, photographers prefer to have the foreground, middle ground, and background all in focus).

A mix of global and local features has been used by C. C. Li and Chen (2009) to model the aesthetics problem for paintings. Feature selection is based on the belief that people use a top-down approach to appreciate art. Prominent factors that determine the choice of features include measuring blur, which is seen as an important artistic effect, and presence and distribution of edges because edges are used by artists for emphasis. The perceptual qualities that differentiate professional pictures from snapshots based on input from professional and amateur photographers are identified by Ke et al. (2006), who found that professional shots are distinguished by (1) a clear distinction between subject and background brought about by choice of complementary colors, higher contrast between subject and background, or a small depth of field, and (2) a surrealism created by the proper choice of camera parameters and appropriate lighting conditions.

While low-level color and texture features capture useful information, modeling spatial characteristics of pixels or regions and spatial relationships among regions in images has also been shown to be very helpful. A computational visual attention model using a face-sensitive saliency map has been proposed by Sun and Yao (2009). A rate of focused attention measure (using the saliency map and the main subject of the image) is proposed as an indicator of aesthetics. The method employs a subject mask generated using several hundreds of manually annotated photos for computation of attention. Yang et al. (2010) propose an interesting pseudogravitational field-based visual attention model in which each pixel is assigned a mass based on its luma and chroma values (YCbCr space) and pixels exert a gravity-like mutual force.

Some recent papers focus on enhancement of images or suggestion of ideal composition based on aesthetically learned rules (Bhattacharya, Sukthankar, & Shah, 2010; Cheng, Ni, Yan, & Tian, 2010). Two distinct recomposition techniques based on key aesthetic principles (rule of thirds and the golden ratio) have been proposed (Bhattacharya et al., 2010). The algorithm performs segmentation of single-subject images into "sky," "support," and "foreground" regions. Two key aesthetically relevant segment-based features are introduced in this work: the first computes the position of the visual attention center with respect to focal stress points in the image (rule of thirds), and the second feature measures the ratio of weights of support and sky regions (expected to be close to the golden ratio). Another interesting work (Cheng et al., 2010) models local and far contexts from aesthetically pleasing pictures to determine rules that are later applied to suggest good composition to new photographers. According to the authors, whereas local context represents visual continuity, far context models the arrangement of objects/regions as desirable by expert photographers. Contextual modeling involves learning a spatial Gaussian mixture model for pairwise visual words. A recent work (Luo, Wang, & Tang, 2011) explores the role of content in image aesthetics by designing specific visual features for different categories (e.g., landscape, plant, animal, night, human, static, and architecture). The work focuses on detecting and extracting local features from the most attractive image region (from among region of focus, vertical standing objects, or human faces).

Several recent papers have emphasized the usability of generic descriptors constructed by local features for image aesthetics. Along this line bag of visual words and Fisher vectors (which encode more local information) have been explored to improve the accuracy of image aesthetics assessment (Marchesotti, Perronnin, Larlus, & Csurka, 2011). Gradient information is extracted through SIFT, and color features and significant improvements (over previous works) have been reported. The influence of the color harmony of photos on their aesthetic quality has been investigated (Nishiyama, Okabe, Sato, & Sato, 2011). By representing photos as a collection of local regions, the work models the color harmony (as predictor of aesthetic quality) of photos through bags of color patterns. Patchwise bag-of-aesthetics-preserving features that encode contrast information have been explored (Su, Chen, Kao, Hsu, & Chien, 2011). Donovan and colleagues model the quality of color themes that refer to a five-color palette by learning from a large-scale data set with a regression method (Donovan, Agarwala, & Hertzmann, 2011).

Although there exists some concrete rationalization for feature design with respect to the aesthetics inference problem, the design of features that capture emotions is still a challenge. Yanulevskaya et al. (2008) diverge from the common codebook approach to a methodology in which similarity to all vocabulary elements is preserved for emotion category modeling. Bressan et al. (2008) extract low-level local visual features including SIFT and color histograms, and a Fisher kernel-based image similarity is used to construct a graph of artists to discover mutual and collective artistic influence. Associating low-level image features with human feelings and impressions can also be achieved by using ideas from Kansei engineering (Bianchi-Berthouze, 2003) with sets of neural networks to learn mappings between low-level image features and high-level impression words.

Concepts from psychological studies and art theory are used to extract image features for emotion recognition in images and art (Machajdik & Hanbury, 2010). Among other features, Machajdik and Hanbury (2010) adopt the standardized pleasure-arousal-dominance transform color space, composition features such as low-depth-of-field indicators, and rule of thirds (which have been found to be useful for aesthetics), and proportion of skin pixels in images. Eye gaze analysis yields an affective model for objects or concepts in images (Ramanathan, Katti, Huang, Chua, & Kankanhalli, 2009). More specifically, eye fixation and movement patterns learned from labeled images are used to localize affective regions in unlabeled images. Affective responses in the form of facial expressions have also been explored by Arapakis,

Konstas, and Jose (2009) to understand and predict topical relevance. This work models neurological signals and facial expressions of users looking at images as implicit relevance feedback. In order to classify emotions Arapakis et al. (2009) employ a 3D wire-frame model of faces and track the presence and degrees of changes in different facial regions. Similarly, Valenti et al. (2010) also employ face tracking to extract facial motion features for emotion classification.

In recent work Lu et al. (2012) explored the relationship between shape characteristics (such as roundness, angularity, simplicity, and complexity) and emotions. Shape features constitute line segments, continuous lines, angles, and curves to reflect such characteristics. In an interesting diversion, inferring aroused emotions from images in social networks has been studied by Jia et al. (2012). The work represents the emotion by 16 discrete categories that cover the affective space. Color features (e.g., saturation, brightness, and HSV) and social features (e.g., uploading time and user ID) were extracted as image descriptors.

Finally, psychological theories of perception of beauty (discussed previously) also aid researchers who design features for facial attractiveness modeling using a mix of facial geometry features (Davis & Lazebnik, 2008; Eisenthal, Dror, & Ruppin, 2006) as well as nongeometric ones such as hair color and skin smoothness (Eisenthal et al., 2006).

Modeling and Learning

Aesthetics and emotion modeling literature reports use of both discriminative learning methods such as SVM and CART (Datta et al., 2006; C. C. Li & Chen, 2009; Louis et al., 2008; Yanulevskaya et al., 2008) and generative learning techniques such as naive Bayes, Bayesian networks, and Gaussian mixture models (Cheng et al., 2010; J. Luo, Savakis, Etz, & Singhal, 2000; Machajdik & Hanbury, 2010; Valenti et al., 2010). Although two-class or multiclass classification paradigms seem to be the norm, support vector and kernel regression methods have also been explored (Bhattacharya et al., 2010; Davis & Lazebnik, 2008). An adapted regression approach to map visual features extracted from photos to a distribution has been presented (Wu et al., 2011). A dimensional approach to represent emotions (to capture correlations between emotional words) has been explored (Lu et al., 2012). Jia et al. (2012) present a partially labeled factor graph model to infer the emotions aroused from images within a social network setting. A bilayer sparse representation is proposed to encode similarities among global images, local regions, and the regions' co-occurrence property (B. Li, Xiong, Hu, & Ding, 2012). The proposed context-aware classification model with the bilayer sparse representation shows a higher accuracy in predicting categorized emotions on the IAPS data set. In conclusion we can state that although learning lies at the heart of every computational inference problem that we consider here, choices of the modeling and learning strategies vary with the nature of the task and features.

Data Resources

Data from Controlled Studies

Methods for experimental investigation of aesthetic perception and preferences and associated emotional experience vary from traditional collection of verbal judgments along aesthetic dimensions, to multidimensional scaling of aesthetic value and other related attributes, to measuring behavioral, psychophysiological, and neurophysiological responses to art pieces and images in controlled and free viewing conditions. The arsenal of measured response is vast, a few instances being reaction time, various electrophysiological responses that capture activity of the central and autonomic nervous systems, such as an electroencephalogram (EEG), electrooculogram, heart rhythm, pupillary reactions, and, more recently, neural activity in various brain areas obtained using functional magnetic resonance imaging (fMRI) (Doherty et al., 2003; Kirk et al., 2009). Recording eye movements is also a valuable technique that helps detect where the viewers are looking when evaluating aesthetic attributes of art compositions (Nodien, Locher, & Krupinski, 1993).

Certain efforts have resulted in the creation of a specialized database for emotion studies known as the International Affective Picture Systems (IAPS) database (figure 12.5; Lang, Bradley, & Cuthbert, 1997). This collection contains a diverse set of pictures that depict animals, people, activities, and nature and has been categorized mainly in valences (positive, negative, no emotions) along various emotional dimensions (Yanulevskaya et al., 2008).

Data from Community-Contributed Resources

Obtaining controlled experimental data is expensive in time and cost. At the same time, converting user response (captured as described above) to categorical or numerical aesthetics or emotional parameters is another challenge. One should also note that controlled studies are not scalable in nature and can yield only limited human response in a given time. Researchers increasingly turn to the Web, a potentially boundless resource for information. In the last few years a growing phenomenon called *crowd sourcing* has hit the Web. By definition, crowd sourcing is the process by which Web users contribute collectively to the useful information on the Web (Howe, 2006). Several Web photo resources take advantage of these contributions to make their content more visible, searchable, and open to public discussions and feedback. Tapping such resources has proven useful for research in our discussion domain. Here we briefly describe some Web-based data resources.

Flickr (www.flickr.com) is one of the largest online photo-sharing sites in the world. Besides being a platform for photography, tagging, and blogging, Flickr captures contemporary community interest in the form of an interestingness feature. According to Flickr, interestingness of a picture is dynamic and depends on a plurality of criteria

Figure 12.5
(Top) Pictures of Yosemite National Park from Terragalleria.com. (Bottom) Example images from the International Affective Picture System (IAPS) data set. Images with a more positive affect from left to right and higher arousal from bottom to top.

including its photographer, who marks it as a favorite, comments, and tags given by the community.

Photo.Net (www.photo.net) is a platform for photography enthusiasts to share and have their pictures peer-rated on a 1–7 scale of aesthetics. The photography community also provides discussion forums, reviews on photos and photography products, and galleries for members and casual surfers.

DPChallenge (www.dpchallenge.com) allows users to participate and contest in theme-based photography on diverse themes such as life and death, portraits, animals, geology, street photography. Peer-rating on overall quality, on a 1–10 scale, determines the contest winners.

Terragalleria (www.terragalleria.com) showcases travel photography of Quang-Tuan Luong (a scientist and a photographer) and is one of the finest resources for U.S. national park photography on the Web (Figure 12.5). All photographs here have been taken by one person (unlike Photo.Net), but multiple users have rated them on overall quality on a 1–10 scale.

ALIPR–Automatic Photo Tagging and Visual Image Search (www.alipr.com) is a Web-based image search and tagging system that also allows users to rate photographs along 10 different emotional categories such as surprising, amusing, pleasing, exciting, and adorable.

Besides this, certain research efforts have created their own collections of data from the above sources, notably (1) a manually labeled data set with over 17,000 photos covering seven semantic categories (J. Luo, Boutell, & Brown, 2006), and (2) the AVA data set to facilitate aesthetics visual analysis (Murray, Marchesotti, & Perronnin, 2012), consisting of about 250,000 images from DPChallenge.

Data Analysis

Feature Plots of Aesthetics Ratings

We performed a preliminary analysis of the above data sources to compare and contrast the different rating patterns. A collection of images (14,839 images from Photo.net, 16,509 images from DPChallenge, 14,449 images from Terragalleria, and 13,010 emotion-tagged images from ALIPR) was formed, drawing at random, to create real-world data sets. These can be used to compare competing algorithms in the future. Here we present plots of features of the data sets, in particular the nature of user ratings received in each case (not necessarily comparable across the data sets).

We first describe the nature of the plots. Figure 12.6 shows the distribution of mean aesthetics. Figure 12.7 shows the distribution of the number of ratings each photo received. In figure 12.8, the number of ratings per photo is plotted against the average score received by it in an attempt to visualize possible correlation between the number of ratings and the average ratings each photo received. In figure 12.9, we plot

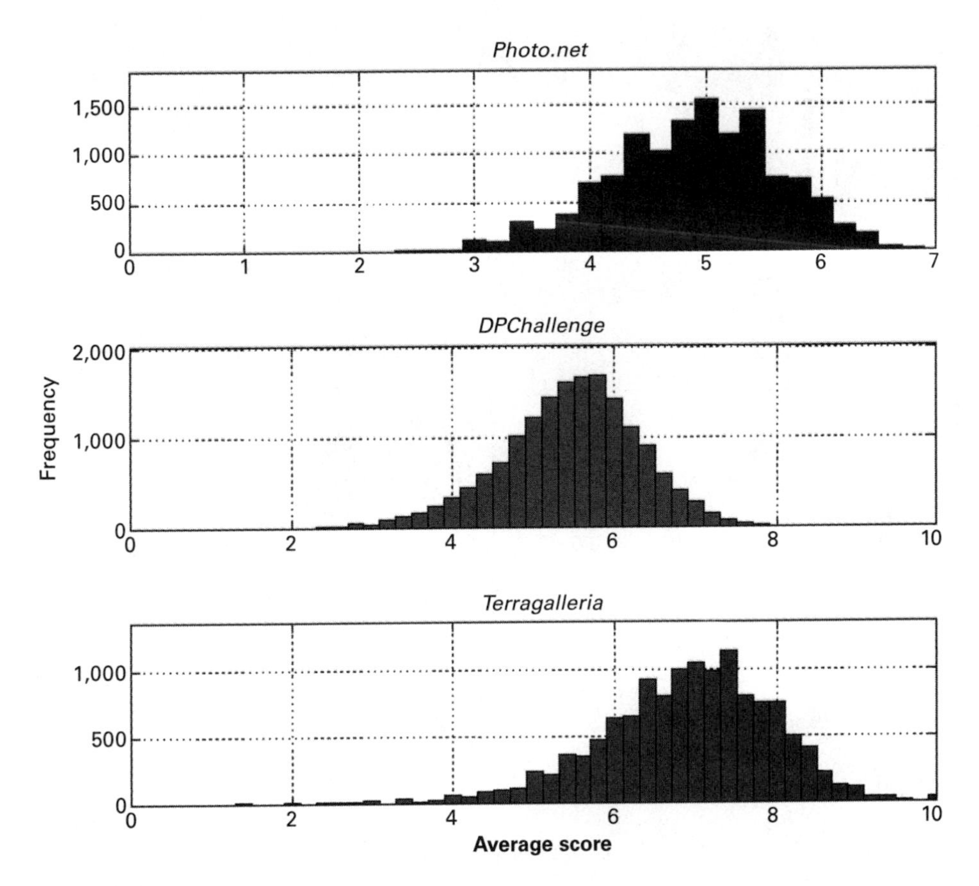

Figure 12.6
Distributions of average aesthetics scores from three different data collections.

the distribution of the fraction of ratings received by each photo within ± 0.5 of its own average. In other words we examine every score received by a photo, find the average, count the number of ratings that are within ± 0.5 of this average, and take the ratio of this count to the total number of ratings this photo received. This is the ratio whose distribution we plot. Each of the aforementioned figures comprises this analysis separately for each collection (Photo.net, Terragalleria, and DPChallenge). Finally, in figure 12.10 we plot the distribution of emotions votes in the dataset sampled from ALIPR. In the following section we analyze each of these plots separately and share with readers the insights drawn from them.

Analysis of Feature Plots

When we look closely at each of the plots in figures 12.6–12.10, we obtain insights about the nature of human ratings of aesthetics. Broadly speaking, we note that this

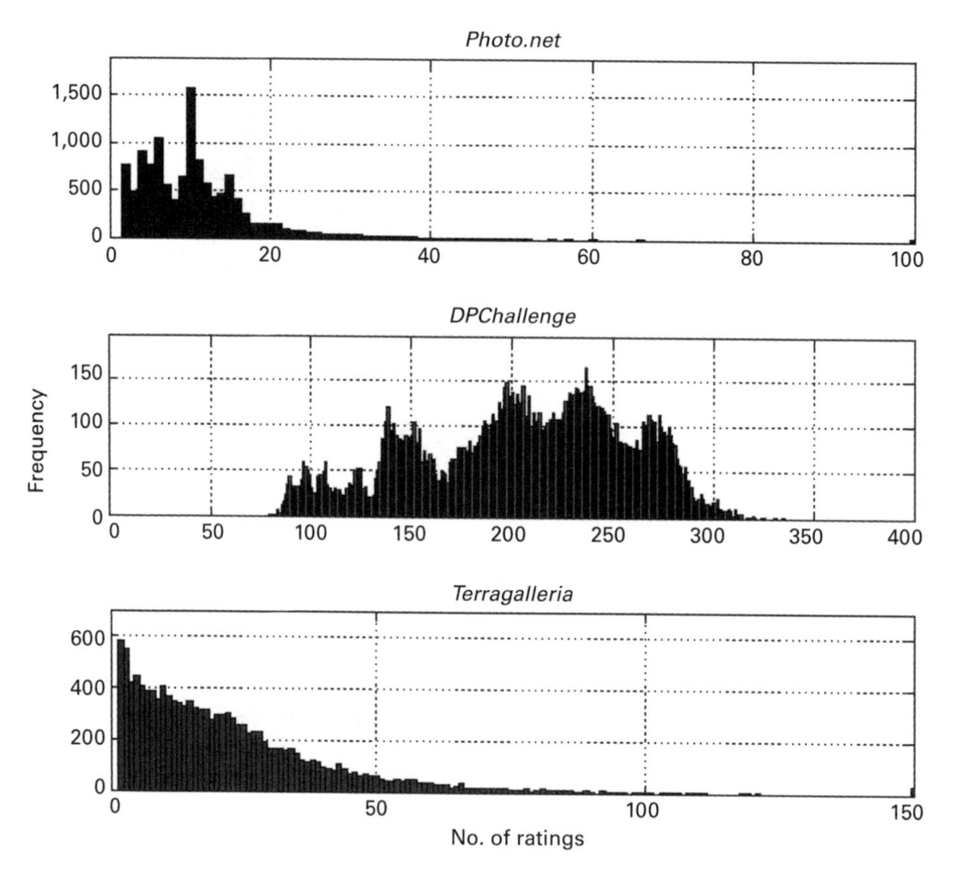

Figure 12.7
Distributions of number of ratings from three different data collections.

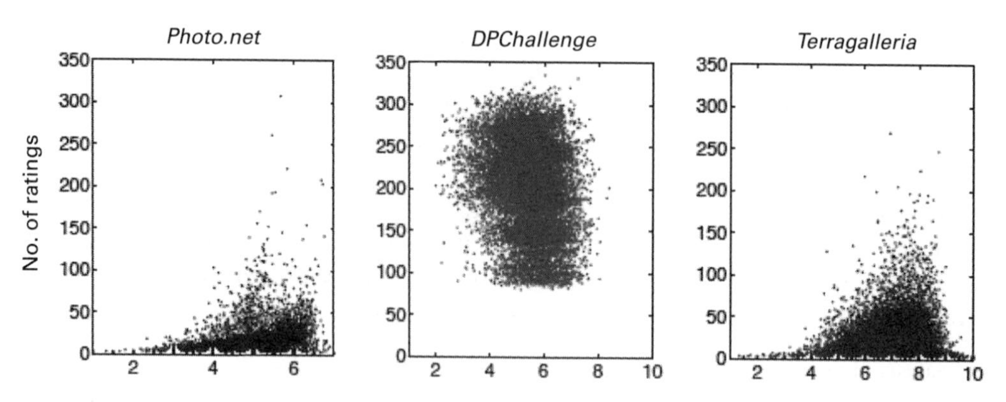

Figure 12.8
Correlation plot of (average score, number of ratings) pairs.

analysis pertains to the overall social phenomenon of peer rating of photographs rather than the true perception of photographic aesthetic quality by individuals. In Photo.net, for example, users (at least at the time of data collection) could see who rated their photographs. This naturally makes the rating process a social rather a true scientifically unbiased test or process. Another side effect of the rating process is that the photos that people upload for others to rate are generally not drawn at random from a person's broad picture collection. Rather, it is more likely that users select to share what they consider their best shots. This introduces another kind of bias. Models and systems trained on these data therefore learn how people rate each other's photos in a largely nonblind social setting and learn this for only a subset of the images that users consider worthy of being posted publicly, which helps to explain the inherent bias found in the distributions. Conversely, the bias corroborates the assumption that collection of aesthetics ratings in public social forums is primarily a social experiment rather than a principled scientific one.

In figure 12.6 we see that for each data set the peak of the average score distribution lies to the right of the mean position in the rating scale. For example, the peak for Photo.net is approximately 5, which is a full point above the midpoint 4. There are two possible explanations for this phenomenon:

- Users tend to post only those pictures that they consider to be their best shots.
- Because public photo rating is a social process, peers tend to be lenient or generous by inflating the scores that they assign to others' photos as a means of encouragement and also particularly when the Web site reveals the rater's identity.

Another observation we make from figure 12.6 is that the distribution is smoother for DPChallenge than for the other two. This may simply be because this data set has the largest sample size. In figure 12.7, we consider the distribution of the number of ratings each photo received. This graph looks dramatically different for each source. This feature almost entirely reflects the social nature of public ratings rather than anything intrinsic to photographic aesthetics. The most well-balanced distribution is found in DPChallenge, in part because of the incentive structure (it is a time-critical, peer-rated competitive platform). The distribution almost resembles a mixture of Gaussians with means at well-spaced locations. The nature of the social phenomenon on DPChallenge.com that these peaks might be associated with is unclear to the authors. Photos on Photo.net are much rarer, mainly because the process is noncompetitive and voluntary, and the system of soliciting ratings is not designed to attract many ratings per photo. The distribution looks heavy-tailed in the case of Terragalleria, which much more resembles typical rating distribution plots.

The purpose of the plots in figure 12.8 is to determine if there exists a correlation between the number of ratings a photo receives and the average of those ratings. The plots for Photo.net as well as Terragalleria most clearly demonstrate what can be

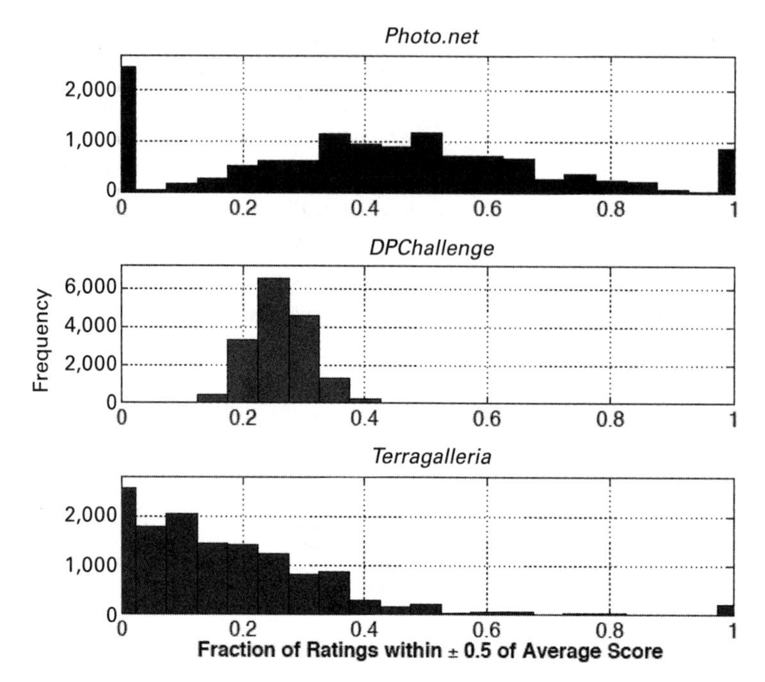

Figure 12.9
Distribution of the level of consensus among ratings.

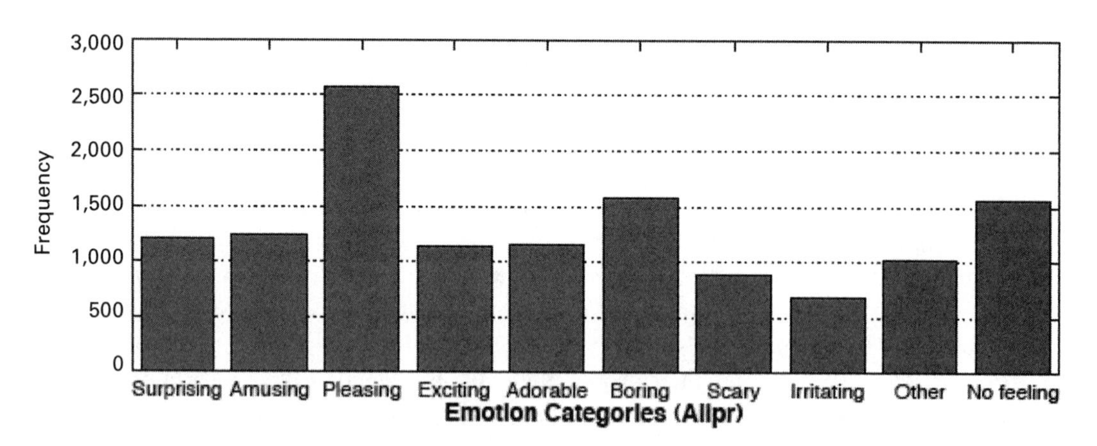

Figure 12.10
Distribution of emotion votes given to images (ALIPR).

anticipated about social peer-rating systems: people rate inherently positively, and they tend to highly rate photos that they like, and not rate at all those they consider to be poor. This phenomenon is not peculiar to photo-rating systems or even social systems: we also observe this clearly in movie-rating systems found in Web sites such as IMDb. Associated with the issue that people tend to explicitly rate mainly things they like is the fact that the Web sites also tend to surface highly rated entities to newer audiences (through top-k lists and recommendations). Together, these two forces help generate much data on good-quality entities while other candidates are left with sparse amounts of feedback and rating. Conversely, DPChallenge, because it is a competitive site, attempts to gather feedback from all candidate photos fairly. Therefore, we see a less biased distribution of its scores, making it unclear whether the correlation is at all significant or not.

In figure 12.9 we plot the distribution of the fraction of ratings received by each photo within $\pm$ 0.5 of its own average. What we expect to see is whether or not most ratings are closer to the average score. In other words do most raters roughly agree with each other for a given photo, or is the variance per photo high for most photos? The observation for Photo.net is that there is a wide and healthy distribution of the fraction of rater agreement, and then there are the boundary conditions. A small but significant fraction of the photos had everyone essentially give the photo the same rating $\pm$ 0.5 (this corresponds to $x = 1$ in the plot). These photos have high consensus or rater agreement. However, three times larger is the fraction of photos to which nearly no one has given a rating close to the average (this corresponds to $x = 0$ in the plot). This occurs primarily when there are two groups of raters: one group that likes the photo and another group that does not. This way, the average lies somewhere between the sets of scores given by the two camps of raters. The distribution looks quite different for DPChallenge: roughly one third of the ratings tend to lie close to the average value while the rest of the ratings lie further apart on either side of average. For Terragalleria, users tend to be less in agreement with each other on ratings. Nearly all of the raters are in agreement on only a small fraction of the photos (corresponding to $x = 1$ in the plot).

Note than the graphs in figure 12.9 are particularly unfit for an apples-to-apples comparison: an absolute difference of 0.5 implies different things for the different Web sites, especially when the score ranges are different. Furthermore, DPChallenge receives so many ratings per photo that it is improbable that all raters would agree on the same score (hence $y = 0$ at $x = 1$ in that graph). Finally, in figure 12.10, we observe that the dominant emotion expressed by Web users while viewing pictures is "pleasing," followed by "boring" and "no feeling." Conversely, "irritating" and "scary" are relatively rare responses. The reason for this may well be what emotions people find easy to attribute to the process of looking at a picture. On the Web we are accustomed to expressing ourselves on like-dislike scales of various kinds. Hence, it is convenient to refer to what one likes as "pleasing" and what one does not like as "boring."

Future Research Directions

Understanding Sociocultural, Personal, and Psychology-Induced Preferences from Data

Social and cultural backgrounds can affect one's judgment of aesthetics or influence one's emotions in a particular scenario. An important future research direction would be to incorporate cultural, social, and personal differences into the learning methodologies. An important starting point can be to determine how many distinct "preference groups" (cultural or social) there are in a population. This could be followed by discovering characteristic rating distributions of scores that differ across different preference groups. Additional personalization can be achieved by understanding tastes of individuals, which will, however, require a significant amount of personalized data for model building.

The emotional and aesthetic impact of art and visual imagery is also linked to the emotional state of the viewer, who, according to the *emotional congruence theory*, perceives his or her environment in a manner congruent with his/her current emotional state (Bower, 1981). Studies have also shown that art preferences and art judgment can vary significantly across expert and nonexpert subjects (Hekkert, 1997). Artists and experienced art viewers tend to prefer artworks that are challenging and emotionally provocative (Winston & Cupchik, 1992), which is in contrast to the majority of people who prefer art that makes them happy and feel relaxed (Wypijewski, 1997). The results reported by Axelsson (2007), Cerosaletti and Loui (2009), and Fedorovskaya, Neustaedter, and Hao (2008) demonstrate that such differences are significant and can be explained on the basis of common mechanisms, as suggested by Berlyne (1971).

Understanding and Modeling Context

Context plays an important role in semantic image understanding (Luo et al., 2006). Context within the purview of images has been explored as spatial context (leveraging spatial arrangement of objects in images), temporal context (leveraging the time and date information when pictures were taken), geographical context (leveraging information about geographical location of pictures) (Kennedy & Naaman, 2007, 2008), and social context (leveraging information about the social circle of a person or social relationship reflected in pictures) (Gallagher & Chen, 2009; Shevade & Sundaram, 2007; Zunjarwad & Sundaram, 2007). For example, people may well associate special emotions with pictures taken on special occasions or about special people in their lives. Similarly, pictures taken during one's trip to a national park may be aesthetically more pleasing than pictures taken in a local park, purely because of their content and opportunities for high-quality shots. Determining the extent to which such factors affect the aesthetic or emotional value of pictures will be a potent future research direction.

Developing Real-World Usable Research Prototypes

Perhaps one of the most important steps in the life cycle of a research idea is its incorporation into a usable and testable system open to the scrutiny of common people. This is important for two reasons: (1) it provides a realistic test bed for evaluating the research machinery, and (2) user reaction and feedback can be very useful in helping the design of future prototypes. In light of this a key future direction could be to take some of the proposed ideas in the current research domain to the next level in their life cycle. We briefly describe Aesthetic Quality Inference Engine (ACQUINE), www.acquine.alipr.com), an attempt in this direction. ACQUINE is a machine-learning-based online system that showcases computer-based prediction of aesthetic quality for color natural photographic pictures (figure 12.1, plate 24). Labeled images from Photo.net have been obtained to achieve supervised learning of aesthetic quality rating models. A number of visual features that are assumed to be correlated with aesthetic quality are extracted from images, and an SVM-based classifier is used to obtain the aesthetic rating of a given picture. Users can upload their own images, use links to images that exist on the Web, or simply browse photographs uploaded by others. They are also able to look at the ratings that were machine given and optionally add their own ratings. This is a valuable source of feedback and labeled data for future iterations of the system. As of May 2011, nearly 250,000 images from nearly 32,000 different users have been uploaded to ACQUINE for automatic rating. Over 65,000 user ratings of photos have also been provided. Another recently developed system, On-Site Composition and Aesthetics Feedback (OSCAR), aims at helping photographers to generate high-quality photos (Yao, Suryanarayan, Qiao, Wang, & Li, 2012). OSCAR provides on-site analyses of photos in terms of the composition and aesthetics quality and generates feedback through high-quality examples.

We envision a future where consumer cameras and smartphones are equipped with an automated personal assistant that can provide aesthetics judgment so that only the highest quality photos are taken and stored. Such a module can be a post-photography filter or a real-time filter (such as a real-time aesthetics meter). A recent effort in this direction is the Nadia camera, which uses ACQUINE to offer a real-time aesthetics score (Nadia Camera, n.d.). Real-time photography feedback is not a stranger to today's photographers (face detection, smile detection, etc.). Hence, the dream of aesthetics feedback in cameras may not be that distant.

Conclusion

In this chapter we have looked at key aspects of aesthetics, emotions, and associated computational problems with respect to natural images and artwork. We discussed these problems in relation to philosophy, photography, paintings, visual arts, and psychology. Computational frameworks and representative approaches proposed to address problems in this domain were outlined, followed by a discussion of available data sets for research use. An analysis of the nature of data and ratings among the available resources was also presented. In conclusion, we laid out a few intriguing directions for future research in this area. We hope that this tutorial significantly increases the visibility of this research area and serves to foster dialogue and collaboration among artists, photographers, and researchers in signal processing, computer vision, pattern recognition, and psychology.

Acknowledgments

The authors acknowledge the constructive comments of the anonymous reviewers. J. Z. Wang and J. Li would like to thank the Van Gogh and Kröller-Müller Museums for providing the photographs of paintings for their study. National Science Foundation Grants IIS-0347148, CCF-0936948, and EIA-0202007 provided partial funding for their research. Part of the work of J. Z. Wang was done while working at the National Science Foundation. Any opinion, findings, and conclusions or recommendations expressed in this material are those of the authors and do not necessarily reflect the views of the Foundation. The work of D. Joshi was done while working at Eastman Kodak Research Labs.

Portions of this chapter appeared in D. Joshi, R. Datta, Q.-T. Luong, E. Fedorovskaya, J. Z. Wang, J. Li, and and J. Luo, Aesthetics and emotions in images: A computational perspective, *IEEE Signal Processing Magazine*, 28(5) (2011), 94–115.

References

Arapakis, I., Konstas, I., & Jose, J. M. (2009). Using facial expressions and peripheral physiological signals as implicit indicators of topical relevance. *Proceedings of ACM Multimedia* (pp. 461–470).

Axelsson, O. (2007). Towards a psychology of photography: Dimensions underlying aesthetic appeal of photographs. *Perceptual and Motor Skills*, *105*, 411–434.

Berezhnoy, I. E., Postma, E. O., & van den Herik, H. J. (2005). Computerized visual analysis of paintings. *Proceedings of 16th International Conference Association for History and Computing* (pp. 28–32).

Berezhnoy, I. E., Postma, E. O., & van den Herik, H. J. (2007). Computer analysis of Van Gogh's complementary colors. *Pattern Recognition Letters*, *28*(6), 703–709.

Berlyne, D. E. (1971). *Aesthetics and psychobiology*. New York: Appleton-Century-Crofts.

Bhattacharya, S., Sukthankar, R., & Shah, M. (2010). A framework for photo-quality assessment and enhancement based on visual aesthetics. *Proceedings of ACM Multimedia* (pp. 271–280).

Bianchi-Berthouze, N. (2003). K-dime: An affective image filtering system. *IEEE MultiMedia*, *10*(3), 103–106.

Bower, G. H. (1981). Mood and memory. *American Psychologist*, *36*, 129–148.

Bressan, M., Cifarelli, C., & Perronnin, F. (2008). An analysis of the relationship between painters based on their work. *Proceedings of IEEE International Conference on Image Processing* (pp. 113–116).

Cerosaletti, C., & Loui, A. (2009). Measuring the perceived aesthetic quality of photographic images. *Proceedings of 1st International Workshop on Quality of Multimedia Experience* (pp. 47–52).

Cheng, B., Ni, B., Yan, S., & Tian, Q. (2010). Learning to photograph. *Proceedings of ACM Multimedia* (pp. 291–300).

Daly, S. (1993). The visible differences predictor: An algorithm for the assessment of image fidelity. In A. B. Watson (Ed.), *Digital images and human vision* (pp. 179–206). Cambridge, MA: MIT Press.

Datta, R., Joshi, D., Li, J., & Wang, J. Z. (2006). Studying aesthetics in photographic images using a computational approach. *Computer Vision ECCV*, *3953*, 288–301.

Datta, R., Joshi, D., Li, J., & Wang, J. Z. (2008). Image retrieval: Ideas, influences, and trends of the New Age. *ACM Computing Surveys*, *40*(2), 1–60.

Datta, R., Li, J., & Wang, J. Z. (2007). Learning the consensus on visual quality for next generation image management. *Proceedings of ACM Multimedia* (pp. 533–536).

Datta, R., Li, J., & Wang, J. Z. (2008). Algorithmic inferencing of aesthetics and emotion in natural images: An exposition. *Proceedings of International Conference on Image Processing* (pp. 105–108).

Davis, B. C., & Lazebnik, S. (2008). Analysis of human attractiveness using manifold kernel regression. *Proceedings of International Conference on Image Processing* (pp. 109–112).

Doherty, J. O., Winston, J., Critchley, H., Perrett, D., Burt, D. M., & Dolan, R. J. (2003). Beauty in a smile: The role of medial orbitofrontal cortex in facial attractiveness. *Neuropsychologia*, *41*, 147–155.

Donovan, P. O., Agarwala, A., & Hertzmann, A. (2011). Color compatibility from large datasets. *ACM Transactions on Graphics*, *30*(4), 1–12.

Eisenthal, Y., Dror, G., & Ruppin, E. (2006). Facial attractiveness: Beauty and the machine. *Neural Computation*, *18*(1), 119–142.

Falco, C. M. (2007). Computer vision and art. *IEEE MultiMedia*, *14*(2), 8–11.

Fang, Y., Geman, D., & Boujemaa, N. (2005). An interactive system for mental face retrieval. *Proceedings of ACM SIGMM International Workshop on Multimedia Information Retrieval* (pp. 193–200).

Fechner, G. T. (1871). Zur experimentalen Ästhetik (On experimental aesthetics). *Abhandlungen der Königlich Sächsischen Gesellschaft der Wissenschaften*, *9*, 555–635.

Fechner, G. T. (1876). *Vorschule der Aesthetik*. Leipzig: Breitkopf und Härtel.

Fedorovskaya, E. A., Neustaedter, C., & Hao, W. (2008). Image harmony for consumer images. *Proceedings of International Conference on Image Processing* (pp. 121–124).

Freeman, M. (2007). *The photographer's eye*. Waltham, MA: Focal Press.

Gallagher, A., & Chen, T. (2009). Using context to recognize people in consumer images. *IPSJ Transactions on Computer Vision and Applications*, *1*, 115–126.

Geng, B., Yang, L. J., Xu, C., Hua, X.-S., & Li, S. P. (2011). The role of attractiveness in web image search. *Proceedings of ACM Multimedia* (pp. 63–72).

Hekkert, P. (1997). Beauty in the eye of expert and non-expert beholders: A study in the appraisal of art. *American Journal of Psychology*, *109*(3), 389–407.

Howe, J. (2006). The rise of crowdsourcing. *Wired Magazine, 14*.

Jia, J., Wu, S., Wang, X. H., Hu, P. Y., Cai, L. H., & Tang, J. (2012). Can we infer Van Gogh's mood? Learning to infer affects from images in social networks. *Proceedings of ACM Multimedia* (pp. 857–860).

Johnson, C. R. J., Hendriks, E., Berezhnoy, I. J., Brevdo, E., Hughes, S. M., Daubechies, I., et al. (2008). Image processing for artist identification: Computerized analysis of Vincent van Gogh's painting brushstrokes. *IEEE Signal Processing Magazine*. *Special Issue on Visual Cultural Heritage*, *25*(4), 37–48.

Kawabata, H., & Zeki, S. (2004). Neural correlates of beauty. *Journal of Neurophysiology*, *91*, 1699–1705.

Ke, Y., Tang, X., & Jing, F. (2006). The design of high-level features for photo quality assessment. *Proceedings of IEEE International Conference on Computer Vision and Pattern Recognition, 1*, 419–426.

Kennedy, L., & Naaman, M. (2007). How Flickr helps us make sense of the world: Context and content in community-contributed media collections. *Proceedings of ACM Multimedia* (pp. 631–640).

Kennedy, L., & Naaman, M. (2008). Generating diverse and representative image search results for landmarks. *Proceedings of the 17th International Conference on World Wide Web* (pp. 297–306).

Kirk, U., Skov, M., Hulme, O., Christensen, M. S., & Zeki, S. (2009). Modulation of aesthetic value by semantic context: An fMRI study. *NeuroImage*, *44*, 1125–1132.

Koffka, K. (1935). *Principles of Gestalt psychology*. New York: Harcourt Brace.

Kroner, S., & Lattner, A. (1998). Authentication of free hand drawings by pattern recognition methods. *Proceedings of IEEE International Conference on Pattern Recognition, 1*, 462–464.

Kushki, A., Androutsos, P., Plataniotis, K., & Venetsanopoulos, A. (2004). Retrieval of images from artistic repositories using a decision fusion framework. *IEEE Transactions on Image Processing*, *13*(3), 277–292.

Lang, P. J., Bradley, M. M., & Cuthbert, B. N. (1997). *International Affective Picture System (IAPS): Technical manual and affective ratings*. Bethesda, MD: NIMH Center for the Study of Emotion and Attention.

Lang, P. J., Greenwald, M. K., Bradley, M. M., & Hamm, A. O. (1993). Looking at pictures: Affective, facial, visceral, and behavioral reactions. *Psychophysiology*, *30*, 261–273.

Li, B., Xiong, W. H., Hu, W. M., & Ding, X. M. (2012). Context-aware affective images classification based on bilayer sparse representation. *Proceedings of ACM Multimedia* (pp. 721–724).

Li, C. C., & Chen, T. (2009). Aesthetic visual quality assessment of paintings. *IEEE Journal of Selected Topics in Signal Processing*, *3*(2), 236–252.

Li, J., & Wang, J. Z. (2004). Studying digital imagery of ancient paintings by mixtures of stochastic models. *IEEE Transactions on Image Processing*, *13*(3), 340–353.

Liu, Y., Schmidt, K. L., Cohn, J. F., & Mitra, S. (2003). Facial asymmetry quantification for expression invariant human identification. *Proceedings of IEEE International Conference on Computer Vision and Pattern Recognition, 91*(1–2), 138–159.

Louis, A., Wood, M. D., Scalise, A., & Birkelund, J. (2008). Multidimensional image value assessment and rating for automated albuming and retrieval. *Proceedings of International Conference on Image Processing* (pp. 97–100).

Lu, X., Suryanarayan, P., Adams, R. B., Li, J., Newman, M. G., & Wang, J. Z. (2012). On shape and computability of emotions. *Proceedings of ACM Multimedia* (pp. 229–238).

Luo, J., Boutell, M., & Brown, C. (2006). Exploiting context for semantic scene content understanding. *IEEE Signal Processing Magazine. Special Issue on Semantic Retrieval of Multimedia*, *23*(2), 101–114.

Luo, J., Savakis, A., Etz, S., & Singhal, A. (2000). On the application of Bayes networks to semantic understanding of consumer photographs. *Proceedings of International Conference on Image Processing, 3*, 512–515.

Luo, W., Wang, X. G., & Tang, X. O. (2011). Content-based photo quality assessment. *Proceedings of IEEE International Conference on Computer Vision* (pp. 2206–2213).

Lupsa, C. (2006). What if your laptop knew how you felt? *USA Today*, 18 December.

Machajdik, J., & Hanbury, A. (2010). Affective image classification using features inspired by psychology and art theory. *Proceedings of ACM Multimedia* (pp. 83–92).

Marchesotti, L., Perronnin, F., Larlus, D., & Csurka, G. (2011). Assessing the aesthetic quality of photographs using generic image descriptors. *Proceedings of IEEE International Conference on Computer Vision* (pp. 1784–1791).

Murray, N., Marchesotti, L., & Perronnin, F. (2012). AVA: A large-scale database for aesthetic visual analysis. *Proceedings of IEEE International Conference on Computer Vision and Pattern Recognition* (pp. 2408–2415).

Nadia Camera. (n.d.). Retrieved from http://www.wired.com/gadgetlab/2010/07/nadia-camera-offers-opinion-of-your-terrible-photos/

Nishiyama, M., Okabe, T., Sato, I., & Sato, Y. (2011). Aesthetic quality classification of photographs based on color harmony. *Proceedings of IEEE International Conference on Computer Vision and Pattern Recognition* (pp. 33–40).

Nodien, C. F., Locher, P. J., & Krupinski, E. A. (1993). The role of formal art training on perception and aesthetic judgment of art compositions. *Leonardo, 26*(3), 219–227.

Palmer, S. E. (2009). Aesthetic science [keynote address]. *Proceedings of IS&T/SPIE Electronic Imaging Conference*.

Perrett, D. I., May, K. A., & Yoshikawa, S. (1994). Facial shape and judgments of female attractiveness. *Nature, 368*, 239–242.

Peters, G. (2007). Aesthetic primitives of images for visualization. *Proceedings of IEEE International Conference on Information Visualization* (pp. 316–325).

Ramachandran, V. S., & Hirstein, W. (1999). Science of art: A neurological theory of aesthetic experience. *Journal of Consciousness Studies, 6*, 15–51.

Ramanathan, S., Katti, H., Huang, R., Chua, T.-S., & Kankanhalli, M. (2009). Automated localization of affective objects and actions in images via caption text-cum-eye gaze analysis. *Proceedings of ACM Multimedia* (pp. 729–732).

Reber, R. N., Schwarts, N., & Winkielman, P. (2004). Processing fluency and aesthetic pleasure: Is beauty in the perceiver's processing experience? *Personality and Social Psychology Review, 8*(4), 364–382.

Redi, M., & Merialdo, B. (2012). Enhancing semantic features with compositional analysis for scene recognition. *Computer Vision ECCV, 7585*, 446–455.

Rockmore, D., Lyu, S., & Farid, H. (2006). A digital technique for authentication in the visual arts. *International Foundation for Art Research, 8*(2), 12–23.

Savakis, A., Etz, S., & Loui, A. (2000). Evaluation of image appeal in consumer photography. *Proceedings of SPIE Human Vision and Electronic Imaging, 3959*, 111–120.

Scheib, J. E., Gangestad, S. W., & Thornhill, R. (1999). Facial attractiveness, symmetry, and cues of good genes. *Proceedings of the Royal Society of London, 266*, 1913–1917.

Sheikh, H. R., Bovik, A. C., & Cormack, L. (2005). No-reference quality assessment using natural scene statistics: JPEG2000. *IEEE Transactions on Image Processing, 14*(11), 1918–1927.

Shevade, B., & Sundaram, H. (2007). Modeling personal and social network context for event annotation in images. *Proceedings of Joint Conference on Digital Libraries* (pp. 127–134).

Smeulders, A. W. M., Worring, M., Santini, S., Gupta, A., & Jain, R. (2000). Content-based image retrieval at the end of early years. *IEEE Transactions on Pattern Analysis and Machine Intelligence, 22*(12), 1349–1380.

Solso, R. L. (2003). *The psychology of art and the evolution of the conscious brain*. Cambridge, MA: MIT Press.

Stork, D. (2009). Computer vision and computer graphics analysis of paintings and drawings: An introduction to the literature. *Proceedings of the International Conference on Computer Analysis of Images and Patterns* (pp. 9–24).

Su, H.-H., Chen, T.-W., Kao, C.-C., Hsu, W. H., & Chien, S.-Y. (2011). Scenic photo quality assessment with bag of aesthetics-preserving features. *Proceedings of ACM Multimedia* (pp. 1213–1216).

Sun, X., & Yao, H. (2009). Photo assessment based on computational visual attention model. *Proceedings of ACM Multimedia* (pp. 541–544).

Swaddle, J. P., & Cuthill, I. C. (1995). Asymmetry and human facial attractiveness: Symmetry may not always be beautiful. *Proceedings. Biological Sciences, 261*(1360), 111–116.

Taylor, R. (2004). Pollock, Mondrian and the nature: Recent scientific investigations. *Chaos and Complexity Letters, 1*(3), 265–277.

Taylor, R., Micolich, A. P., & Jones, D. (1999). Fractal analysis of Pollock's drip paintings. *Nature, 399*, 422.

Valenti, R., Jaimes, A., & Sebe, N. (2010). Sonify your face: Facial expressions for sound generation. *Proceedings of ACM Multimedia* (pp. 1363–1372).

Valenti, R., Sebe, N., & Gevers, T. (2007). Facial expression recognition: A fully integrated approach. *Proceedings of Intlernational Workshop on Visual and Multimedia Digital Libraries* (pp. 125–130).

Valentine, C. W. (1962). *The experimental psychology of beauty*. London: Methuen and Co.

Wang, W., & He, Q. (2008). A survey on emotional semantic image retrieval. *Proeedings of. International Conference on Image Processing* (pp. 117–120).

Watson, A. B. (1998). Toward a perceptual video quality metric. *Proceedings of SPIE Human Vision and Electronic Imaging*, *3299*, 139–147.

Winston, A. S., & Cupchik, G. C. (1992). The evaluation of high art and popular art by naive and experienced viewers. *Visual Arts Research*, *18*(1), 1–14.

Wu, O., Hu, W. M., & Gao, J. (2011). Learning to predict the perceived visual quality of photos. *Proceedings of IEEE International Conference on Computer Vision* (pp. 225–232).

Wypijewski, J. E. (1997). *Painting by the numbers: Komar and Melamid's scientific guide to art*. New York: Farrar, Straus & Giroux.

Yang, Y., Song, M., Li, N., Bu, J., & Chen, C. (2010). Visual attention analysis by pseudo gravitational field. *Proceedings of ACM Multimedia* (pp. 553–556).

Yanulevskaya, V., van Gemert, J. C., Roth, K., Herbold, A. K., Sebe, N., & Geusebroek, J. M. (2008). Emotional valence categorization using holistic image features. *Proceedings of International Conference on Image Processing* (pp. 101–104).

Yao, L., Suryanarayan, P., Qiao, M., Wang, J. Z., & Li, J. (2012). OSCAR: On-site composition and aesthetics feedback through exemplars for photographers. *International Journal of Computer Vision*, *96*(3), 353–383.

Zaidel, D. W., & Cohen, J. A. (2005). The face, beauty, and symmetry: Perceiving asymmetry in beautiful faces. *International Journal of Neuroscience*, *115*, 1165–1173.

Zeki, S. (1999). *Inner vision: An exploration of art and the brain*. Oxford: Oxford University Press.

Zunjarwad, A., & Sundaram, H. (2007). Contextual wisdom: Social relations and correlations for multimedia event annotation. *Proceedings of ACM Multimedia* (pp. 615–624).

13

Emotion and Motivation in the Perceptual Processing of Natural Scenes

Margaret M. Bradley, Dean Sabatinelli, and Peter J. Lang

Because human cognition, including perception, has evolved in the context of a fundamental drive to survive, it is useful to consider the role of motivation in perceptual processing. In this chapter, we focus on perceptual processing of natural scenes that humans describe as emotionally arousing—both pleasant and unpleasant scenes. It is proposed that these scenes engage motivational circuits that have evolved in the mammalian brain to promote the survival of individuals and their progeny. Motive circuit activation prompts enhanced perceptual processing and information intake in the service of selecting and implementing effective coping actions in both appetitive and defensive contexts.

We begin with a brief description of this survival circuitry as it has been elucidated in research with animals. Evidence for motive circuit activation in humans and its effects on perceptual processing during natural scene viewing is then presented, based on research with a large set of picture stimuli defined by standardized evaluative ratings and varying widely in arousal and pleasant/unpleasant affect. Studies monitoring eye movements provide evidence of enhanced information intake for emotionally arousing scenes, and both functional neural imaging (fMRI) and electroencephalographic recording of event-related potentials (ERP) confirm defensive and appetitive circuit engagement and illuminate timing and activation patterns in the visual cortex. Subsequent studies explore the hypothesis that motivational engagement facilitates recognition in rapid serial visual processing (RSVP). Overall, the data strongly support the view that the perceptual processing of natural scenes is significantly enhanced when the array includes a motivationally significant (emotional) cue, independent of the contributions of perceptual complexity, familiarity, color, and other perceptual characteristics of the scene.

Motivational Circuits in the Brain

To begin simply it is useful to consider the brain's survival circuit as comprised of two fundamental motive systems, defensive and appetitive, that are evolutionarily old and shared across mammalian species (e.g., Lang, Bradley, & Cuthbert, 1990). The defense system is primitively activated in contexts involving physical threat, with a potential output repertoire of withdrawal, escape, and counterattack. Conversely, the appetitive system is activated in contexts that promote survival and well-being, including sustenance, procreation, and nurturance, with a basic behavioral repertoire of ingestion, copulation, and caregiving. Perceptual and learned memorial cues activate these motivational circuits through existing associations to the subcortical and cortical structures that mediate the expressive, autonomic, and somatic changes in emotion.

Our understanding of survival networks owes much to the study of animal subjects (e.g., Davis, 2000; Fanselow & Poulos, 2005; Kapp, Supple, & Whalen, 1994; LeDoux, 2003). Key regions implicated in both the appetitive and defensive circuits include the bilateral amygdalae—two small, almond shaped bundles of nuclei in the temporal lobe. The basolateral nucleus of each amygdala receives input from the thalamus (sensory), hippocampus (memory), and other regions, subsequently engaging, via the central nucleus and extended amygdala (basal nucleus of the stria terminalis), downstream outputs that include regions that modulate sensory processing (vigilance), facilitate associated information processing, and activate the autonomic and somatic structures that mobilize the organism for defensive or appetitive action (see Davis & Lang, 2003; Lang & Bradley, 2010; Lang & Davis, 2006).

Importantly, when an animal perceives a motivationally significant cue, the first output of the survival circuit includes activities whose goal is to enhance sensory processing. Thus, for example, if a foraging prey animal perceives a predator in the distance, its first response is to orient toward the predator, immobile ("freezing"), fixedly regarding the potential attacker, with a slowing pulse rate (sometimes called "fear bradycardia": Campbell, Wood, & McBride, 1997). Although the motivational aim differs, a similar information-gathering response characterizes the predator's initial reaction on first espying potential prey. A secondary response, mobilization for action (sympathetic arousal), is tuned to the prey's perception of the imminence (increasing danger) of attack and the necessity of fight or flight, and, for the predator, the optimal probability of a successful strike. In the investigations of scene processing we describe, the human research participant is considered, metaphorically, to be at an early stage of motivational activation—in a predominantly observational posture, viewing a series of natural scenes that may contain aversive or appetitive cues that activate mammalian circuitry similar to that previously elucidated in animal research.

The International Affective Picture System

Photographs of natural scenes are particularly good cues for associative activation in humans, as they share perceptual features with the actual object. For example, a picture of an attacking animal vividly portrays its staring eyes, snarling mouth, teeth, among other signs, all increasing the probability that this highly symbolic cue will activate (through feature matching) the aggressor animal's mental representation, which includes associative connections to the neural circuit mediating defensive behavior. Other media cues, such as narrative language ("the dog growls menacingly"), also engage the survival circuit. However, activation is less immediate, as words do not share the sensory/perceptual features with the objects they represent. Indeed, because primates in general are highly visual creatures, many subspecies are known to respond meaningfully to motivationally significant pictures of natural scenes (e.g., Fagot, 2000; Itakura, 1994). In consideration of these factors we developed a standardized set of photographic pictures to be used as a research tool. Our aims were (1) to include images representing a wide range of emotional events experienced in the natural world by human participants and (2) to make them available to the larger research community with the goal of improving the reliability of research findings in the study of emotion and to provide for better replication of test results between laboratories.

Thus, the *International Affective Picture System* (IAPS) (Bradley & Lang, 2007; Lang, Bradley, & Cuthbert, 1997a; Lang, Bradley, & Cuthbert, 2008) was developed as a catalog of natural scenes that vary in emotional valence and arousal and are freely available to not-for-profit researchers studying emotion and cognition. The IAPS includes over 1000 scenes depicting a wide range of events in human experience—threatening, attractive, dressed and undressed people, art objects, household objects, housing projects, funerals, pollution, dirty toilets, cityscapes, seascapes, landscapes, sports events, photojournalism from wars and disasters, medical treatments, sick patients, mutilated bodies, baby animals, attacking animals, insects, loving families, waterfalls, children playing—a virtual world of pictures.

The IAPS pictures are initially standardized with a large sample of participants who provide evaluative ratings using two pictographic (nonlinguistic) scales whose emotional anchors range from pleasant to unpleasant and from calm to aroused—two dimensions consistently identified as fundamental in factor analyses of language judgments (e.g., Bradley & Lang, 1994; Osgood, Suci, & Tannenbaum, 1957; Russell & Mehrabian, 1977). For instance, dimensions of pleasure and arousal reliably organize judgments of words rated on 50 different bipolar scales (e.g., hot-cold, white-black, fast-slow) and hold equally well for judgments of nonverbal stimuli that range from sonar signals to aesthetic paintings (Osgood et al., 1957). Mehrabian (1970) found that the same dimensions underlie judgments of facial expressions, hand and body movements, and postural position. The fact that these two dimensions, pleasure and arousal, account for significant variance in human judgments of so many different signal stimuli suggests their primacy in organizing human experience, both semantic and affective. It is of further significance that they mirror the two factors that define the concept of motivation, which includes the *direction* of behavior (e.g., approach/avoidance) and behavioral *intensity* (Hebb, 1949).

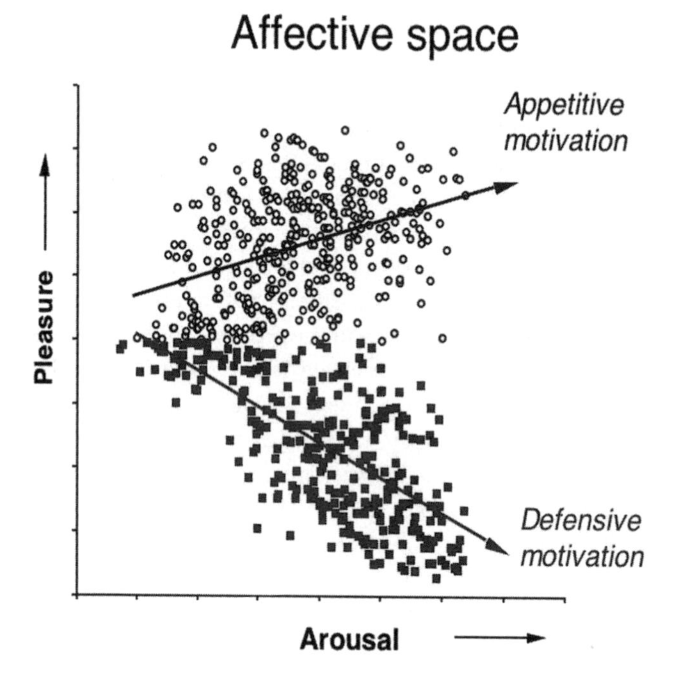

Figure 13.1
When each picture is plotted in a two-dimensional space defined by its mean pleasure and arousal rating, the resulting scatterplot is consistent with a biphasic organization of emotion in appetitive and defensive systems that varies in the intensity of activation.

Figure 13.1 illustrates individual IAPS pictures in a two-dimensional *affective space* defined by their pleasure and arousal ratings. The distribution of these images is consistent with an underlying motivational organization—appetitive and defensive—varying in the intensity of activation. Significant support for this view is provided by research that assesses the somatic and autonomic reflex responses measured during picture viewing (e.g., Cuthbert, Schupp, Bradley, Birbaumer, & Lang, 2000). In these studies physiological measures are combined with evaluative judgments in a principal components analysis. Again, two primary factors of hedonic valence and arousal organize the pattern of reactivity with (1) pleasure ratings, facial muscle activity—zygomatic (smiling) and corrugator EMG (frowning)—and a decelerative heart rate response loading on a first factor, identifying the scene as defensively (unpleasantly) or appetitively (pleasantly) engaging, and (2) heightened skin conductance activity, increased viewing time, and degree of interest ratings together with ratings of *affective arousal*, loading on a second factor reflecting the intensity of motivational activation in each system.

Scanning Natural Scenes: Motivated Behavior

When processing natural scenes the eye is the initial organ that gathers information regarding threat and safety, sexual opportunity and successful foraging. Visual information is initially processed by neuroectodermal photoreceptors at the back of the eye, and the optical system as a whole—lenses, receptor bed, and pupillary and orienting musculature—is under fine nervous control. Studies measuring eye movements therefore provide a sensitive measure of initial orienting and information intake during visual scene processing, as the information available for further neural processing is determined almost solely by eye fixations, their duration, and their extent.

In general, eye movements during scene perception follow stereotypical patterns with periods of repose (fixations) interspersed with rapid movement (saccades). However, eye movements are affected by the stimulus, the task, and, importantly, by semantic relevance. For example, Henderson and Hollingworth (1998) found longer fixation durations as well as greater saccade lengths during free-viewing of scenes compared to when reading text. Not surprisingly, eye movements are also influenced by characteristics of a visual array, including contrast, edges, complexity, and other sensory features (e.g., Itti & Koch, 2001; Parkhust, Law, & Niebur, 2002; Peters, Iyer, Itti, & Koch, 2005). Nonetheless, information relevant to scene interpretation (meaning) is salient and also strongly affects eye movement behavior (e.g., Loftus & Mackworth, 1978), with informative cues receiving more fixations than those that are semantically irrelevant (Henderson, Weeks, & Hollingworth, 1999).

Considering these factors, we undertook research assessing the effects of both perceptual composition and hedonic content on scanning behavior (Bradley, Houbova, Miccoli, Costa, & Lang, 2011). Participants viewed 192 color photographs selected from the IAPS that varied systematically in ratings of affect and compositional complexity. Pictures were presented for several seconds with no other instruction than to look at the picture. This free viewing context is similar to the way scenes are scanned in natural environments, without either a specific task or explicit a priori goal driving perceptual processing (e.g., see Einhäuser, Rutishauser, & Koch, 2008). As figure 13.2 illustrates, the resulting scan pattern included fixations that were initially brief but then lengthened in duration across the viewing interval, with saccade amplitude first rapidly increasing and then decreasing, suggesting an initial rapid, broad scan of the scene, followed by an increasing focus or zeroing in on the most interesting or informative information (see also Castelhano, Wieth, & Henderson, 2007).

Specific scan parameters were affected by both perceptual complexity and the emotional intensity of the picture stimuli. Viewing complex scenes prompted more fixations of briefer duration and overall longer scan paths than simple figure-ground compositions, regardless of hedonic content. And emotionally engaging cues (whether pleasant or unpleasant) also prompted more fixations of briefer duration and a longer

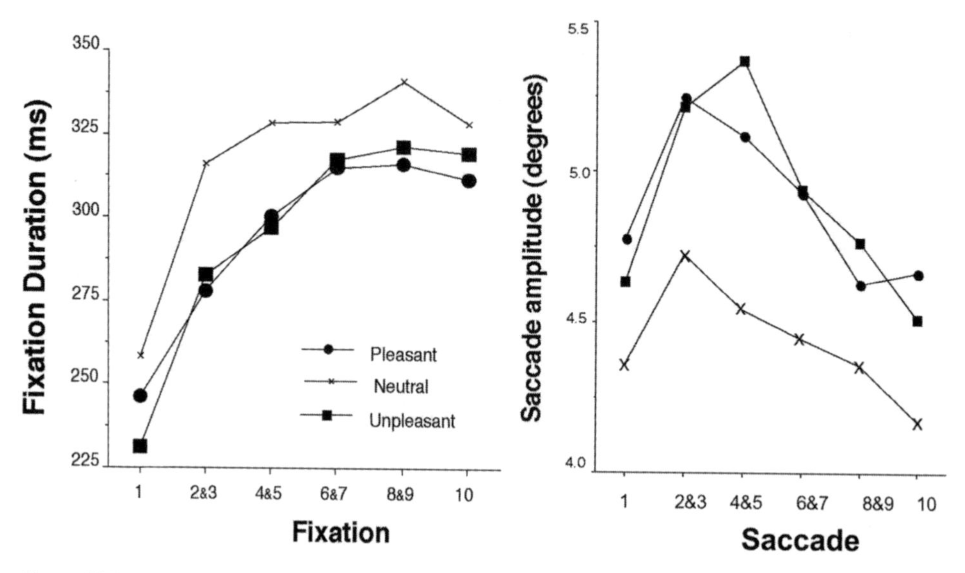

Figure 13.2
Fixation duration (left panel) and saccade amplitude (right panel) when viewing novel pleasant, neutral, and unpleasant scenes suggest a general scan pattern in which fixations are initially brief and spatially broad but lengthen in duration and spatially focus over the course of the viewing interval.

total scan path compared to neutral pictures. Thus, for either emotional or complex scenes, an initial flurry of brief fixations covered a larger portion of the scene, suggesting enhanced information intake. For perceptually complex scenes enhanced information seeking reflects the demands of the sensory array, and previous research has shown that fixations rarely land on spatial locations in which the information is constant or uniform (e.g., Henderson, 2003), which was a defining feature of the simple, figure-ground compositions in our studies.

Information seeking is otherwise determined for emotional scenes. When perceptual differences between pictures, including brightness, contrast, spatial frequency, and rated picture composition were excluded using stepwise regression, analysis of the residuals continued to show highly significant modulation by emotional arousal of both fixation duration and scan path length. This finding is consistent with the view that enhanced information seeking and intake are consequent on activation of the evolved appetitive/defensive motivational systems, as suggested by the animal model (Bradley, 2009; Lang, Bradley, & Cuthbert, 1997b). Interestingly, effects of motivational arousal on scanning behavior are found even when emotional scenes are simple in perceptual composition. The sensory array is repeatedly and broadly scanned: *Is a predator attack imminent? Is prey approaching or receding?* Despite both the perceptual simplicity and the symbolic nature of these cues, the evolved heightened attention and information intake to motivational cues persist in human perception of natural scenes that are emotionally evocative.

A further question addressed in this research concerns whether the scan patterns elicited by emotional arousal are based on stimulus novelty; that is, do they reflect initial orienting that would be absent if the scene were familiar? To address this issue scenes were repeated four times either massed (contiguous) or distributed across the experimental session. As expected, novel pictures elicited more discrete fixations and longer scan paths than repeated pictures. Despite increased familiarity, however, repeated emotional pictures continued to prompt more fixations and longer scan paths than repeated neutral pictures. Interestingly, there were very few differences in eye movement indices as a function of whether repetitions were massed or distributed.

These data reaffirm the view that formal properties of the images (brightness, complexity, etc.) are not the only determinant of enhanced scanning for emotionally evocative scenes. After the same picture is viewed four times in a row, effects of physical features on scanning patterns should be attenuated, given that recent memory will dampen effects primarily due to sensory differences (Hollingworth, 2004). Rather, the data suggest that emotional cues continue to reactivate the brain's motive system, reflexively persisting in information-seeking behavior, even though the scene offers no new input.

Neuroimaging of Emotional Scenes

The view that emotional scenes activate the brain's motivational circuits, appetitive and defensive, is strongly supported by functional magnetic resonance imaging (fMRI) research. Thus, it has been shown repeatedly that both pleasant and unpleasant IAPS pictures prompt enhanced activation in the amygdala and that reactivity is greatest for those scenes rated most arousing (e.g., Sabatinelli, Bradley, Fitzsimmons, & Lang, 2005). Furthermore, significant increases in amygdala activity are accompanied by enhanced, more widespread activation in visual cortex (e.g., Lang et al., 1998; Sabatinelli et al., 2005). A recent meta-analysis (Sabatinelli et al., 2011) found that, compared to the viewing of neutral scenes, viewing emotionally evocative scenes—whether pleasant or unpleasant—is associated specifically with greater amplitude and more extensive changes in the extrastriate visual cortex, particularly the fusiform and inferotemporal (TE) regions, as well as in the lateral occipital and posterior parietal cortex—all regions that are critical to the perceptual processing of natural scenes (see figure 13.3A, plate 28).

Detailed studies with nonhuman primates (Amaral, Price, Pitkanen, & Carmichael, 1992) suggest that enhanced activation of the posterior visual system for motivational cues may result from reentrant projections from the amygdala. That is, initially, visual information proceeds from the eye to the posterior, striate cortex. It is then passed forward in a hierarchical fashion along the ventral temporal lobe to successively higher-level processing areas, finally proceeding from the inferotemporal area to the lateral nucleus of the amygdala. Subsequently, the amygdala's basal nucleus

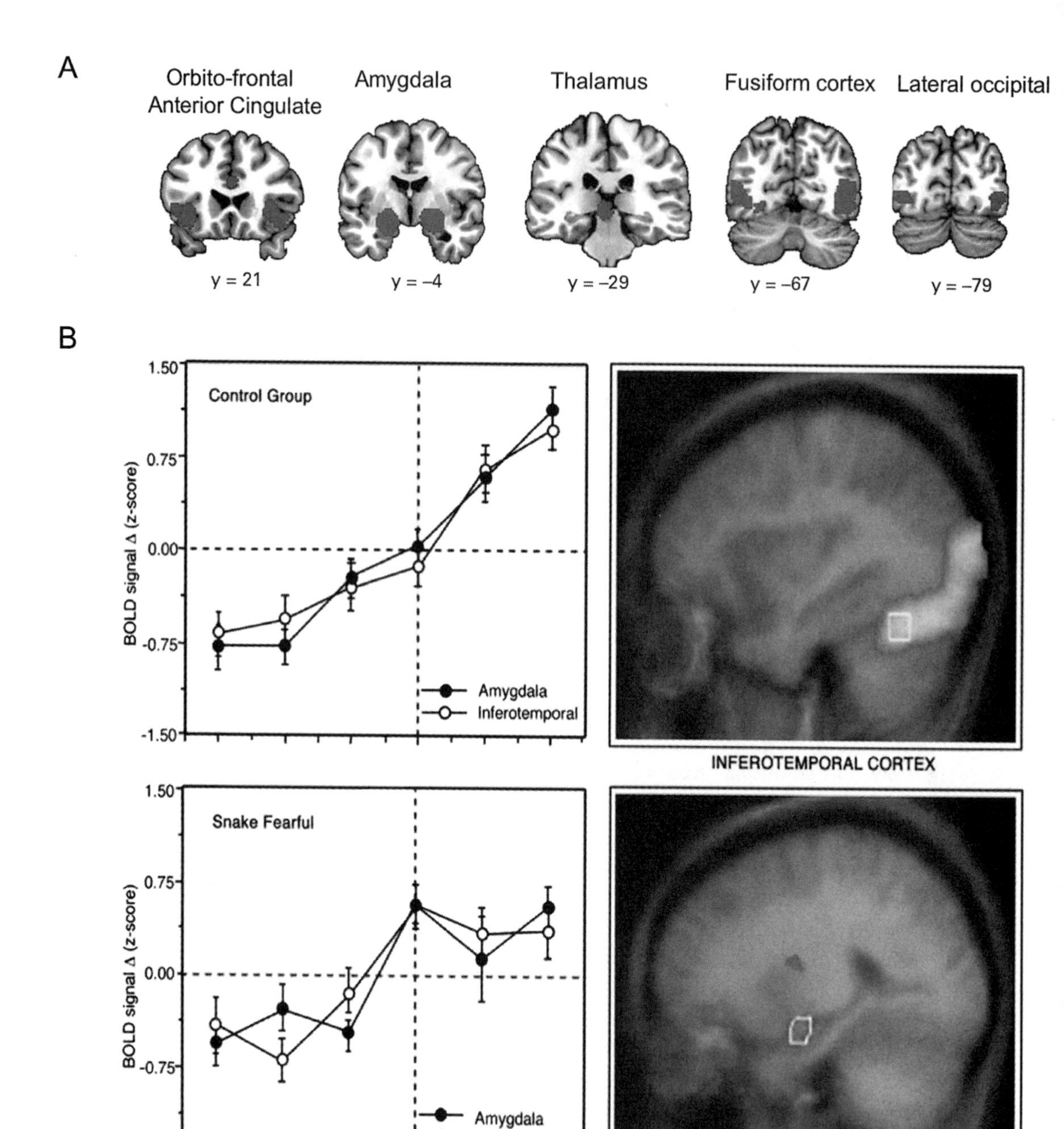

Figure 13.3 (plate 28)

(A) A functional network in which enhanced activation is found when viewing emotionally arousing (pleasant or unpleasant), compared to neutral, scenes includes cortical and subcortical regions. (B) Enhanced activation levels in the amygdala and fusiform cortex covary closely when emotional scenes are viewed, and coactivation when pictures of snakes are viewed is enhanced only for participants reporting high snake fear.

back-projects “to virtually all levels of the visual cortex” (Amaral et al., 1992, p. 44). In this way, cues signaling threat or reward could receive heightened, more sustained perceptual processing mediated by a positive feedback loop that is specifically enhanced for motivationally relevant scenes.

Functional MRI research suggests that this hierarchical path is also activated in humans. To examine this hypothesis we rapidly sampled (500 ms) neural activity in a single axial brain slice, strategically located to assess activation in both the amygdala and critical regions of the TE cortex, including the fusiform and extrastriate occipital cortex (Sabatinelli, Lang, Bradley, Costa, & Keil, 2009). Restricting the functional sampling to a single slice with fast sampling allowed us to determine the relative timing of functional activity in relevant regions. The resulting data were consistent with a hypothesis of reentrant processing based on the animal data (Amaral et al., 1992). Thus, enhanced functional activity when viewing emotionally arousing, compared to neutral, pictures was first apparent in the amygdala and then in the proximal TE cortex and only later in more distant, posterior regions of the visual system (e.g., the medial occipital cortex). Findings based on a Granger causality source analysis of steady-state visual potentials evoked during emotional perception (Keil et al., 2009) are also consistent with this analysis, suggesting, furthermore, a broader reentrant connectivity to the visual processing system from anterior neural structures, both subcortical and cortical.

Individual differences in response to emotional scenes have also indicated that, depending on the participant, the same scene can be either an emotional or a neutral cue, ruling out potential differences due to specific picture characteristics. For example, we presented pictures of snakes to individuals reporting high or low snake fear, as well as pictures of neutral people and objects, nonthreatening animals, erotica, and mutilated human bodies (Sabatinelli et al., 2005). As illustrated in figure 13.3B (plate 28), when people viewed pictures that most people report as highly arousing (e.g., erotica and mutilations), heightened functional brain reactivity was found in the bilateral amygdala and TE cortex for both groups of participants. Only highly fearful individuals viewing pictures of snakes, however, showed enhanced functional activity in these regions that was similar in magnitude to the activity when other highly arousing scenes were viewed. For low-fear participants viewing snake scenes, amygdala activation and activation of its reentrant target, the TE cortex, did not differ from that evoked by pictures with neutral content.

Because enhanced neural activation in the amygdalae and the visual cortex is found when people view either highly arousing unpleasant or pleasant cues, these regions can be considered integral structures in both the defensive and appetitive motivational circuits. Research with animals suggests, however, that reward-seeking behavior uniquely involves heightened activity in the medial prefrontal cortex and the nucleus accumbens (e.g., Ishikawa, Ambroggi, Nicola, & Fields, 2008), which interface with motor areas (Mogenson, 1977). Furthermore, imaging research with human participants has implicated these regions in the perception of positive scenes (e.g., pictures of loved ones: Aron et al., 2005; Bartels & Zeki, 2004; addictive drug cues: David et al., 2005). Another body of research, however, suggests that the striatum (which includes the nucleus accumbens) is activated when one is perceiving any salient stimulus that arouses and prompts attention (Zink, Pagnoni, Chappelow, Martin-Skurski, & Berns, 2006).

We used fMRI to assess whether these mesolimbic circuits are specifically activated when one is perceiving pleasant visual scenes or, alternatively, if activation is determined by stimulus salience (arousal). Participants viewed a variety of different pleasant and unpleasant contents, including erotica, romance, mutilations, threat, as well as pictures of neutral people (Sabatinelli, Bradley, Lang, Costa, & Versace, 2007). Replicating previous data, the response in the bilateral amygdala was significantly enhanced when participants viewed erotic or mutilation pictures compared to quotidian pictures of neutral people, again indicating that the amygdala is equivalently activated by emotionally arousing stimuli, whether appetitive or aversive. In contrast, only pleasant scenes significantly activated medial prefrontal cortex and the nucleus accumbens. Neither unpleasant nor neutral scenes prompted activation of these regions, and activation when subjects were viewing these scenes was often below baseline, particularly for medial prefrontal cortex. These data do not support the hypothesis that the nucleus accumbens and medial prefrontal cortex sites react to any scene that arouses or captures attention. Rather, unlike amygdala and TE cortex, which are activated when subjects view all emotionally arousing scenes, these specific regions in the appetitive subcircuit are silent for aversive cues, perhaps even deactivated.

The Late Positive Potential and Motivational Relevance

When changes in the brain's electrical activity are measured with an EEG during scene viewing, a centroparietal late positive potential is elicited that is a reliable, replicable index of motivational relevance. As recorded from scalp electrodes, the centroparietal positive slow wave begins around 300 ms after the onset of an emotional picture and can persist (using the appropriate filters) for almost the entire duration of a 6-second viewing interval (Cuthbert et al., 2000). The late positive potential is most enhanced for the natural scenes rated highest in emotional arousal, regardless of whether these depict appetitive (e.g., erotica) or aversive (e.g., mutilated bodies) hedonic contents (Schupp et al., 2004).

We originally entertained the hypothesis that sustained positive potentials during emotional scene perception reflected enhanced orienting and attention allocation as a function of motivational activation. Other findings seemed consistent with this early interpretation: emotional scene perception reduced the amplitude of the P3 component of a secondary event-related potential (elicited not by the picture but by a brief acoustic probe) and also slowed probe reaction times (Bradley, Cuthbert, & Lang,

1996; Schupp, Cuthbert, Bradley, Birbaumer, & Lang, 1997). These findings suggested that fewer resources were available for processing the irrelevant cue (i.e., the secondary probe) when subjects are viewing emotional scenes. The finding that later incidental memory performance is reliably enhanced for emotional, compared to neutral, scenes also suggested greater attention to the picture stimulus and, thus, better initial encoding (Bradley, Greenwald, Petry, & Lang, 1992). A more recent series of studies that investigated effects of picture repetition on event-related potentials during scene perception, however, suggest that an attention-capture interpretation may be too narrow. In these studies the same scene was repeatedly presented throughout the experiment—up to as many as 90 repetitions in the same session (Codispoti, Ferrari, & Bradley, 2006, 2007; Ferrari, Bradley, Codispoti, Karlsson, & Lang, 2013). In these experiments the amplitude of the P3 component to the secondary probe was attenuated when the subject was viewing novel scenes but not when viewing repeated, emotional scenes, which is consistent with a resource allocation account of probe P3 modulation.

In contrast, significant modulation of the late positive potential when subjects were viewing emotional, compared to neutral, scenes was found regardless of repetition. Because of the persisting effect of emotion on the LPP, despite massive repetition, the late positive potential of the ERP has been interpreted as a broad index of motivational "significance"—that is, that an eliciting perceptual cue has activated, through existing associations, the defensive or appetitive motivation system (Bradley, 2009). Because the measuring electrodes lay on the surface of the scalp distant from the subcortical regions mediating motivational activation, the LPP presumably reflects enhanced cortical activity that may partially reflect reentrant processing from subcortical regions. Indeed, when the amplitudes of the electrocortical LPP and fMRI estimates of neural reactivity are correlated either within or between participants, the amplitude of the cortical late positive potential correlates highly with fMRI activation not only in inferotemporal and lateral occipital regions of the visual cortex (Sabatinelli et al., 2009) but in the amygdala as well (Sabatinelli, Keil, Frank, & Lang, 2013).

Attributing differences in LPP amplitude (or any measure) between scenes to the emotion-evoking properties of the cue raises a critical question: Could the presumed differences between emotional and neutral scenes be determined by confounding physical differences in the array, such as color, brightness, familiarity, or complexity? Color is not a confounding variable because IAPS pictures, whether presented in color or grayscale, prompt similar differences between emotional and neutral picture contents in measured autonomic (skin conductance) and somatic (startle potentiation) physiological responses (Bradley, Codispoti, Cuthbert, & Lang, 2001), in enhanced activation in sensory cortex (Bradley et al., 2003), and in the amplitude of the LPP during scene perception (Codispoti, De Cesarei, & Ferrari, 2012).

When emotional and neutral pictures are presented that depict either simple figure-ground compositions or more complex scenes, perceptual composition does impact the resulting event-related potential (Bradley, Hamby, Löw, & Lang, 2007). These effects, however, are earlier in time (150–250 ms) than the late positive potential, are maximal over occipital-temporal rather than centroparietal, sensors, and, importantly, are the same for both emotional and neutral scenes. The late positive potential, on the other hand, is significantly enhanced when one is viewing emotional (pleasant or unpleasant), compared to neutral, pictures regardless of perceptual composition. Interestingly, effects of emotion on the late positive potential are somewhat larger for simple figure-ground compositions, contrary to a hypothesis that the late positive potential might reflect the complexity of information processing. Rather, simple figure-ground compositions appear to be better cues for activating motivational systems, a finding consistent with previous data indicating stronger affective modulation of the startle reflex for simple, compared to complex, scenes (Bradley et al., 2007).

Rapid Serial Visual Presentation

Thus far we have considered scene perception in the context of a relatively leisurely (6-second) free viewing interval. A number of studies, however, explore natural scene processing in the context of RSVP, in which scenes are presented at extremely rapid rates without blank intervals, resulting in a perceptual array consisting of fleeting images. RSVP research has determined that natural scenes presented at such rapid rates are indeed perceived but quickly forgotten (Potter, 1976; Potter, Staub, Rado, & O'Connor, 2002; see chapter 9 by Potter in this volume). We have used ERP methodology to investigate the electrophysiological correlates of RSVP as well as to determine whether emotion has an impact on perceptual processing during RSVP.

In an initial study we presented pictures that alternated between those rated high and low in emotional arousal at a rate of either 3 or 5 Hz (Junghöfer, Bradley, Elbert, & Lang, 2001). A pronounced difference was found in which emotional, compared to neutral, scenes prompted an enhanced negative deflection over occipital sensors that was maximal around 250 ms after picture onset. One hypothesis is that this occipital negative deflection reflects the success of perceptual recognition for individual images presented in this rapid stream. If so, one prediction is that these pictures will be remembered better in a subsequent memory test. We tested this hypothesis in a second experiment in which an array of eight pictures was presented, alternating between those rated high and low in emotional arousal at a rate of 5.4 Hz (Versace, Bradley, & Lang, 2010). Because memory performance following RSVP rapidly declines as the number of items tested increases, presumably reflecting interference (e.g., Potter et al., 2002), we tested recognition of a single picture following each brief RSVP sequence, expecting that the hit rate would be above chance but not perfect, permitting assessment of the effect of emotional arousal on recognition performance.

As noted by Potter and Fox (2004), recognition of a briefly glimpsed scene could be due to a match in perceptual information, conceptual information, or both. To evaluate these possible contributions we covaried both the perceptual and conceptual similarity between a picture presented during RSVP and its subsequent test item. *Perceptual similarity* was manipulated by presenting the natural scenes in color during RSVP and varying whether the test picture was presented in color (matching its presentation in the RSVP sequence) or in grayscale (mismatch) during recognition. *Conceptual similarity* was manipulated by varying whether a critical picture and a test item were semantically similar (e.g., both babies) or unrelated in semantic content (e.g., a baby and shoe). These two variables were crossed to produce four different conditions in the immediate recognition test; each participant was tested with only one version of the test picture.

Replicating Junghöfer et al. (2001), during the RSVP encoding phase, emotionally arousing pictures prompted a more negative-going potential over occipital sites compared to less arousing pictures in a window approximately 250 ms after picture onset (Junghöfer et al., 2001; Peyk, Schupp, Keil, Elbert, & Junghöfer, 2009). More importantly, emotionally arousing pictures were also better recognized than low-arousal scenes on the immediate recognition test, with a significantly greater proportion of hits for emotionally engaging pictures following RSVP as well as better discrimination performance. Drawing on research by Irwin and Andrews (1996), Henderson (1997), and others, Potter (2012) has concluded that the memory representation resulting from RSVP is at least partially conceptual. Consistent with this, better discrimination performance for emotionally arousing pictures was dependent, to some extent, on whether a new test picture was similar in semantic content to the RSVP scene. When a new test picture was similar (e.g., both were pictures of erotica), participants were more likely to wrongly classify a "new" picture as "old." When the new test picture was not semantically confusable, however, a clear effect of emotional arousal was found, with better memory discrimination for emotional, compared to neutral, pictures.

Whether a test picture was perceptually similar (i.e., in color) or different (i.e., grayscale) did not affect memory performance, suggesting that color, at least, is not a critical factor in mediating recognition of RSVP stimuli. Visual images of natural scenes, of course, vary widely in other perceptual properties that could affect the ease of processing, especially when presented at rapid speeds. For instance, according to Gestalt psychologists, one of the major principles in perceptual organization is segmentation of a scene into figure and ground (Palmer, 1999). When people process the high perceptual loads imposed by the rapidly changing visual scenes in RSVP, simple figure-ground pictures may seem to pop out of the fleeting array. In addition, because of the social significance of faces (Adolphs, 1999) as well as human expertise in processing faces and people (Gauthier et al., 2000), face processing might also be facilitated at rapid speeds of presentation. It is possible that differences attributed to emotional arousal during RSVP are instead mediated by differences in perceptual composition or the presence of people, which both may pop out of the rapid stream enhancing the early occipital potential.

Thus, in a third experiment (Löw, Bradley, & Lang, 2013) using RSVP we presented emotional and neutral pictures that were either simple figure-ground compositions or more complex scenes that either did or did not include people to assess the contribution of these three factors to the ease of perceptual processing. All three factors influenced perceptual processing as indexed by ERPs: enhanced occipital negativity was found for perceptually simple figure-ground compositions compared to more complex scenes; for pictures depicting people, compared to those that did not; and for emotionally arousing compared to neutral scenes. When occipital negativity was computed individually for each picture presented using RSVP, regression analysis indicated that perceptual composition accounted for the most variance, with enhanced negativity for figure-ground compositions compared to more complex scenes regardless of whether these depicted emotional or neutral content or included people or did not, confirming that figure-ground composition is a key variable that facilitates the ease of perceptual processing during RSVP of natural scenes.

Whether a picture included people (or not) also accounted for significant variance, consistent with hypotheses and data documenting that humans, experts at face detection, show facilitated processing for these salient stimuli (Bentin, Allison, Puce, Perez, & McCarthy, 1996; Kanwisher & Yovel, 2006). Interestingly, however, the presence of people affected occipital negativity only when scenes were simple figure-ground in composition and therefore tended to depict faces; for perceptually complex scenes that included a variety of people and objects, the mere presence of a person did not facilitate perceptual processing.

Whether a scene depicted emotionally arousing or neutral content affected early occipital negativity most for scenes that included people, whether these were simple or complex in perceptual composition. Previous studies have clearly determined that viewers' emotional engagement when viewing pictures of natural scenes is strongest for pictures that depict people in sexual, violent, and other affective contexts (Bradley et al., 2001; Lang & Bradley, 2010; Schupp et al., 2004; Weinberg & Hajcak, 2010), with pictures of objects less able to strongly engage emotional reactions.

Taken together, the data suggest that in the rapid visual processing stream of natural scenes, simple figure-ground segmentation, the presence of faces, and emotional features facilitate perceptual recognition. Intraub (1984) reported that the probability of correctly recognizing an item within a rapid stream is enhanced when the viewer is intentionally looking for it. We have suggested that in the absence of specific tasks or explicit instructions, emotional stimuli *naturally* engage attention by activation of appetitive or defensive motivational circuits that are the foundation of human emotion (Bradley, 2009; Lang, Bradley, & Cuthbert, 1997b). Both the heightened occipital negativity in the ERP found for emotional pictures during RSVP and enhanced postviewing memory performance support the view that emotional features draw attention and facilitate perception, even when scenes are presented at rapid rates.

Summary

The data reviewed in this chapter are consistent with the hypothesis that perceptual processing of emotionally arousing scenes activates motivational circuits in the mammalian brain that originally promoted the organism's survival. Our studies of eye movements during picture processing indicate enhanced scanning and information intake for emotional pictures that is independent of physical features in the scene's visual array. ERP studies suggest that heightened attention and enhanced recognition and memory are hallmarks of emotional scene processing at both slow and rapid rates of presentation. Animal studies and human neuroimaging research provide a detailed picture of the neural circuits that might mediate enhanced perceptual processing. Taken together, studying the perception of scenes that activate the fundamental motivational systems of appetite and defense highlights critical information-processing mechanisms that operate during natural scene recognition.

Acknowledgments

This work was supported in part by NIMH grants MH098078 and MH094386 to Peter J. Lang.

References

Adolphs, R. (1999). Social cognition and the human brain. *Trends in Cognitive Sciences*, *3*(12), 469–479.

Amaral, D. G., Price, J. L., Pitkanen, A., & Carmichael, S. T. (1992). Anatomical organization of the primate amygdaloid complex. In J. P. Aggleton (Ed.), *The amygdala: Neurobiological aspects of emotion, memory, and mental dysfunction* (pp. 1–66). New York: Wiley.

Aron, A., Fisher, H., Mashek, D. J., Strong, G., Li, H., & Brown, L. L. (2005). Reward, motivation, and emotion systems associated with early-stage intense romantic love. *Journal of Neurophysiology*, *94*, 327–337.

Bartels, A., & Zeki, S. (2004). The neural correlates of maternal and romantic love. *NeuroImage, 21*, 1155–1166.

Bentin, S., Allison, T., Puce, A., Perez, E., & McCarthy, G. (1996). Electrophysiological studies of face perception in humans. *Journal of Cognitive Neuroscience*, *8*(6), 551–565.

Bradley, M. M. (2009). Natural selective attention: Orienting and emotion. *Psychophysiology*, *46*(1), 1–11.

Bradley, M. M., Codispoti, M., Cuthbert, B. N., & Lang, P. J. (2001). Emotion and motivation I: Defensive and appetitive reactions in picture processing. *Emotion*, *1*(3), 276–298.

Bradley, M. M., Cuthbert, B. N., & Lang, P. J. (1996). Picture media and emotion: Effects of a sustained affective context. *Psychophysiology*, *33*(6), 662–670.

Bradley, M. M., Greenwald, M. K., Petry, M., & Lang, P. J. (1992). Remembering pictures: Pleasure and arousal in memory. *Journal of Experimental Psychology: Learning, Memory, and Cognition*, *18*(2), 379–390.

Bradley, M. M., Hamby, S., Löw, A., & Lang, P. J. (2007). Brain potentials in perception: Picture complexity and emotional arousal. *Psychophysiology*, *44*(3), 364–373.

Bradley, M. M., Houbova, P., Miccoli, L., Costa, V. D., & Lang, P. J. (2011). Scan patterns when viewing natural scenes: Emotion, complexity, and repetition. *Psychophysiology*, *48*(11), 1543–1552.

Bradley, M. M., & Lang, P. J. (1994). Measuring emotion: The Self-Assessment Manikin and the semantic differential. *Journal of Behavioral Therapy and Experimental Psychiatry*, *25*, 49–59.

Bradley, M. M., & Lang, P. J. (2007). The International Affective Picture System (IAPS) in the study of emotion and attention. In J. A. Coan & J. B. Allen (Eds.), *Handbook of emotion elicitation and assessment* (pp. 29–46). Oxford: Oxford University Press.

Bradley, M. M., Sabatinelli, D., Lang, P. J., Fitzsimmons, J. R., King, W. M., & Desai, P. (2003). Activation of the visual cortex in motivated attention. *Behavioral Neuroscience*, *117*(2), 369–380.

Campbell, B. A., Wood, G., & McBride, T. (1997). Origins of orienting and defensive responses: An evolutionary perspective. In P. J. Lang, R. F. Simons, & M. T. Balaban (Eds.), *Attention and orienting: Sensory and motivational processes* (pp. 41–67). Hillsdale, NJ: Lawrence Erlbaum Associates.

Castelhano, M., Wieth, M., & Henderson, J. (2007). I see what you see: Eye movements in real-world scenes are affected by perceived direction of gaze. In L. Paletta & E. Rome (Eds.), *Attention in cognitive systems. Theories and systems from an interdisciplinary viewpoint* (pp. 251–262). Berlin: Springer-Verlag.

Codispoti, M., De Cesarei, A., & Ferrari, V. (2012). The influence of color on emotional perception of natural scenes. *Psychophysiology*, *49*(1), 11–16.

Codispoti, M., Ferrari, V., & Bradley, M. M. (2006). Repetitive picture processing: Autonomic and cortical correlates. *Brain Research*, *1068*(1), 213–220.

Codispoti, M., Ferrari, V., & Bradley, M. M. (2007). Repetition and event-related potentials: Distinguishing early and late processes in affective picture perception. *Journal of Cognitive Neuroscience*, *19*(4), 577–586.

Cuthbert, B. N., Schupp, H. T., Bradley, M. M., Birbaumer, N., & Lang, P. J. (2000). Brain potentials in affective picture processing: Covariation with autonomic arousal and affective report. *Biological Psychology*, *52*(2), 95–111.

David, S. P., Munafo, M. R., Johansen-Berg, H., Smith, S. M., Rogers, R. D., Matthews, P. M., & Walton, R. T. (2005). Ventral striatum/nucleus accumbens activation to smoking—related pictorial cues in smokers and nonsmokers: A functional magnetic resonance imaging study. *Biological Psychiatry*, *58*(6), 488–494.

Davis, M. (2000). The role of the amygdala in conditioned and unconditioned fear and anxiety. In J. P. Aggleton (Ed.), *The amygdala* (Vol. 2, pp. 213–287). Oxford: Oxford University Press.

Davis, M., & Lang, P. J. (2003). Emotion. In M. Gallagher & R. J. Nelson (Eds.), *Handbook of Psychology, Volume 3: Biological Psychology* (pp. 405–439). New York: Wiley.

Einhäuser, W., Rutishauser, U., & Koch, C. (2008). Task-demands can immediately reverse the effects of sensory-driven saliency in complex visual stimuli. *Journal of Vision*, *8*(2), 1–19.

Fagot, J. (2000). *Picture perception in animals*. Philadelphia: Taylor & Francis.

Fanselow, M. S., & Poulos, A. M. (2005). The neuroscience of mammalian associative learning. *Annual Review of Psychology*, *56*, 207–234.

Ferrari, V., Bradley, M. M., Codispoti, M., Karlsson, M., & Lang, P. J. (2013). Repetition and brain potentials when recognizing natural scenes: Task and emotion differences. *Social Cognitive and Affective Neuroscience*, *8*(8), 847–854.

Gauthier, I., Tarr, M. J., Moylan, J., Skudlarski, P., Gore, J. C., & Anderson, A. W. (2000). The fusiform "face area" is part of a network that processes faces at the individual level. *Journal of Cognitive Neuroscience*, *12*(3), 495–504.

Hebb, D. (1949). *The organization of behavior: A neuropsychological theory*. New York: Wiley.

Henderson, J. M. (1997). Transsaccadic memory and integration during real-world object perception. *Psychological Science*, *8*(1), 51–55.

Henderson, J. M. (2003). Human gaze control during real-world scene perception. *Trends in Cognitive Sciences*, *7*(11), 498–504.

Henderson, J. M., & Hollingworth, A. (1998). Eye movements during scene viewing: An overview. In G. Underwood (Ed.), *Eye guidance in reading and scene perception* (pp. 269–293). Oxford: Elsevier.

Henderson, J. M., Weeks, P. A., & Hollingworth, A. (1999). The effects of semantic consistency on eye movements during complex scene viewing. *Journal of Experimental Psychology: Human Perception and Performance*, *25*(1), 210–228.

Hollingworth, A. (2004). Constructing visual representations of natural scenes: The roles of short- and long-term visual memory. *Journal of Experimental Psychology: Human Perception and Performance*, *30*(3), 519–537.

Intraub, H. (1984). Conceptual masking: The effects of subsequent visual events on memory for pictures. *Journal of Experimental Psychology: Learning, Memory, and Cognition*, *10*(1), 115–125.

Irwin, D. E., & Andrews, R. V. (1996). Integration and accumulation of information across saccadic eye movements. In T. Inui & J. L. McClelland (Eds.), *Attention and performance XVI: Information integration in perception and communication* (pp. 125–155). Cambridge, MA: MIT Press.

Ishikawa, A., Ambroggi, F., Nicola, S. M., & Fields, H. L. (2008). Dorsomedial prefrontal cortex contribution to behavioral and nucleus accumbens neuronal responses to incentive cues. *Journal of Neuroscience*, *28*(19), 5088–5098.

Itakura, S. (1994). Recognition of line-drawing representations by a chimpanzee (*Pan troglodytes*). *Journal of General Psychology*, *121*, 189–197.

Itti, L., & Koch, C. (2001). Computational modelling of visual attention. *Nature Reviews. Neuroscience*, *2*(3), 194–203.

Junghöfer, M., Bradley, M. M., Elbert, T. R., & Lang, P. J. (2001). Fleeting images: A new look at early emotion discrimination. *Psychophysiology*, *38*(2), 175–178.

Kanwisher, N., & Yovel, G. (2006). The fusiform face area: A cortical region specialized for the perception of faces. *Philosophical Transactions of the Royal Society of London. Series B, Biological Sciences*, *361*(1476), 2109–2128.

Kapp, B. S., Supple, W. F., & Whalen, P. J. (1994). Effects of electrical stimulation of the amygdaloid central nucleus on neocortical arousal in the rabbit. *Behavioral Neuroscience*, *108*(1), 81–93.

Keil, A., Sabatinelli, D., Ding, M., Lang, P. J., Ihssen, N., & Heim, S. (2009). Re-entrant projections modulate visual cortex in affective perception: Evidence from Granger causality analysis. *Human Brain Mapping*, *30*(2), 532–540.

Lang, P. J., & Bradley, M. M. (2010). Emotion and the motivational brain. *Biological Psychology*, *84*(3), 437–450.

Lang, P. J., Bradley, M. M., & Cuthbert, B. N. (1990). Emotion, attention, and the startle reflex. *Psychological Review*, *97*(3), 377–395.

Lang, P. J., Bradley, M. M., & Cuthbert, B. N. (1997a). *International Affective Picture System (IAPS): Technical manual and affective ratings*. Bethesda, MD: NIMH Center for the Study of Emotion and Attention.

Lang, P. J., Bradley, M. M., & Cuthbert, M. M. (1997b). Motivated attention: Affect, activation and action. In P. J. Lang, R. F. Simons, & M. T. Balaban (Eds.), *Attention and orienting: Sensory and motivational processes* (pp. 97–135). Hillsdale, NJ: Lawrence Erlbaum Associates.

Lang, P. J., Bradley, M. M., & Cuthbert, B. N. (2008). *International affective picture system (IAPS): Affective ratings of pictures and instruction manual. Technical Report A-8.* Gainesville, FL: University of Florida.

Lang, P. J., Bradley, M. M., Fitzsimmons, J. R., Cuthbert, B. N., Scott, J. D., Moulder, B., & Nangia, V. (1998). Emotional arousal and activation of the visual cortex: An fMRI analysis. *Psychophysiology*, *35*(2), 199–210.

Lang, P. J., & Davis, M. (2006). Emotion, motivation, and the brain: Reflex foundations in animal and human research. *Progress in Brain Research*, *156*, 3–29.

LeDoux, J. (2003). The emotional brain, fear, and the amygdala. *Cellular and Molecular Neurobiology*, *23*(4/5), 727–738.

Loftus, G. R., & Mackworth, N. H. (1978). Cognitive determinants of fixation location during picture viewing. *Journal of Experimental Psychology: Human Perception and Performance*, *4*(4), 565–572.

Löw, A., Bradley, M. M., & Lang, P. J. (2013). Perceptual processing of natural scenes at rapid rates: Effects of complexity, content, and emotional arousal. *Cognitive, Affective, & Behavioral Neuroscience, 13*(4), 860–868.

Mehrabian, A. (1970). A semantic space for nonverbal behavior. *Journal of Consulting and Clinical Psychology*, *35*(2), 248-257.

Mogenson, G. J. (1977). *The neurobiology of behavior: An introduction.* Hillsdale, NJ: Lawrence Erlbaum Associates.

Osgood, C., Suci, G., & Tannenbaum, P. (1957). *The measurement of meaning*. Urbana, IL: University of Illinois.

Palmer, S. E. (1999). *Vision Science: Photons to Phenomenology* (Vol. 1). Cambridge, MA: MIT Press.

Parkhust, D., Law, K., & Niebur, E. (2002). Modeling the role of salience in the allocation of overt visual attention. *Vision Research*, *42*(1), 107–123.

Peters, R. J., Iyer, A., Itti, L., & Koch, C. (2005). Components of bottom-up gaze allocation in natural images. *Vision Research*, *45*(18), 2397–2416.

Peyk, P., Schupp, H. T., Keil, A., Elbert, T., & Junghöfer, M. (2009). Parallel processing of affective visual stimuli. *Psychophysiology*, *46*(1), 200–208.

Potter, M. C. (1976). Short-term conceptual memory for pictures. *Journal of Experimental Psychology. Human Learning and Memory*, *2*(5), 509–522.

Potter, M. C. (2012). Recognition and memory for briefly presented scenes. *Frontiers in Psychology*, *3*(32), 1–9.

Potter, M. C., & Fox, L. F. (2004). Perceiving and remembering multiple pictures in RSVP. *Journal of Vision*, *4*(8), 868.

Potter, M. C., Staub, A., Rado, J., & O'Connor, D. H. (2002). Recognition memory for briefly-presented pictures: The time course of rapid forgetting. *Journal of Experimental Psychology. Human Perception and Performance*, *28*(5), 1163–1175.

Russell, J. A., & Mehrabian, A. (1977). Evidence for a three-factor theory of emotions. *Journal of Research in Personality, 11*, 273–294.

Sabatinelli, D., Bradley, M. M., Fitzsimmons, J. R., & Lang, P. J. (2005). Parallel amygdala and inferotemporal activation reflect emotional intensity and fear relevance. *NeuroImage*, *24*, 1265–1270.

Sabatinelli, D., Bradley, M. M., Lang, P. J., Costa, V. D., & Versace, F. (2007). Pleasure rather than salience activates human nucleus accumbens and medial prefrontal cortex. *Journal of Neurophysiology*, *98*(3), 1374–1379.

Sabatinelli, D., Fortune, E. E., Li, Q., Siddiqui, A., Krafft, C., Oliver, W. T., …, & Jeffries, J. (2011). Emotional perception: Meta-analyses of face and natural scene processing. *NeuroImage*, *54*(3), 2524–2533.

Sabatinelli, D., Keil, A., Frank, D. W., & Lang, P. J. (2013). Emotional perception: Correspondence of early and late event-related potentials with cortical and subcortical functional MRI. *Biological Psychology*, *92*(3), 513–519.

Sabatinelli, D., Lang, P. J., Bradley, M. M., Costa, V. D., & Keil, A. (2009). The timing of emotional discrimination in human amygdala and ventral visual cortex. *Journal of Neuroscience*, *29*(47), 14864–14868.

Schupp, H. T., Cuthbert, B. N., Bradley, M. M., Birbaumer, N., & Lang, P. J. (1997). Probe P3 and blinks: Two measures of affective startle modulation. *Psychophysiology*, *34*(1), 1–6.

Schupp, H. T., Cuthbert, B. N., Bradley, M. M., Hillman, C. H., Hamm, A. O., & Lang, P. J. (2004). Brain processes in emotional perception: Motivated attention. *Cognition and Emotion*, *18*(5), 593–611.

Versace, F., Bradley, M. M., & Lang, P. J. (2010). Memory and event-related potentials for rapidly presented emotional pictures. *Experimental Brain Research*, *205*(2), 223–233.

Weinberg, A., & Hajcak, G. (2010). Beyond good and evil: The time-course of neural activity elicited by specific picture content. *Emotion*, *10*(6), 767-782.

Zink, C. F., Pagnoni, G., Chappelow, J. C., Martin-Skurski, M. E., & Berns, G. S. (2006). Human striatal activation reflects degree of stimulus saliency. *NeuroImage*, *29*(3), 977–983.

14

Threat Perception in Visual Scenes: Dimensions, Action, and Neural Dynamics

Kestutis Kveraga

Efficient recognition of threat is necessary for survival. Identifying threats in natural environments can be a difficult task for which humans have evolved a finely tuned visual recognition and action system. Although threat is a type of negative stimulus that typically initiates defensive reactions and mobilizes fight-or-flight systems in the brain, nonthreatening negative stimuli can produce exploratory behavior and engage associative processing. In this chapter I describe a number of studies that have explored visual discrimination of different types of negative stimuli in real-world scene images. I discuss the role of spatial and temporal properties of threat, the neural systems mediating the processing of different types of negative stimuli, and the visual pathways that may subserve them. In addition, I go over some findings of the latency and connectivity in the brain networks involved in threat assessment. I finish by discussing how our brain may deal with natural versus culturally learned sources of threat from other humans.

I put my face close to the thick glass-plate in front of a puff-adder in the Zoological Gardens, with the firm determination of not starting back if the snake struck at me; but, as soon as the blow was struck, my resolution went for nothing, and I jumped a yard or two backwards with astonishing rapidity. My will and reason were powerless against the imagination of a danger which had never been experienced.

—Charles Darwin, The Expression of the Emotions in Man and Animals (1872)

Picture yourself slogging through thick jungle on some ill-considered tropical adventure. The region in which you are traveling is remote, lawless, and is rife with natural and human predators. Big cats with a taste for human flesh are known to roam here, and venomous snakes slither underfoot. Even worse, this area is swarming with various rebel factions who make their living by kidnapping tourists and demanding a ransom for their release. Although everyone in your group is carrying a machete and your guide has a hunting rifle slung across his back, these weapons are helpful only against the natural predators—if you spot them quickly enough. If confronted by rebels armed with AK-47s, the guide warns, it is best to cooperate fully. The guerillas are ruthless, and any signs of resistance will have consequences far worse than being taken hostage.

The forest is alive with sounds of the local fauna and the crackling of twigs broken by feet, hands, and machetes, so your sense of hearing is of little use. Constantly scanning the dense foliage for signs of potential predators has kicked your visual system into overdrive. You are mulling alternate responses in your mind—hacking at four-legged or legless attackers with your machete, immediately surrendering to the two-legged ones, or perhaps freezing on sight of predators and blending into the brush. Selecting the correct response could literally save your life. After a while your guide spies a half-torn-apart carcass of some unfortunate ungulate. Your group feels impelled to investigate it for any signs of what might have happened to the poor creature. Did this happen recently? Had it been attacked by a leopard? Or did some rebels shoot the animal and then hack away the edible parts for roasting?

While the situation I describe here is perhaps overly dramatized, it provides several examples of the decisions our visual brain is tasked with in appropriately responding to negative stimuli. The brain must react rapidly to imminent danger because to delay action can be deadly. Therefore, it evolved to be sensitive to things that appear suddenly and/or are highly salient. Camouflage employed by many predators loses its effectiveness as soon as they move quickly because certain types of visual neurons are very sensitive to sharp transitions evoked by movement. However, not all negative or suddenly appearing stimuli represent threats. Automatically treating the latter as threat can also be extremely costly, both in high metabolic costs of initiating survival-related responses and in consequences for nearby conspecifics, who are likely to be close kin. For example, you do not want to strike out at close genetic relatives who suddenly appear in your peripheral view or immediately flee the area if they are seeking your help. Merely negative stimuli that do not present current danger, such as a dead animal or an accident site, can offer valuable clues about what events caused their present state, whether they present potential threats to you and, if so, how to stay out of danger.

Therefore, the brain must have mechanisms for both identifying imminent threats and quickly suppressing potentially harmful threat-related responses when the stimuli that are initially perceived as threats turn out to be innocuous. Although, aside from risky jungle adventures, most of us face such life-and-death decisions fairly rarely, law enforcement and military personnel in war zones are expected to make such snap judgments quickly and accurately, lest they kill an innocent person or be killed themselves. What information does the brain use to make decisions about what is and is not a threat? The studies reviewed in this chapter represent some of the initial attempts at answering this question using ecologically realistic images—photographs of scenes as well as derivations of these images to delve more deeply into the processing mechanisms.

Spatial and Temporal Dimensions of Threat

Negative stimuli can be classified along certain threat properties, both spatial and temporal. The most salient of different types of negative stimuli is a direct, immediate threat to one's person. This can be a wild animal, a human attacker, or something far more common (though still deadly) such as an impending crash. A *direct threat*, as the term implies, is harm aimed toward you, and it is about to happen or happening right now. Therefore, its spatial orientation (defined in egocentric coordinates) is *direct* or toward the observer, and its temporal direction is current or in the very near future. Moreover, in most cases, direct threat tends to be located in close proximity to you (effective long-range weapons are a relatively recent invention). Another type of negative stimulus is *indirect threat,* which involves an impending or in-progress attack on another human or animal. Therefore, its spatial orientation is averted, or away from the observer, and it has the temporal property of happening currently or in the very near future. Yet another type of negative stimulus is one in which harm has already occurred and no extant threat is visible, which I call a *merely negative* stimulus. In these stimuli, the threat direction is averted from the observer, and its temporal arrow points back in time. Negative stimuli can be classified on other grounds, such as by presence or absence of intent, whether the stimuli evoke feelings of disgust, even whether the potential harm is physical or "only" mental, but in this chapter I focus on the types of negative stimuli described above. Likewise, there are other, more complex types of negative stimuli—think of images of someone about to commit suicide (self-inflicted current, direct threat), images of harm done to you in the past (direct, past threat), or even viewing the very morally fraught images of your own (obviously unsuccessful) suicide attempt. However, we have not yet studied the latter classes of negative stimuli and therefore focus on impending direct and indirect threats, and merely negative (past threat) stimuli. These three types of stimuli cover the most common negative situations that humans have encountered during their evolutionary history and which many still live with in their daily lives.

How Well Do We Discriminate Different Types of Negative Stimuli?

Dimensional theories of affective perception (those mapping all stimuli along the axes of valence and arousal) predict that all negative stimuli should be perceived as aversive and threatening (Barrett & Russell, 1999; Larsen & Diener, 1992; Russell, 1980; Watson & Tellegen, 1985; see Barrett & Bliss-Moreau, 2009, for a review).

However, the introductory example and even our everyday experiences suggest that perhaps not all negative stimuli are perceived the same way. Imminent danger from a predator in the jungle or from an out-of-control car careening toward us on the street will cause us to act immediately, either to avoid the threat or to brace for the inevitable encounter. But seeing the carcass of an animal or an auto accident by the side of the road evokes the desire to know what happened, to search for clues that may be relevant to our own survival, or perhaps to help if it is not too late. If the attack or an accident is unfolding nearby but does not involve us, we are torn among fleeing for safety, watching it to learn its causes and outcome, or perhaps rushing to someone's aid.

Given these experiences, we predicted that observers would have sharply different reactions to different types of negative images.

We wanted to test this hypothesis by presenting photographs of natural and urban scenes and asking observers to rate the amount of harm depicted in the images that (1) the observer might suffer personally, were he or she in that situation; (2) someone else might suffer; or (3) someone else has already suffered. These questions were designed to elicit ratings corresponding to direct threat, indirect threat, and merely negative situations, respectively. We employed over 500 photographic scene images that we had obtained from the web and a priori classified as direct threat, indirect threat, merely negative, or neutral (figure 14.1A, plate 29). We found that subjects discriminated the different types of negative stimuli quickly and accurately (figure 14.1B, plate 29). In a follow-up experiment in which participants had to detect the presence or absence of a threat rather than explicitly make a decision about the presence and direction of harm, we found that direct threats were responded to most quickly and merely negative stimuli most slowly, slower than even the neutral scene images we employed as control stimuli (figure 14.1C, plate 29). These findings reveal evidence about different responses to visual stimuli that are negative *and* threatening versus those that are merely negative. Moreover, they show that merely negative scene images engage exploration processes that take longer than those of neutral or threatening scene images.

Brain Networks Engaged by Threat and Merely Negative Stimuli

Threat has been shown to activate brain structures whose functions include detection of salient and potentially aversive stimuli, evaluating their proximity and appraisal of threat potential. These regions include the periaqueductal gray (PAG) (e.g., Bandler & Shipley, 1994; Mobbs et al., 2007; Zhang, Bandler, & Carrive, 1990), locus coeruleus (Berridge & Waterhouse, 2003; Isbell, 2006), and the amygdala (e.g., Bliss-Moreau, Toscano, Bauman, Mason, & Amaral, 2011; Fanselow, 1994; Feinstein, Adolphs, Damasio, & Tranel, 2011; Larson et al., 2006; LeDoux, 2000, 2012; Sabatinelli, Lang, Bradley, Costa, & Keil, 2009; Whalen, 2007). But what brain networks differentiate among different types of negative stimuli—direct and indirect threat and merely negative stimuli? To answer this question, we employed a subset of the photographic scene images from our initial studies that did not differ in their low-level qualities (such as spatial frequency, average brightness) in an event-related fMRI experiment. We asked the participants to pay close attention to the scene images and report whether the scene repeated on a subsequent trial (a one-back task). Thus, they were not required to detect or report threats explicitly.

What we found was that direct threat, indirect threat, and merely negative images engaged distinct, although somewhat overlapping, sets of regions. The direct threat

A.

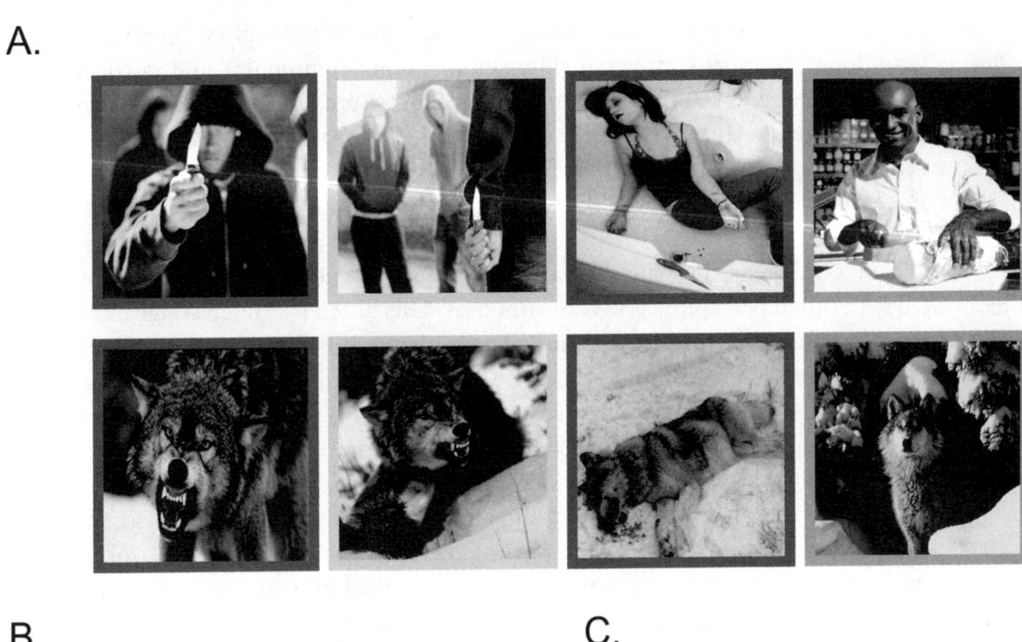

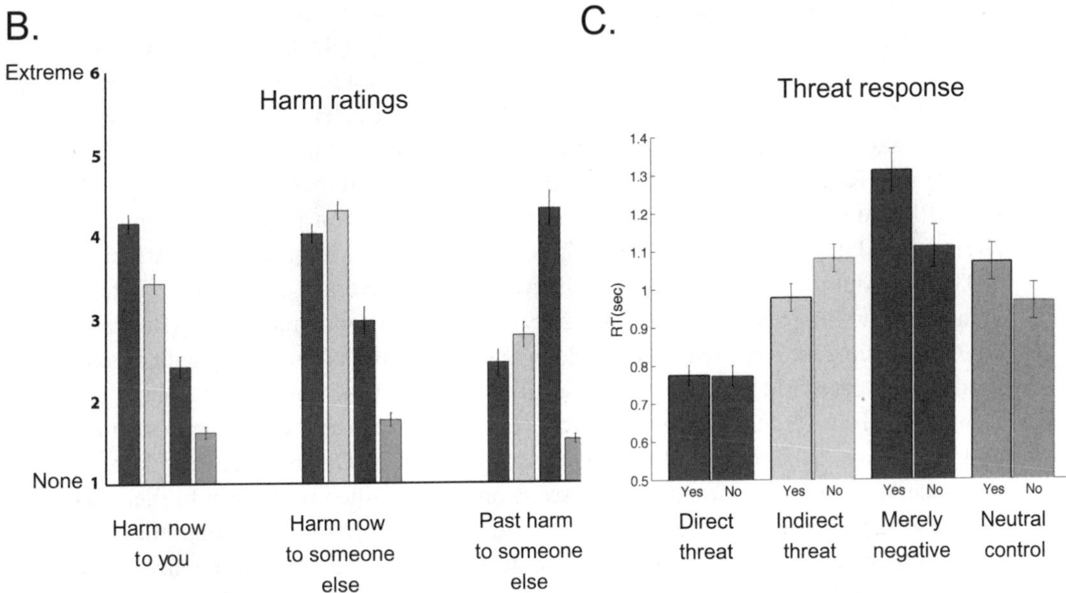

Figure 14.1 (plate 29)
Stimuli and behavioral results. (A) Four types of color photographic scene stimuli. From left to right: direct threat, indirect threat, merely negative, and neutral. (B) Behavioral rating results. Three groups rated the scene stimuli on questions designed to elicit measures of the scenes' direct threat (first bar in each subpanel), indirect threat (second bar), and past threat (third bar). Neutral control scenes (fourth bar) were rated implicitly. (C) Behavioral responses from an experiment in which participants had to make a "threat present or absent" decision as quickly and accurately as possible and report via a key press. Bars with solid borders are "yes-threat" responses; bars with dotted borders (second bar in each subgrouping) are "no-threat" responses.

images activated a set of regions associated with fear: subcortical regions (periaqueductal gray, locus coeruleus, pulvinar, hippocampus, and amygdala) and cortical regions in the orbital and ventrolateral prefrontal cortices (premotor cortex, fusiform and posterior cingulate cortex) (Kveraga et al., 2014). Direct threat stimuli activated these regions most strongly, followed by activations evoked by indirect threat stimuli. Direct threat orientation has been found to affect behavioral (e.g., electrodermal) responses to threat stimuli (Flykt, Esteves, & Öhman, 2007), and our findings demonstrate that the difference in spatial threat orientation is reflected in the brain regions associated with threat perception and fear responses. This difference, interestingly, was not driven by more negative valence and higher evoked arousal of the direct threat stimuli, as the indirect threat stimuli were actually rated as slightly more negative and more arousing than direct threat stimuli (Kveraga et al., 2014). Therefore, the essence of threat is not fully captured by the valence and arousal dimensions but rather involves factors such as spatial and temporal threat directions.

Compared with the direct and indirect threat stimuli, the merely negative scenes evoked strongest activations in a different set of regions (although they still evoked activations in many of the same regions as the threat scenes, but with a delayed peak). Merely negative scene images strongly activated the parahippocampal and retrosplenial cortices (PHC and RSC), and parts of the medial prefrontal cortex (MPFC) (figure 14.2, plate 30). Note that PHC and RSC are consistently implicated in scene perception (Aguirre, Detre, Alsop, & D'Esposito, 1996; Epstein & Kanwisher, 1998; also see chapter 6 by Epstein, chapter 3 by Park and Chun, chapter 1 by Intraub, and chapter 4 by Rajimehr, Nasr, and Tootell in this volume), and these regions along with the MPFC are also heavily involved in contextual association processing (Aminoff, Kveraga, & Bar, 2013; also see chapter 7 by Aminoff in this volume; Bar & Aminoff, 2003; Bar, Aminoff, & Schacter, 2008; Davachi, 2006; Kveraga et al., 2011; Peters, Daum, Gizewski, Forsting, & Suchan, 2009) and default-mode mentation (Buckner, Andrews-Hanna, & Schacter, 2008; Buckner & Carroll, 2007). This, along with behavioral results described in the previous section, suggests that when scene images are merely negative without an obvious current threat, observers may engage in extraction and processing of the context of the events depicted therein. During poststudy debriefing, observers uniformly noted that the scene images classified (unknown to them) as merely negative required more thought "to figure out what is going on" than the clear (direct threat) images. It is important to note that these activation differences were not due to low-level differences such as luminance or spatial frequency spectra, which were not different between the conditions (Kveraga et al., 2014). Furthermore, they were compared with neutral control scenes, which were matched on general context but different in affective content. Last, activation differences between the conditions were generally observed only in higher-level visual, affective, and associative regions, and no differences were found in low-level visual regions. Thus, the activation of quite

A.

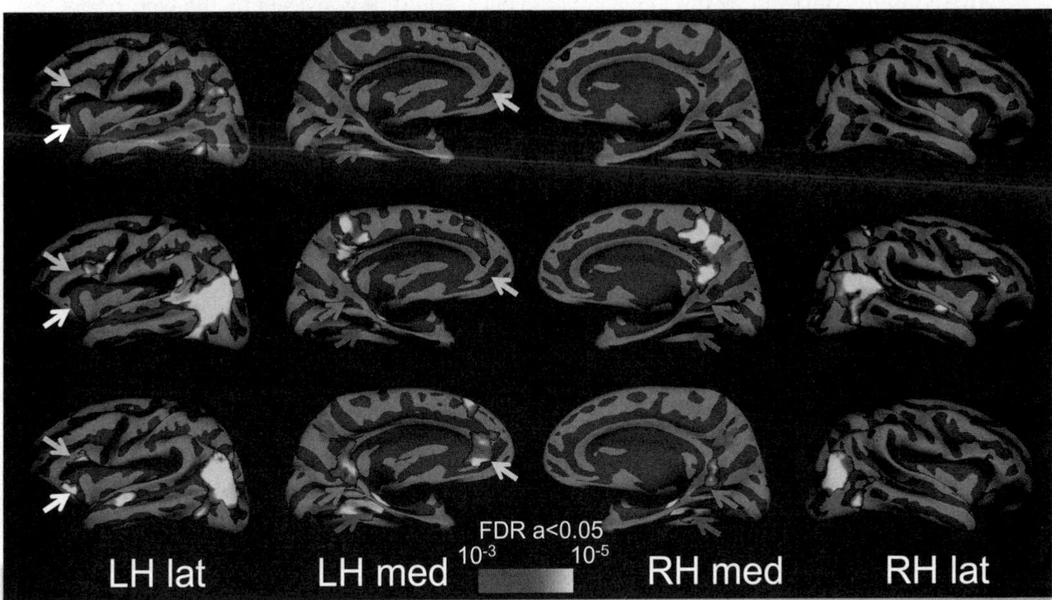

B.

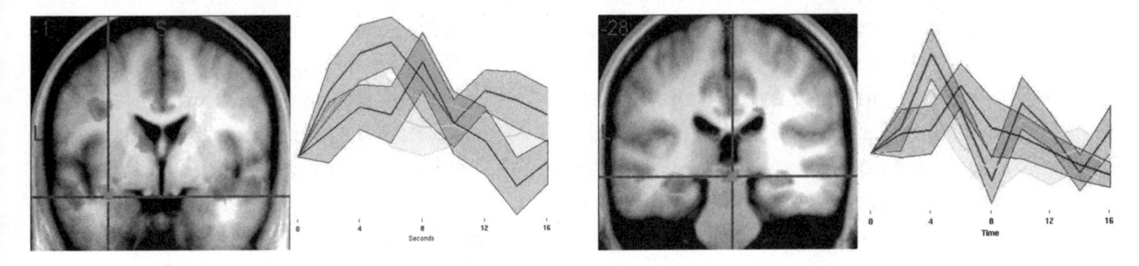

Figure 14.2 (plate 30)
fMRI activations evoked by photographic scene stimuli. (A) Statistical parametric maps for the direct threat, indirect threat, and merely negative conditions shown on the inflated group average brain. The BOLD activations evoked by the direct threat > neutral images contrast (top row), indirect threat > neutral images contrast (middle row), and merely negative > neutral images contrast (bottom row). The arrows indicate some of the ROIs with notable activation differences for the different scene categories: PHC (blue), RSC (magenta), PCC (red), OFC/vmPFC (white), vlPFC (cyan). (B) Subcortical activations and time courses in the amygdala (left panels) and periaqueductal gray (right panels).

different networks for the threat and merely negative scenes must be ascribed to affective content conveyed by threat cues. These cues would be conveyed by the face, body, and posture of human or animal threats, presence and use of weapons or weapon-like objects, as well as presence of signs of tissue damage, such as blood, wounds, or unnatural body positions.

Cortical Dynamics Mediating Perception of Threat and Merely Negative Scenes

Decisions about threat stimuli must be made quickly, and rapid brain dynamics are not well captured by hemodynamics-based techniques such as fMRI. To investigate rapidly changing neurodynamics during threat perception, we thus employed magnetoencephalography (MEG) using the same stimuli as in the previous fMRI study. Although MEG is not well suited for recording subcortical activity, particularly in the amygdala, it also does not suffer from tissue-air interaction artifacts that result in significant fMRI signal dropout in certain regions, such as OFC, particularly in posteromedial OFC (Deichmann, Gottfried, Hutton, & Turner, 2003). OFC is involved in resolving ambiguity (Kringelbach & Rolls, 2004), including ambiguous visual stimuli (Chaumon, Kveraga, Barrett, & Bar, 2013) early during perception (Bar et al., 2006), likely using magnocellular projections (Kveraga, Boshyan, & Bar, 2007). We thus wanted to investigate its activity with MEG and characterize its interactions with the cortical prefrontal and temporal regions that were activated in our fMRI study (Kveraga et al., 2014). We recorded data from 20 participants viewing our affective and neutral scene stimuli and found that posterior OFC was activated earlier (at around 130 ms) and more strongly for direct threat than for merely negative stimuli. However, ventrolateral prefrontal cortex (VLPFC), which is involved in threat appraisal and reassessment (Wager et al., 2008), was activated more strongly by the merely negative scenes at the later processing stage beginning at around 220 ms. Given the evident lack of current threat in merely negative scenes, this VLPFC activity might be involved in suppression of threat-related response during the reflective, reiterative-stage processing of negative stimuli (Cunningham & Zelazo, 2007; Lieberman, Gaunt, Gilbert, & Trope, 2002; Lieberman, 2003).

Strong bilateral connections between OFC and anterior temporal cortex (Barbas & De Olmos, 1990; Ghashghaei & Barbas, 2002; Kondo, Saleem, & Price, 2003), and between OFC and VLPFC (Barbas, 1995) are thought to mediate emotional processing. Therefore, given our MEG latency findings, we wanted to investigate functional connectivity between these regions. Furthermore, our earlier studies had shown early functional interactions between PHC and RSC during contextual association processing (Kveraga et al., 2011), the same regions that were strongly activated by the merely negative scene images. To accomplish these goals, we computed phase locking, a measure of communication between neural regions (Lachaux, Rodriguez, Martinerie,

& Varela, 1999), of MEG activity evoked by direct threat and merely negative scenes in those regions. We found greater early phase locking in the theta band (4–7 Hz) between anterior temporal cortex and OFC, beginning at around 130 ms for the direct threat versus merely negative scenes. Between OFC and VLPFC, there was stronger phase locking for merely negative versus direct threat scenes, which began at around 150 ms in the theta band and intensified at around 220 ms. Last, between PHC and RSC we found stronger phase locking in the alpha (8–12 Hz) for direct threat versus merely negative stimuli. Increases in the alpha band are usually indicative of active blocking or suppression of regional activity when resources are needed elsewhere (Haegens, Nácher, Luna, Romo, & Jensen, 2011; Klimesch, Sauseng, & Hanslmayr, 2007). Therefore, this finding suggests that activity in the association processing regions PHC and RSC is attenuated during threat perception. This makes sense in terms of dealing with direct, imminent threats—the brain needs to engage rapid threat-related response, not time-consuming deliberations.

Visual Pathways Mediating Threat and Merely Negative Scene Processing

Detecting threat rapidly confers a significant evolutionary advantage. This ability serves us well to this day even in environments (e.g., avoiding being hit by a car on the street) that are quite far removed from the ones in which humans evolved. But how does the brain implement visual threat detection? Our visual system comprises several parallel but strongly interacting visual pathways (Ungerleider & Mishkin, 1982). The two major, most extensively studied pathways are the magnocellular (M) pathway, whose outputs predominantly form the so-called dorsal visual stream (Goodale & Milner, 1992), and the parvocellular (P) pathway, which comprises much, though not all, of the ventral visual pathway (for reviews, see Nassi & Callaway, 2009). The faster M pathway is involved in action guidance, ambiguity resolution based on low spatial frequencies (Chaumon et al., 2013), and triggering top-down facilitation by the orbitofrontal (OCFC) cortex in the slower P pathway in the ventral temporal cortex (Kveraga et al., 2007). The slower P pathway is involved in processing color and fine details but, contrary to conventional wisdom, is also able to process low spatial frequencies in the color dimension and global shape (Thomas, Kveraga, Huberle, Karnath, & Bar, 2012). Extensive connections between the dorsal and ventral streams regions (Kravitz, Saleem, Baker, & Mishkin, 2011) enable interactions that shape scene perception in ways that are thus far little understood.

In this study we wanted to determine how the M and P pathways contribute to scene perception and how those contributions may vary depending on scene content. To accomplish this, the color photographs employed in the previous studies were converted to line drawings, and M-biased (low-luminance-contrast grayscale stimuli) and P-biased (isoluminant, chromatic contrast) scene stimuli were created dynamically, based on individual calibration procedures described by Kveraga et al. (2007) and Thomas et al. (2012). We also presented the intact line drawings (black and white; figure 14.3A, plate 31) as the control condition that engaged both M and P pathways. Scene line drawings in direct threat, indirect threat, merely negative, and neutral conditions were randomly assigned to M, P, or unbiased conditions and created "on the fly," based on individual pretest values obtained inside the scanner from each participant. The task was to report the presence or absence of a threat.

The contrast of M > P scene images (of all affective types) revealed activations in OFC, whereas P > M contrast yielded activations in the ventral occipital and temporal cortex. This was consistent with the M and P results obtained by Kveraga et al. (2007), which used line drawings of single objects rather than scenes. Moreover, the direct threat > merely negative contrast also revealed activations in OFC that were adjacent and overlapping with those evoked by M > P contrast (figure 14.2, plate 30). Focusing further on only M-biased direct threat > P-biased merely negative comparison (and the reverse contrast) revealed the same pattern. Thus, it seems that threat scene images may preferentially engage M pathways to trigger OFC activation, whereas merely negative scene images activate the ventral stream P pathway and activate PHC more strongly. These results also replicate the main activation patterns of the fMRI study with photographic scene images (Kveraga et al., 2014) with line-drawing stimuli, a different (explicit recognition) paradigm, and a new cohort of participants scanned in a different (1.5 T vs. 3 T) scanner (Kveraga, Boshyan, Ward, Adams, Barrett, Bar, unpublished data). Finally, it suggests that our findings of fMRI activations during threat and merely negative stimuli are quite robust (Kveraga et al., 2014).

How Does Threat Information Reach OFC?

These findings raise a question: How is threat information identified, and by which route is it sent to the OFC? We know that the OFC is activated quite early in visual perception, on the order of 100 ms (Thorpe, Rolls, & Maddison, 1983; Tomita, Ohbayashi, Nakahara, Hasegawa, & Miyashita, 1999). The earliest activation is likely due to low-spatial-frequency information carried by the M pathway (Bar et al., 2006). The exact path by which M-biased information reaches the OFC has not been resolved. One possibility is a cortical projection through the dorsal stream, perhaps involving the parietal lobe and the frontal eye fields, regions that mediate spatial attention and identification of objects of interest in visual scenes (see chapter 5 by Brooks, Sigurdardottir, and Sheinberg in this volume). Another possibility is a magnocellular projection through the ventral stream, although regions in the ventral stream mediating scene (PHC) (e.g., Kveraga et al., 2011) or face perception (including angry or fearful faces) typically have longer activation latencies than the OFC (Hadjikhani, Kveraga, Naik, & Ahlfors, 2009). Yet another intriguing possibility is a projection through the pulvinar nucleus of the thalamus, either via cortical projections from the primary

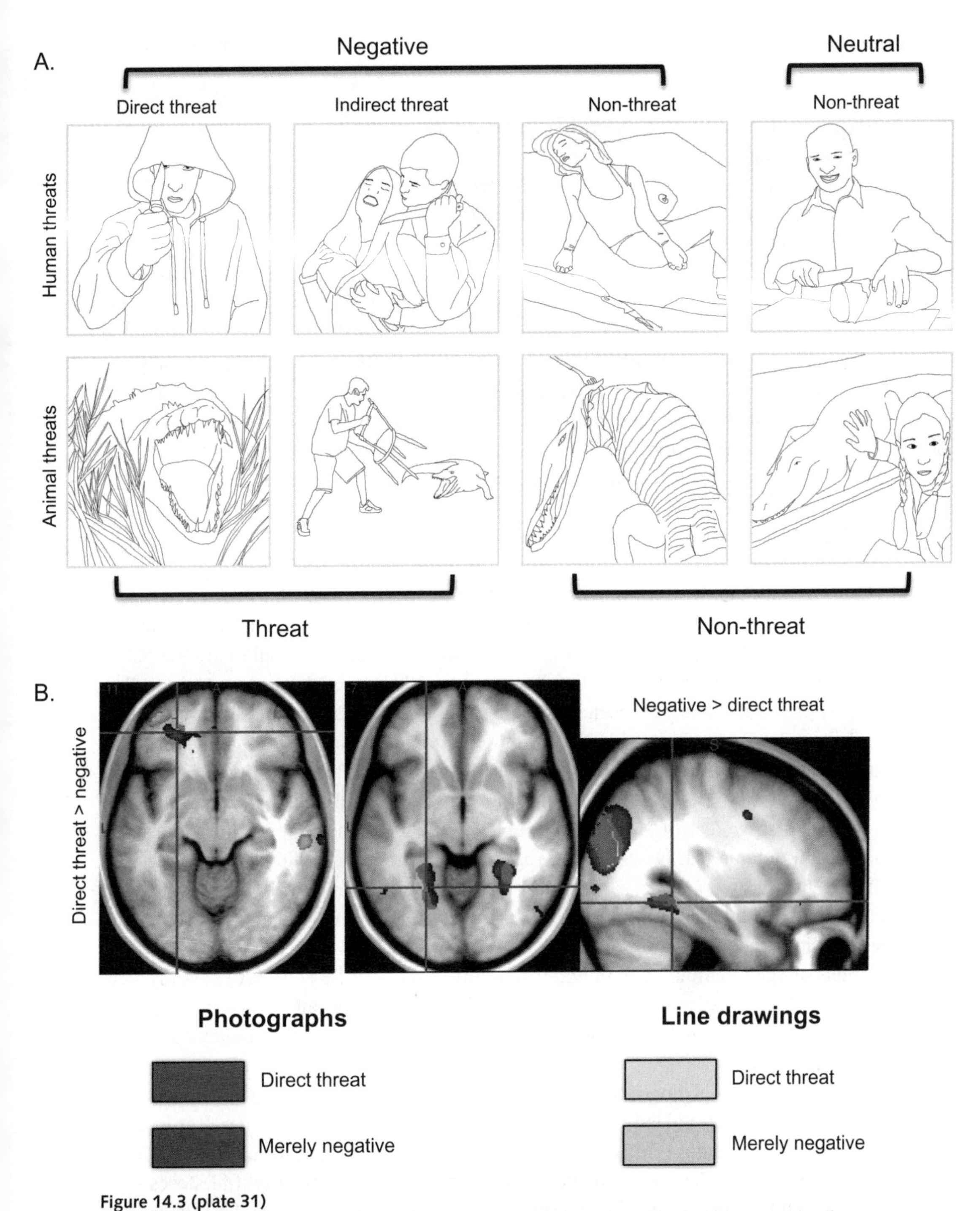

Figure 14.3 (plate 31)
Line drawing scenes and fMRI activations evoked by those images. (A) Line drawing scene stimuli depicting human and animal contexts: from left to right, direct threat, indirect threat, merely negative, and neutral scenes. (B) Comparison of results from two fMRI studies using color photo scenes and line drawing scenes. (Left panel) Direct threat > merely negative scene activations in fMRI study with color photo scenes (lighter color map) and study with line drawings (darker color map). (Middle and right panels) Negative > direct threat activations for color photo images (darker color map) and study line drawings (lighter color map).

visual cortex or from subcortical structures such as the superior colliculus or the lateral geniculate nucleus.

A recent study by Van Le et al. (2013) recorded from medial and dorsolateral pulvinar neurons in monkeys and found that pulvinar neurons were very responsive to natural threat stimuli—particularly images of snakes. However, the neurons were responsive to threat images only when they contained LSFs (i.e., intact and low-pass-filtered images). The pulvinar receives input from the superior colliculus and has heavy projections to the prefrontal cortex, including the OFC. It also has undergone dramatic expansion in nonhuman primates and hominids. Therefore, it is a good candidate to serve as an intermediate processing stage of coarse (that is, M/LSF) visual information. This stage may include phylogenetic templates for certain natural threats, which play a role in initiating survival-related responses before passing the information on to the prefrontal cortex for further disambiguation.

Are Some Threats Innate?

Responses to some types of threat stimuli may be innate and shaped by evolution. The so-called snake detection hypothesis (Isbell, 2006) proposes that particular visual regions of the brain unique to primates evolved in the early primates because of the need to detect snakes. Both constrictor and venomous species of snakes are the oldest predator of small primates (Isbell, 2006; Ohman, 2007), predating big cats and other natural predators. The very earliest drawings by hominids discovered in caves consist of diagonal cross-hatch and ladder-like patterns, which occur in nature only on dorsal and ventral body surfaces of snakes (Henshilwood et al., 2002; Van Le et al., 2013). Naive infant monkeys raised in artificial environments and never exposed to snakes nonetheless show fear of snake stimuli unless critical regions of their threat detection network are damaged. One of these regions is the superior colliculus, whose lesions disrupt normal fear reactions in infant monkeys (Maior et al., 2011). The superior colliculus is a laminar midbrain structure involved in coarse perception of visual stimulation and programming of automatic saccades; it projects to the pulvinar nucleus of the thalamus. The pulvinar neurons in the Van Le et al. (2013) study were particularly responsive to snakes, which evoked higher firing rates and lower latencies than other threat stimuli relevant to monkeys (such as threatening monkey faces) and neutral stimuli.

Although some phylogenetic threat stimuli (snakes, perhaps also spiders, carnivorans, raptors) may indeed be recognized more efficiently because of selective pressures on early primates during evolution, this can not be the case for so-called cultural or ontogenetically learned threats, such as guns. A study by Flykt, Esteves, and Öhman (2007) investigated skin conductance responses (SCRs) evoked by snakes and guns in human participants. The stimuli were conditioned with electrical shock (for phylogenetic stimuli) or noise bursts (ontogenetic stimuli) and were backward masked to prevent conscious recognition. SCRs for both snakes and guns were preserved (that is, they resisted backward masking), but only when the stimuli were oriented toward the observers (i.e., direct threat stimuli). Thus, it seems both innate and overlearned threats can be effective in evoking preattentive responses, at least when they are imminent and directed toward the observer. Future studies will need to delve more deeply into this interesting question: whether we are innately predisposed to perceive certain threats (and if so, which), and whether overlearned threats, such as guns, can become as efficiently recognizable as phylogenetic threats.

Conclusions

The findings described in this chapter show that humans are keenly sensitive to the spatial and temporal direction of threats in visual scenes. This perceptual discrimination results in clear differences in overt appraisal, as well as action toward, visual scenes with varying threat loading. The behavioral differences reflect the activation patterns in fMRI, where threat scenes engaged regions implicated in avoiding danger, whereas merely negative scenes activated networks known to be involved in associative thinking. The timing and functional connectivity of the cortical regions likewise reflected early engagement of areas involved in ambiguity resolution, such as the OFC, and blocking of associative thinking during threat perception in scene images. Later stages of threat reappraisal and assessment triggered activity in the VLPFC and interactions between the OFC and VLPFC, consistent with the role of the VLPFC. This research is just the beginning in developing our understanding of the neural processes underlying the perception of the spatial and temporal aspects of affective scenes. Future efforts will need to consider how the ingredients of complex affective images influence the larger percepts of threat and negativity in visual scenes. These component parts include facial expression and identity cues, body posture and appearance, display of natural and manufactured weapons, signs of tissue damage, as well as more elusive influences such as the general affective tone of a scene.

Acknowledgments

This work would not have been possible without many valuable discussions with Jasmine Boshyan, Lisa Feldman Barrett, Reginald Adams, Moshe Bar, and Nouchine Hadjikhani, their input and support. I also would like to thank Jasmine Boshyan and Jasmine Mote for their help with stimulus preparation, and Jasmine Boshyan, Noreen Ward, Jasmine Mote, and Nicole Betz for their help with data collection. I also want to thank Nao Suzuki, Naoro Tanaka, Sheraz Khan, Seppo Alhfors, and Matti Hämäläinen for technical support in conducting MEG research, and Thomas Witzel for fMRI technical support. Finally, this research was supported by NIH grant

K01MH084011 and MGH ECOR ISF grant to the author and by the Athinoula A. Martinos Center for Biomedical Imaging.

References

Aguirre, G. K., Detre, J. A., Alsop, D. C., & D'Esposito, M. (1996). The parahippocampus subserves topographical learning in man. *Cerebral Cortex*, *6*(6), 823–829.

Aminoff, E. M., Kveraga, K., & Bar, M. (2013). The role of parahippocampal cortex in cognition. *Trends in Cognitive Sciences*, *17*(8), 379–390.

Bandler, R., & Shipley, M. T. (1994). Columnar organization in the midbrain periaqueductal gray: Modules for emotional expression? *Trends in Neurosciences*, *17*(9), 379–389.

Bar, M., & Aminoff, E. M. (2003). Cortical analysis of visual context. *Neuron*, *38*, 347–358.

Bar, M., Aminoff, E. M., & Schacter, D. L. (2008). Scenes unseen: The parahippocampal cortex intrinsically subserves contextual associations, not scenes or places per se. *Journal of Neuroscience*, *28*(34), 8539–8544.

Bar, M., Kassam, K. S., Ghuman, A. S., Boshyan, J., Schmidt, A. M., Dale, A. M., et al. (2006). Top-down facilitation of visual recognition. *Proceedings of the National Academy of Sciences of the United States of America*, *103*(2), 449–454.

Barbas, H. (1995). Anatomic basis of cognitive-emotional interactions in the primate prefrontal cortex. *Neuroscience and Biobehavioral Reviews*, *19*(3), 499–510.

Barbas, H., & De Olmos, J. (1990). Projections from the amygdala to basoventral and mediodorsal prefrontal regions in the rhesus monkey. *Journal of Comparative Neurology*, *300*(4), 549–571.

Barrett, L. F., & Bliss-Moreau, E. (2009). Affect as a psychological primitive. *Advances in Experimental Social Psychology*, *41*, 167–218.

Barrett, L. F., & Russell, J. A. (1999). Structure of current affect. *Current Directions in Psychological Science*, *8*, 10–14.

Berridge, C. W., & Waterhouse, B. D. (2003). The locus coeruleus-noradrenergic system: Modulation of behavioral state and state-dependent cognitive processes. *Brain Research. Brain Research Reviews*, *42*, 33–84.

Bliss-Moreau, E., Toscano, J. E., Bauman, M. D., Mason, W. A., & Amaral, D. G. (2011). Neonatal amygdala lesions alter responsiveness to objects in juvenile macaques. *Neuroscience*, *178*, 123–132.

Buckner, R. L., Andrews-Hanna, J. R., & Schacter, D. L. (2008). The brain's default network: Anatomy, function, and relevance to disease. *Annals of the New York Academy of Sciences*, *1124*, 1–38.

Buckner, R. L., & Carroll, D. C. (2007). Self-projection and the brain. *Trends in Cognitive Sciences*, *11*(2), 49–57.

Chaumon, M., Kveraga, K., Barrett, L. F., & Bar, M. (2013). Visual predictions in the orbitofrontal cortex rely on associative content. *Cerebral Cortex*. Epub ahead of print.

Cunningham, W. A., & Zelazo, P. D. (2007). Attitudes and evaluations: A social cognitive neuroscience perspective. *Trends in Cognitive Sciences*, *11*, 97–104.

Davachi, L. (2006). Item, context and relational episodic encoding in humans. *Current Opinion in Neurobiology*, *16*(6), 693–700.

Deichmann, R., Gottfried, J. A., Hutton, C., & Turner, R. (2003). Optimized EPI for fMRI studies of the orbitofrontal cortex. *NeuroImage*, *19*(2 Pt 1), 430–441.

Epstein, R. A., & Kanwisher, N. (1998). A cortical representation of the local visual environment. *Nature*, *392*(6676), 598–601.

Fanselow, M. S. (1994). Neural organization of the defensive behavior system responsible for fear. *Psychonomic Bulletin & Review*, *1*(4), 429–438.

Feinstein, J. S., Adolphs, R., Damasio, A., & Tranel, D. (2011). The human amygdala and the induction and experience of fear. *Current Biology*, *21*(1), 34–38.

Flykt, A., Esteves, F., & Öhman, A. (2007). Skin conductance responses to masked conditioned stimuli: Phylogenetic/ontogenetic factors versus direction of threat? *Biological Psychology*, *74*, 328–336.

Ghashghaei, H., & Barbas, H. (2002). Pathways for emotion: Interactions of prefrontal and anterior temporal pathways in the amygdala of the rhesus monkey. *Neuroscience*, *115*, 1261–1279.

Goodale, M. A., & Milner, A. D. (1992). Separate visual pathways for perception and action. *Trends in Neurosciences*, *15*(1), 20–25.

Hadjikhani, N., Kveraga, K., Naik, P., & Ahlfors, S. P. (2009). Early (M170) activation of face-specific cortex by face-like objects. *Neuroreport*, *20*(4), 403–407.

Haegens, S., Nácher, V., Luna, R., Romo, R., & Jensen, O. (2011). Alpha oscillations in the monkey sensorimotor network influence discrimination performance by rhythmical inhibition of neuronal spiking. *Proceedings of the National Academy of Sciences of the United States of America*, *108*(48), 19377–19382.

Henshilwood, C. S., d'Errico, F., Yates, R., Jacobs, Z., Tribolo, C., Duller, G. A. T., et al. (2002). Emergence of modern human behavior: Middle stone age engravings from South Africa. *Science*, *295*(5558), 1278–1280.

Isbell, L. A. (2006). Snakes as agents of evolutionary change in primate brains. *Journal of Human Evolution*, *51*, 1–35.

Klimesch, W., Sauseng, P., & Hanslmayr, S. (2007). EEG alpha oscillations: The inhibition-timing hypothesis. *Brain Research. Brain Research Reviews*, *53*, 63–88.

Kondo, H., Saleem, K. S., & Price, J. L. (2003). Differential connections of the temporal pole with the orbital and medial prefrontal networks in macaque monkeys. *Journal of Comparative Neurology*, *465*(4), 499–523.

Kravitz, D. J., Saleem, K. S., Baker, C. I., & Mishkin, M. (2011). A new neural framework for visuospatial processing. *Nature Reviews. Neuroscience*, *12*(4), 217–230.

Kringelbach, M. L., & Rolls, E. T. (2004). The functional neuroanatomy of the human orbitofrontal cortex: Evidence from neuroimaging and neuropsychology. *Progress in Neurobiology*, *72*(5), 341–372.

Kveraga, K., Boshyan, J., Adams, R. B. J., Mote, J., Betz, N., Ward, N., et al. (2014). If it bleeds, it leads: Separating threat from mere negativity. *Social Cognitive and Affective Neuroscience*, 4 Apr.

Kveraga, K., Boshyan, J., & Bar, M. (2007). Magnocellular projections as the trigger of top-down facilitation in recognition. *Journal of Neuroscience*, *27*(48), 13232–13240.

Kveraga, K., Ghuman, A. S., Kassam, K. S., Aminoff, E. A., Hämäläinen, M. S., Chaumon, M., et al. (2011). Early onset of neural synchronization in the contextual associations network. *Proceedings of the National Academy of Sciences of the United States of America*, *108*(8), 3389–3394.

Lachaux, J. P., Rodriguez, E., Martinerie, J., & Varela, F. J. (1999). Measuring phase synchrony in brain signals. *Human Brain Mapping*, *8*, 194–208.

Larsen, R. J., & Diener, E. (1992). Promises and problems with the circumplex model of emotion. In M. S. Clark (Ed.), *Review of personality and social psychology: Emotion* (Vol. 13, pp. 25–59). Newbury Park, CA: Sage.

Larson, C. L., Schaefer, H. S., Siegle, G. J., Jackson, C. A., Anderle, M. J., & Davidson, R. J. (2006). Fear is fast in phobic individuals: Amygdala activation in response to fear-relevant stimuli. *Biological Psychiatry*, *60*(4), 410–417.

LeDoux, J. (2000). Emotion circuits in the brain. *Annual Review of Neuroscience*, *23*, 155–184.

LeDoux, J. (2012). Rethinking the emotional brain. *Neuron*, *73*(4), 653–676.

Lieberman, M., Gaunt, R., Gilbert, D., & Trope, Y. (2002). Reflexion and reflection: A social cognition neuroscience approach to attributional inference. *Advances in Experimental Social Psychology*, *34*, 199–249.

Lieberman, M. D. (2003). Reflective and reflexive judgment processes: A social cognitive neuroscience approach. In J. P. Forgas, K. R. Williams, & W. von Hippel (Eds.), *Social judgments: Implicit and explicit processes* (pp. 44–67). New York: Cambridge University Press.

Maior, R. S., Hori, E., Barros, M., Teixeira, D. S., Tavares, M. C., Ono, T., et al. (2011). Superior colliculus lesions impair threat responsiveness in infant capuchin monkeys. *Neuroscience Letters*, *504*(3), 257–260.

Mobbs, D., Petrovic, P., Marchant, J. L., Hassabis, D., Weiskopf, N., Seymour, B., et al. (2007). When fear is near: Threat imminence elicits prefrontal-periaqueductal gray shifts in humans. *Science*, *317*(5841), 1079–1083.

Nassi, J. J., & Callaway, E. M. (2009). Parallel processing strategies of the primate visual system. *Nature Reviews. Neuroscience*, *10*(5), 360–372.

Ohman, A. (2007). Has evolution primed humans to "beware the beast"? *Proceedings of the National Academy of Sciences of the United States of America*, *104*(42), 16396–16397.

Peters, J., Daum, I., Gizewski, E., Forsting, M., & Suchan, B. (2009). Associations evoked during memory encoding recruit the context-network. *Hippocampus*, *19*(2), 141–151.

Russell, J. A. (1980). A circumplex model of affect. *Journal of Personality and Social Psychology*, *39*, 1161–1178.

Sabatinelli, D., Lang, P. J., Bradley, M. M., Costa, V. D., & Keil, A. (2009). The timing of emotional discrimination in human amygdala and ventral visual cortex. *Journal of Neuroscience*, *29*(47), 14864–14868.

Thomas, C., Kveraga, K., Huberle, E., Karnath, H.-O., & Bar, M. (2012). Enabling global processing in simultanagnosia by psychophysical biasing of visual pathways. *Brain*, *135*(5), 1578–1585.

Thorpe, S. J., Rolls, E. T., & Maddison, S. (1983). The orbitofrontal cortex: Neuronal activity in the behaving monkey. *Experimental Brain Research*, *49*, 93–115.

Tomita, H., Ohbayashi, M., Nakahara, K., Hasegawa, I., & Miyashita, Y. (1999). Top-down signal from prefrontal cortex in executive control of memory retrieval. *Nature*, *401*, 699–703.

Ungerleider, L. G., & Mishkin, M. (1982). Two cortical visual systems. In D. J. Ingle, M. A. Goodale, & R.J.W. Mansfield (Eds.), *Analysis of visual behavior* (pp. 549–586). Cambridge, MA: MIT Press.

Van Le, Q., Isbell, L. A., Matsumoto, J., Nguyen, M., Hori, E., Maior, R. S., et al. (2013). Pulvinar neurons reveal neurobiological evidence of past selection for rapid detection of snakes. *Proceedings of the National Academy of Sciences of the United States of America*, *110*(47), 19000–19005.

Wager, T. D., Davidson, M. L., Hughes, B. L., Lindquist, M. A., & Ochsner, K. N. (2008). Prefrontal-subcortical pathways mediating successful emotion regulation. *Neuron*, *59*(6), 1037–1050.

Watson, D., & Tellegen, A. (1985). Toward a consensual structure of mood. *Psychological Bulletin*, *98*, 219–223.

Whalen, P. J. (2007). The uncertainty of it all. *Trends in Cognitive Sciences*, *11*(12), 499–500.

Zhang, S. P., Bandler, R., & Carrive, P. (1990). Flight and immobility evoked by excitatory amino acid microinjection within distinct parts of the subtentorial midbrain periaqueductal gray of the cat. *Brain Research*, *520*(1–2), 73–82.

Contributors

Elissa M. Aminoff
Center for the Neural Basis of Cognition, Carnegie Mellon University

Moshe Bar
Gonda Multidisciplinary Brain Research Center, Bar-Ilan University, Israel, and Martinos Center for Biomedical Imaging, Massachusetts General Hospital, Harvard Medical School

Margaret M. Bradley
Center for the Study of Emotion and Attention, University of Florida

Daniel I. Brooks
Department of Neuroscience, Brown University

Marvin M. Chun
Department of Psychology, Yale University

Ritendra Datta
Senior Engineer, Google

Russell A. Epstein
Department of Psychology, University of Pennsylvania

M. Fabre-Thorpe
Centre de Recherche Cerveau et Cognition (CerCo), CNR Université Paul Sabatier

Elena Fedorovskaya
Paul and Louise Miller Professor, School of Media Sciences, Rochester Institute of Technology

Jack L. Gallant
Vision Science Program and Department of Psychology, University of California, Berkeley

Helene Intraub
Department of Psychology, University of Delaware

Dhiraj Joshi
Senior Research Scientist, FX Palo Alto Lab (previously with Eastman Kodak Research Labs)

Kestutis Kveraga
Department of Radiology, Harvard Medical School; Martinos Center for Biomedical Imaging, Massachusetts General Hospital

Peter J. Lang
Center for the Study of Emotion and Attention, University of Florida

Jia Li
Professor, Department of Statistics, Penn State University

Xin Lu
Ph.D. student, College of Information Sciences & Technology, Penn State University.

Jiebo Luo
Department of Computer Science, University of Rochester

Quang-Tuan Luong
Terra Galleria Photography

George L. Malcolm
Department of Psychology, George Washington University

Shahin Nasr
Martinos Center for Biomedical Imaging, Massachusetts General Hospital

Soojin Park
Assistant Professor, Department of Cognitive Science, Johns Hopkins University

Mary C. Potter
Department of Brain and Cognitive Sciences, Massachusetts Institute of Technology

Reza Rajimehr
McGovern Institute for Brain Research, Massachusetts Institute of Technology

Dean Sabatinelli
Departments of Psychology & Neuroscience, Bioimaging Research Center, University of Georgia

Philippe G. Schyns
Institute of Neuroscience and Psychology, University of Glasgow

David L. Sheinberg
Department of Neuroscience, Brown University

Heida Maria Sigurdardottir
Department of Neuroscience, Brown University and The University of Iceland

Dustin E. Stansbury
Vision Science Program, University of California, Berkeley

Simon Thorpe
Centre de Recherche Cerveau & Cognition, CNRS-Université Toulouse

Roger Tootell
Martinos Center for Biomedical Imaging, Massachusetts General Hospital

Quang-Tuan Luong
Freelance Nature and Travel Photographer and owner of Terragalleria.com

James Z. Wang
Professor, College of Information Sciences & Technology (IS&T), Penn State University

Index

Action, 17, 45, 49, 56, 89, 106, 146, 192, 229, 247, 273–274, 291–292, 299, 303
Affect, 253, 256–257, 259, 273, 275–277, 284, 286, 293, 296, 298, 300, 303
Affective Picture Systems. *See* IAPS
Attention, 1, 3, 77, 85, 89–95, 97–98, 126, 181, 185, 187, 191, 225, 237, 244–245, 255, 278, 282–283, 286–287, 294
  attentional blink, 186
  attentional selection, 91
  attentional set, 185, 192
  attentional tracking paradigm, 77
  divided, 13
  focused, 255
  spatial, 92, 300
  visual, 89, 92, 255
  visuospatial, 90, 94
  voluntary, 181

Boundary extension, 5–18, 20–23, 47, 62–65, 144, 148, 177

Cardinal orientation bias, 79, 117
Coarse-to-fine (CtF), 29–30, 32–33
Conceptual short-term memory (CSTM), 183

Darwin, Charles, 291
Detection, 79, 90, 156–157, 173, 177, 184–187, 189–193, 294
  category detection task, 178
  edge, 28
  face, 192, 267, 286
  object, 158, 166
  snake detection hypothesis, 302
  threat, 299, 302
Dynamics, 149, 157–161, 167, 173, 248, 252, 291, 298

Emotion, 1, 3, 17, 241–243, 245, 247, 249–254, 256–258, 260–261, 264–268, 273–287, 291, 298
  emotional congruence theory, 266
  distribution, 249–250
  dominant, 249–250, 265
  prediction, 248–249
  emotional scene perception, 282
Event-related potentials (ERP), 178, 200, 273, 283–284, 286–287
Expression
  emotional, 247
  facial, 36, 250–252, 256–257, 275, 291, 303
Eye movements, 1, 3, 5, 39, 45–46, 60, 90, 237, 258, 273, 277, 287
  fixations, 5, 8–9, 39, 60, 177–178, 180, 185, 192, 277–279
  saccades, 5, 8, 13, 29, 60, 90, 92, 94–95, 97, 156, 171–173, 200, 277–278, 302

Feedforward processing, 169, 177, 188–189, 191, 193, 199, 201–202
Fight-or-flight, 291
Fine-to-coarse (FtC), 32
fMRI, 20–21, 51–53, 61–63, 68, 73–75, 79–81, 107–108, 110, 112–114, 116, 118, 121–124, 126, 128, 137, 139–141, 144–145, 164, 227–228, 233–234, 237, 258, 273, 279, 282–283, 294, 297–298, 300–301, 303
Fourier's theorem, 28

Gaze-contingent, 38, 40
Gaussian
  likelihood function, 237
  mixture, 255, 257, 263
  noise, 217–218
  statistics, 227
  window, 36
Gist, 2, 39, 41–42, 45, 67, 106, 126, 146, 158–159, 168–170, 173–174, 177, 183–185, 187, 192, 237

Haptic, 9–10, 13, 16, 18, 21, 62, 117
High spatial frequency bias, 77. *See also* Spatial frequency
Hippocampus, 21–23, 60–63, 65, 111, 124, 274, 296

IAPS, 257–259, 274–277, 279, 283
Imagination, 8, 16, 18, 111, 291

Lateral occipital cortex (LOC), 47–49, 51–55, 63–64, 66, 112
Level
basic, 37–42, 49, 107, 159
subordinate, 35, 37–40
superordinate, 159–160, 170–171
Low-level visual
characteristics of scene category, 2
cues, 73
data, 253
differences, 116, 119
features, 73, 79, 116–117, 251
properties of an image, 117
regions, 296
systems, 68

Magnetoencephalography (MEG), 149–150, 201, 298–299, 303
Memory, 1–3, 5–6, 8–13, 16, 18, 20–22, 41, 45–49, 57, 62–63, 65–66, 87, 91, 109–110, 137–138, 143–145, 147–148, 150, 157, 163, 179–180, 182–187, 189–192, 217, 226, 274, 279, 283–285, 287
long-term, 7, 14–15, 23, 142, 177–178, 181
Monkey homologue of the human PPA, 75

Natural environments, 27, 166, 277, 291. *See also* Real-world scenes
Natural threats, 302. *See also* Threat
Navon stimuli, 29–30
Negative stimuli, 291–294, 298–300
Neurodynamics, 298. *See also* Dynamics

Occipital place area (OPA), 59, 74, 108, 114

Parahippocampal cortex (PHC), 22, 77, 107, 110, 112, 137–142, 145–147, 149–150, 296–300
Parahippocampal place area (PPA), 18, 20–23, 46–61, 63–67, 74–75, 77–81, 98, 105, 107–128, 137, 228–229, 236. *See also* Parahippocampal cortex (PHC)
Processing of
contextual associations/information, 21, 164, 167, 296
different dimensions, 138
image statistics, 166, 173
key features, 170
meaningless shapes, 140
natural scenes, 273, 279
negative stimuli, 291, 298
objects, 55, 155–157, 161, 168, 174, 220–221
parvocellular information, 169
pictures, 179
reward, 148
scenes, 73, 118, 123, 168–169, 171, 220, 287
single dimension, 136
spatial information, 41, 55, 97, 123, 138–139
surrounding environment, 156
visual information, 123

Rapid serial visual presentation (RSVP), 49, 177, 180–183, 185–191, 193, 201, 273, 284–286
Real-world scenes, 45, 50, 87, 95–96, 99, 106
Retinotopy, 75. *See also* Retinotopic organization
Retinotopic organization, 109–110
Retrosplenial complex (RSC), 3, 20–23, 46–49, 51, 53–54, 56–61, 63, 65–66, 74–75, 80, 108–109, 111–114, 116–117, 122–124, 137–147, 149–150, 236, 296–299. *See also* Occipital place area (OPA); Parahippocampal cortex (PHC); Parahippocampal place area (PPA)

Saccade. *See* Eye movements
Scene categorization, 2, 28, 40–42, 52, 55, 67, 137, 155, 158, 237
Scene perception, 2, 5–6, 10, 12, 14, 21–23, 29, 45, 55, 57–58, 60, 62, 66–68, 85, 99, 111, 128, 137, 277, 282–284, 296, 299
Search, 3, 39, 40, 50, 85, 89, 91, 93, 95, 96, 97, 98, 99, 163, 183, 186, 187, 191, 242, 253, 260, 293. *See also* Visual search
Snakes, 280–281, 291, 302–303
Spatial frequency, 2, 28, 35, 37, 40, 77, 116, 228, 230, 278, 294, 296
high, 29, 40, 77–79, 106, 117, 168
low, 29, 40, 78, 117, 168–169, 300

Threat, 2–3, 274, 277, 281–282, 291–303
culturally learned, 291
direct, 293–301, 303
indirect, 293–297, 300–301
past, 293, 295
spatial and temporal properties of, 291
Transverse occipital sulcus (TOS), 47–49, 59–60, 74–75, 77, 80, 108, 111, 114, 116, 127, 236

Vision
computer, 155, 199, 201–203, 207–208, 219, 252, 268
human, 1, 5–6, 9–10, 12–13, 15–16, 28, 62, 79, 99, 117, 135, 177, 180, 199, 201–202, 220–221, 232–233, 241
spatial, 1
Visual pathways, 169–170, 291, 299
Visual search, 89, 91, 93, 95–97, 163, 183

Wavelet, 41, 228, 254
Gabor, 228, 233